Emanuele Curotto

Theoretical and Computational Chemistry

Also of Interest

Computational Chemistry Methods. Applications
Ponnadurai Ramasami (Ed.), 2020
ISBN 978-3-11-062906-4, e-ISBN (PDF) 978-3-11-063162-3,
e-ISBN (EPUB) 978-3-11-062992-7

Scientific Computing. For Scientists and Engineers
Timo Heister, Leo G. Rebholz, 2023
ISBN 978-3-11-099961-7, e-ISBN (PDF) 978-3-11-098845-1,
e-ISBN (EPUB) 978-3-11-098875-8

Maths in Chemistry. Numerical Methods for Physical and Analytical Chemistry
Prerna Bansal, 2024
ISBN 978-3-11-133392-2, e-ISBN (PDF) 978-3-11-133444-8,
e-ISBN (EPUB) 978-3-11-133450-9

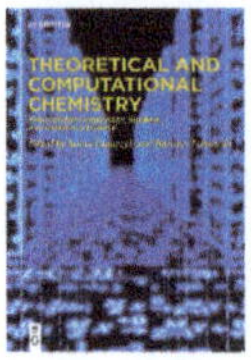

Theoretical and Computational Chemistry. Applications in Industry, Pharma, and Materials Science
Iwona Gulaczyk, Bartosz Tylkowski (Eds.), 2021
ISBN 978-3-11-067815-4, e-ISBN (PDF) 978-3-11-067821-5,
e-ISBN (EPUB) 978-3-11-067828-4

Emanuele Curotto

Theoretical and Computational Chemistry

Gauge Theories for Chemical Physics and Excited State Methods

DE GRUYTER

Author
Prof. Dr. Emanuele Curotto
Arcadia University
Department of Chemistry and Physics
450 South Easton Road
Glenside, PA 19038-3295
USA
curottom@arcadia.edu

ISBN 978-3-11-161008-5
e-ISBN (PDF) 978-3-11-161020-7
e-ISBN (EPUB) 978-3-11-161048-1

Library of Congress Control Number: 2025951223

Bibliographic information published by the Deutsche Nationalbibliothek
The Deutsche Nationalbibliothek lists this publication in the Deutsche Nationalbibliografie; detailed bibliographic data are available on the Internet at http://dnb.dnb.de.

Cover image: Irina Devaeva / iStock / Getty Images Plus
Typesetting: VTeX UAB, Lithuania

www.degruyterbrill.com
Questions about General Product Safety Regulation:
productsafety@degruyterbrill.com

To Katheleen

Introduction

The proclamation that Max Born and Robert J. Oppenheimer are the fathers of "modern" theoretical chemistry would not be considered by many inaccurate. The Born–Oppenheimer approximation simplifies the treatment of a group of point-like nuclei and electrons by replacing the degrees of freedom of the latter set of particles with the adiabatic force field that their (approximately instantaneous) ground state creates. The resulting potential energy surface U is a function of the position of all of the nuclei and its gradient is the adiabatic force vector, the components being the forces that govern the dynamic behaviors, and shape the quantum states of the nuclei.

Much of modern theoretical and computational chemistry research deals with finding increasingly accurate analytical models for U. This is no trivial task, research in this field is ongoing, mostly involving electronic structure theory combined with artificial intelligence tools, and more recently, quantum computing. In this book, we assume we have a reasonably accurate Born–Oppenheimer potential energy surface, and we concern ourselves with the intricacies of applying the laws of physics to atoms and molecules at a very basic level. Assuming the adiabatic approximation is sufficiently accurate, one may begin to explore properties of chemical systems by using Molecular Dynamics (MD) [1], or the Metropolis algorithm [2, 3]. That is only the beginning of the story. The nuclei of arguably the most important elements, in the first two or three rows of the periodic table, are too light to be accurately described by the laws of classical physics. Quantum physics has to be used in its place.

The last two chapters of this textbook contain the results of a relatively long and deep exploratory dive into nonrelativistic gauge theory. The aim was to gain insight into a long standing problem in computational chemical physics. Specifically, we wanted to determine if the principles of gauge theory could provide a rigorously convergent and physically meaningful strategy to construct excited state wave functions from a computed ground state. Briefly, gauge theory is the underlying framework that rigorously develops the mathematical model for all forces of nature by imposing symmetry constraints and using functional calculus to find the equations of motion for all the particles involved, barions and carrier bosons alike. The standard model of physics [4–11], the most advanced and most successful theory known to mankind, is built by gauge and quantum field theories [12–14] combined. But gauge theory is a relativistic quantum scattering theory, one where a conservative system of particles and their autogenous fields interact with one another in a manner consistent with the principle of least action. The interactions themselves are described by mathematical objects that remain invariant under certain types of symmetry transformations. The resulting laws of motion are by nature inhomogeneous, and consequently, even harder to solve than the equation Schrödinger wrote over a century ago, and that we are still trying to solve today.

The Diffusion Monte Carlo (DMC) algorithm [15–54] is one of a handful of stochastic methods routinely used to estimate ground state energies and wave functions of both electronic and molecular systems that satisfy the Schrödinger equation. Unlike vector

https://doi.org/10.1515/9783111610207-202

space methods, the computational effort of most stochastic approaches scales the same way as the computation of U. The latter has to be polynomial time at worst in practice, and this is typically achieved by fitting electronic structure theory energies to neural networks [48]. This process allows our community to compute ground state properties of relatively complex systems.

Finding accurate excited state wave functions by stochastic methods, however, is a long standing challenge. It turns out that convergent methods can be found without using gauge theory, but none of them scale with U, and more importantly all of them are based on assumptions about the functional form of the wave function. The functional forms are based on common sense assumptions, experience, and mathematical theorems applied to simpler models. However, none of them seem to be directly motivated by some physical principle. In the last two chapters, we show how two principles, that is, the orthogonality of states, and the presence of external fields suggest specific functional forms.

What sparked our interest in gauge theory was the observation that left to its own devices, the random walk in DMC will always find the exact ground state, if all goes well that is [32]. After observing this phenomenon, and exploiting it for the advancement of knowledge for decades, one cannot help to question if perhaps this result is trying to tell us something more profound; that there is a deeper explanation than just mathematical or computational difficulties. The chief example of such difficulties is the sign problem. Namely, it is not possible to sample a nonpositive distribution, but excited states have nodes, places where the wave function changes sign. The phenomenon that DMC faithfully reproduces could be the result of the fact that the ground state *is the only* physical solution of the "interaction-free" Schrödinger equation.

In such a view, the excited states are not elements of reality, and are only created into existence by the very interaction that forces transitions into them. That is, the excited states are created by the interactions. This may at first seem an unorthodox view, but it is perfectly consistent with all experimental evidence, and such is the interpretation in scattering theory. A collision between a force carrier and a particle creates, by conservation of momentum, an "excited," that is, a more energetic state of the particle. The widely accepted interpretation of the behavior of the primitive DMC random walk is that it is the implementation of a projection operator that causes all exited states to decay exponentially as the random walk evolves.

We demonstrate how the projection operator works to distill a ground state out of a starting mixture in Chapter 5. However, in order to do so, one has to assume that excited states of the radiation-free Schrödinger equation are known and, therefore, they must exist, or be "real." The fact that, for a few simple systems, we can mathematically express or compute the correct excited states for bound systems is completely irrelevant and cannot be used to prove them as elements of reality: One often computes properties that need not be elements of reality. Neither the projection operator explanation, nor such simple expressions offer a solid counterargument. Theoreticians frequently use objects that are not elements of reality, but that are extremely useful nonetheless.

Some examples are the Lagrangian functionals of classical and quantum mechanics. Lagrangians cannot be elements of reality because they are not unique. One can derive the same laws of motion from different Lagrangians. This fact is precisely what is exploited in gauge theory. The vector potential of electrodynamics is, of course, another important example and the very place where gauge theory originated from.

As further evidence of our claim, Maxwell's equations take a different form in "free space," become separable, and can be solved easily. However, when charge densities and currents are present they take on a more complex form. By analogy, the Schrödinger equation must not be the "globally correct" differential equation for excited sates. It just represents a radiation-free version of a more complex reality, where the only physically meaningful solution *is* the ground state. Evidence corroborating this view follows from the fact that DMC has been used to simulate (in some cases exactly [35]) excited states for decades. However, to achieve this result, one has to employ a strategy known as importance sampling. The modified random walk does not solve the Schrödinger equation, it solves an approximate Smoluchowski differential equation. The "approximate" label comes from assuming some unknown form of the excited state one wants to sample. We might rephrase our question as follows then. What is the "exact" Smoluchowski differential equation and can gauge theory suggest such a thing?

Theoretical chemists do not typically engage in the discovery of new laws of physics. This is not our present intention, nor is it an objective of this textbook to train the reader in such endeavors. However, it is worth to briefly mention that the discussion about the radiation-free Schrödinger equation is valid only for spinless particles. That is, the DMC finds the correct ground state for a system of bosons; these are typically spin zero nuclei in chemical applications. For electrons, the DMC random walk, left to its own devises, and here we mean without importance sampling, *does not* find the correct ground state. It follows logically that the Schrödinger equation must not be the correct nonrelativistic limit for fermionic systems. While we do not pursue this idea in the present textbook, finding such limiting differential equation could finally explain some of the most bizarre behavior of fermions like the Pauli exclusion principle, for example.

Returning to our far less lofty goals, if we want the excited states for bosonic systems, then we must in some localized spacetime volume include the electromagnetic radiation. Additional clues that gauge theory could be insightful come from an interesting and well-known result we rediscovered while pursuing unrelated objectives years ago. Implementing DMC, and other approaches like the imaginary time Path Integral (PI), or the more recently developed Ring Polymer Molecular Dynamics (RPMD) is straightforward when Cartesian coordinates are used. Things are more complicated when alternative coordinate systems [55–80] or atlases [49, 62] have to be used. Our research group has been involved in extending stochastic methods to systems whose configuration space must be mapped with either curvilinear or curved spaces [70, 71], and even spaces with torsion [40] for reasons that will become clear when the vast subject of coordinate changes is visited in Chapters 8 through 10. To restore the quadratic convergence of DMC when simulations take place in curved spaces [29], we had to explore computational methods used to solve stochastic differential equations [81]. When transforming to nonlinear coordinates and performing DMC simulations, one derives drift-like terms that are part of the Laplace–Beltrami operator. At that time, we realized that these drift-like terms become singular in certain regions of configuration space. In particular, we noted that the chart boundaries of the 2-sphere can be used to create the node structure of all the states of the particle in a box, while just sampling its ground state [30].

The intriguing feature of gauge theory is that it blurs the line between dependent (or field) and independent coordinate transformations. Both types of transformations are accompanied by a redefinition of the derivative. The terms added to the regular derivative to enforce covariance create the advection or drift-like terms that become singular at the nodes. Could it be that a DMC random walk can "naturally" lend in an excited state when the fields that create the transition to such states are present in the first place? We soon discovered that local transformations of the dependent variables of the Schrödinger–Lagrangian do not result in singular drift terms, and that non- Abelian versions of the theory has to be explored. The proof of these results are found in Chapter 16. In the same chapter, we show how a scalar theory can produce a Smoluchowski propagator, and we derive a nonscalar version of the Green propagator in imaginary time. In Chapter 17, we consider the simplest possible case: the DMC simulation of the interaction between a dipole and a static electric field in the weak limit.

So far, the quest to find a numerical method that can simulate simultaneously the electromagnetic radiation and the Schrödinger equation in imaginary time and produces a combination of the ground and a single excited state has been unsuccessful. Only more complicated mixtures are produced. These can be untangled by methods discussed in detail in Chapter 17 in a convergent manner when the matrix elements evaluated over the random walk are sufficiently accurate. Briefly, the static Stark effect in the long wavelength, weak field limit, combined with the singular Sturm–Liouville theorem tells us that the excited states are of the form of orthogonal polynomials constructed over the ground state probability amplitude as their Borel measure. In Chapter 17, we briefly discuss the possibility and the difficulties of simulating a time dependent electric field-dipole interaction by means of the propagator derived in Chapter 16.

The methods discussed in Chapter 17 can be obtained, just as well, without the machinery of nonrelativistic gauge theory. Moreover, approaches that exploit orthogonal polynomials constructed over the ground state probability amplitudes can be found in the literature [42, 82]. In Chapters 16 and 17, we show that this idea is not merely an elegant vector space result, but that it is the direct outcome of modeling a static electric field-dipole interaction in the weak field regime. Moreover, we believe that the questions we pose here will lead to more insightful results in the future. Our group is currently investigating determinantal approaches that automatically construct orthogonal polynomials from a given Borel measure in many dimensions, and is currently developing strategies to extend these approaches to curved spaces and manifolds.

The purpose for this textbook

We wrote this book for a far more important reason than just reporting on the results of Chapters 16 and 17. We believe that the collection of our notes, proofs, theorems, and algorithms developed during the investigation could be valuable education tools. Therefore, this book presents the steps taken in our exploration in a detailed and cohesive way. It culminates with the main result in Chapter 17. In the exercises at the end, the reader can experience firsthand how the static Stark effect combined with the theoretical and computational niceties mentioned earlier, and sprinkled throughout, yields a convergent representation of excited states. This representation is computable at every step of a DMC walk while sampling the ground state, and it is closely related to the GSPA [42] and VAMPIR [82] methods introduced in Chapter 6. Since we had to begin with the most fundamental mathematical and physical concepts, and since we had developed detailed proofs using the most elementary strategies, the material contained herein can be used to deliver a solid graduate level course in applied quantum theory.

Alternatively, by properly selecting certain chapters it can be used as a one semester advanced physical chemistry course for undergraduate students that have taken one or two semesters of physical chemistry and/or have been introduced to the fundamentals of quantum theory. We are assuming the reader is familiar with multidimensional calculus. Many of the proofs use linear algebra concepts, and Chapters 1 and 2 begin with a solid review, while at the same time, building some of the essential tools needed to explore gauge theory implementations. Orthogonal polynomials in one dimension built over a ground state amplitude are discussed in Chapter 2 after proving a special case of the Sturm–Liouville theorem. Moreover, this book develops some introductory computation tools and is written without assuming any coding background. Python codes are provided and clear instructions on how to modify them are given in the exercises.

We have attempted to balance the theoretical with the computational aspects of quantum theory with the intent to maximize the learning efficiency of a subject filled with abstract ideas. A multitude of powerful techniques that the community has developed over the past decades to unravel chemical physics puzzles are introduced. Among them are the Diffusion Monte Carlo (DMC) in Chapter 5 and its importance sampling versions in Chapter 6. Associated with DMC is a method recently developed to compute excited states of molecular and ionic aggregates called Ground State Probability Amplitude (GSPA). Additionally, the Discrete Variable Representation (DVR) [83–89], the imaginary time Path Integral, and the Ring Polymer Molecular Dynamics (RPMD) are discussed in detail.

The entire manuscript consists of six parts. The first part begins with a robust introduction of vector fields and their applications to quantum mechanics. The computation of matrix elements of the Discrete Variable Representation is a straightforward exercise once the reader has understood the basic principles of vector spaces and subspace convergence. Computational nonrelativistic gauge theory requires a working knowledge of

Group Theory and Lie algebra, the time dependent Schrödinger equation (TDSE), tensor calculus, and Hamilton's principle, and these subjects are introduced in parts two through five. In these parts, DMC and RPMD are introduced. DMC, in particular, is the tool we have selected to carry out the computations in the final part of the book. A solid one semester advanced undergraduate course can be built with the material in Chapters 1 through 4 and 8 through 10, perhaps with the inclusion of one or more exercises from Chapter 5 as a special project. Two important sampling strategies are introduced in Chapter 6, while the GSPA and the VAMPIR methods are discussed in Chapter 7.

Because the book is conceived to be a teaching tool, each chapter contains a number of exercises aimed at providing the necessary practice needed to achieve a basic understanding of each subject. Of course, a single book that contains a number of vast areas of modern mathematics and physics cannot possibly do justice to any of them. For those readers that wish to master these topics, citations to reasonably approachable textbooks are included. For example, more on orthogonal polynomials can be found here [90, 91], whereas vector spaces are treated very well in a number of quantum mechanics texts [92–94]. For group theory, the book by Wigner [95] is a classic, whereas the treatment of the subject in the book by Gallian [96] is very approachable as it was written for undergraduate students in mathematics. Lie algebras are covered in detail by Gilmore [97], whereas tensor calculus is introduced in this textbook by Neuenschwander [98] who also authored this tome [99] where Hamilton's principle, invariance, Noether's theorem, and gauge theory are carefully introduced. The quantum theory of light by R. Loudon [100] is an authoritative textbook on the subject in its title. The best textbooks to introduce the Feynman path integrals are those of Feynmann and Hibbs [101] and Hagen Kleinert [102].

As mentioned earlier, besides building a basic understanding of the theoretical principles, the other intended outcome of the training contained in this book is to provide some experience with computation. Many of the exercises require a moderate amount of python coding with the `numpy` class. These exercises are all designed to run on a personal computer or laptop in a matter of minutes at most. For pedagogical reasons, most exercises focus on simple monodimensional systems, where it is always possible to compute the quantum properties by vector space approaches. The best strategy to teach programming principles, in our experience, is to provide codes to the student and begin by teaching the proper way to test and debug them. The codes provided in the main body of the chapters and exercises have been thoroughly tested and should be "bug-free," but the readers are better served by maintaining a healthy amount of skepticism. The students should be guided to test each computed line by verifying the numerical results with some other mean, such as a spreadsheet application, or in some cases, by writing separate programs.

Computation tools like python are freely available and can accelerate the learning process and deepen the understanding of the theoretical principles without requiring a lot of time spent on tedious derivations. For example, after introducing eigenanalysis in Chapter 1 and providing some old school paper and pen exercises, the code to diagonalize a symmetric matrix of arbitrary size is provided, so that the ideas of diagonal matrices, null spaces, orthogonal matrices, secular determinants, and similarity transformations can be quickly explored and put into practice. Having said all this about the powerful (but not omnipotent) computational tools available today, the student should not rely on computation alone to understand any subject. For this reason, we have attempted to create a good balance of theoretical and computational exercises to achieve the desired outcomes.

Contents

1 Vector spaces

> I should therefore be permitted to draw your attention to this new formula, which I consider to be the simplest possible, apart from Wien's expression, from the point of view of the electromagnetic theory of radiation.
>
> M. Planck

1.1 Some basic definitions

The subjects of this chapter are used heavily in computational chemistry mainly to achieve two outcomes: to find solutions of differential equations and to fit functions. The most frequently solved differential equation is, of course, the time independent Schrödinger equation. It is not uncommon though to find both uses of vector spaces to answers typical research questions. The most commonly fitted functions are *potential energies*, U and *wave-functions* ψ. Therefore, vector spaces are the nuts and bolts of computational and theoretical chemistry research, mainly because for atomic and molecular physics, U is still considered the "holy grail," and in this book, the ground state wave-function ψ_0 plays just as central a role. A strong foundation in the subject of vector spaces is critical for the practitioner and the student alike; therefore, though much of the foundation is perhaps familiar to the average reader, it is a good idea to introduce the subject from scratch and follow a rigorous path rooted not only in the subject of linear algebra but modern abstract algebra as well. Modern algebra concepts are needed to leave what are familiar grounds like the types of vector spaces one finds in a typical linear algebra course, and soar toward other types of vector spaces that helps us to interpret, represent, translate into code, and compute objects like

$$e^{\frac{d}{dx}} f(x),$$

which are central to the theory and algorithms developed in this book. The first useful concept for which we need an operational definition is a **field** $\mathcal{F}$. There are fields of numbers the reader should be familiar with already. The field of real and complex numbers $\mathbb{R}$ and $\mathbb{C}$ are examples that provide a good intuition. The field of integers is denoted with $\mathbb{Z}$.

Definition 1 (Field). A field $\mathcal{F}$ is a set of numbers and two **binary operations**, multiplication and addition, which satisfies the following axioms. Given elements $a, b, c \in \mathcal{F}$:

1. Addition is commutative and associative:

$$a + b = b + a$$
$$(a + b) + c = a + (b + c).$$

2. There is an additive identity (the number 0) such that $a + 0 = a$.

https://doi.org/10.1515/9783111610207-001

3. Every number in $\mathcal{F}$ has an additive inverse $-a + a = 0$.
4. Both binary operations satisfy closure, namely

$$a + b = c \in \mathcal{F} \quad a \times b = d \in \mathcal{F}.$$

5. Multiplication is associative $(a \times b) \times c = a \times (b \times c)$.
6. There is a multiplicative identity 1 such that $1 \times a = a \times 1 = a$.
7. Every element other than zero has a multiplicative inverse, $a^{-1}a = 1$.

By a binary operation, we mean two members in $\mathcal{F}$ are needed as input for the operation. The multiplicative inverse property is key to the usefulness of the concept, because it allows for solutions of equations like $ax = b$ over $\mathcal{F}$ by the so-called cancellation property, namely we can multiply by the inverse of a on both sides to compute the value of x that satisfies this equation, $a^{-1} \times ax = a^{-1} \times b$, so $x = a^{-1} \times b$.

The set of all positive integers, also known as the set of natural numbers, is a useful construct and is denoted with $\mathbb{N}$. However, $\mathbb{N}$ is not a field. Not all fields important in physics and in chemical physics are endowed with a commutative multiplication. For example, over the field of quaternions $\mathbb{Q}$, often used to represent rotation degrees of freedom, multiplication does not satisfy the commutativity property. Namely, for $q_1, q_2 \in \mathbb{Q}$, $q_1 \times q_2$ is generally not equal to $q_2 \times q_1$. By satisfying closure, we can think of the two binary operations as **maps** from $\mathcal{F}$ to $\mathcal{F}$. Often these are expressed using the $\times : \mathcal{F} \to \mathcal{F}$ notation. The closure property for both operations is important for the researcher to ensure that the result is computable and the physical interpretations are valid. For instance, it is well known that $\mathbb{R}$ is not algebraically closed, meaning the solution of equations like $x^2 + 1 = 0$ cannot be computed with $\mathbb{R}$. The algebraist's trick to solve these types of problems is to develop field extensions. Thus, the field of complex numbers $\mathbb{C}$ is born, so that we may interpret expressions like $x = \pm\sqrt{-1}$. Note that introducing a division and subtraction operation is not necessary. These are included in the existence of additive and multiplicative inverses. The vast majority of the content of this book will make use of vector spaces built over the field of reals $\mathbb{R}$, but for the present purposes, any of the fields discussed so far works, and will continue to use $\mathcal{F}$ until we get to some computations later in the chapter, where specificity becomes necessary.

We now introduce vector spaces, and we shall begin with some familiar examples. E. g., $\mathbb{R}^3$ is a vector space that consists of triples and can be represented using the (a, b, c) notation, where $a, b, c \in \mathbb{R}$.

Definition 2 (Vector space). A vector space is a set $\mathcal{V}$ over a field $\mathcal{F}$ together with two binary operations:

1. Addition $(+)$, let $v_1, v_2 \in \mathcal{V}$.

$$v_1 + v_2 = \{a_1, a_2, \ldots, a_n\} + \{b_1, b_2, \ldots, b_n\} = \{(a_1 + b_1), \ldots, (a_n + b_n)\},$$
$$a_1, a_2, \ldots, a_n, b_1, b_2, \ldots, b_n \in \mathcal{F}.$$

2. Multiplication by a scalar, let $v_1 \in \mathcal{V}, c \in \mathcal{F}$,

$$c \times v_1 = c \times \{a_1, a_2, \ldots, a_n\} = \{cv_1, cv_2, \ldots, cv_n\}.$$

The operations must satisfy the following axioms: Let $v_1 v_2, v_3 \in \mathcal{V}, c, d \in \mathcal{F}$.

1. Closure under addition and scalar multiplication:

$$v_1 + v_2 \in \mathcal{V},$$
$$c \times v_1 = cv_1 \in \mathcal{V}.$$

2. Vector addition is commutative:

$$v_1 + v_2 = v_2 + v_1.$$

3. Associativity of the two operations:

$$(v_1 + v_2) + v_3 = v_1 + (v_2 + v_3),$$
$$c(dv_1) = (cd)v_1.$$

4. The scalar multiplication is distributive:

$$c(v_1 + v_2) = cv_1 + cv_2.$$

5. There exist additive and scalar multiplicative identities:

$$0 + v_1 = v_1,$$
$$1 \times v_1 = v_1,$$

   where 0 is the zero vector, a vector with all n components equal to zero.
6. All elements of $\mathcal{V}$ must have an additive inverse, and all elements of $\mathcal{F}$ a multiplicative inverse other than 0:

$$v_1 - v_1 = 0,$$
$$c^{-1}cv_1 = v_1.$$

The number of components of the set $\{a_1, a_2, \ldots, a_n \in \mathcal{F}\}$ is called the **dimension** of $\mathcal{V}$. The sets $\{a_1, a_2, \ldots, a_n \in \mathcal{F}\}$ contains the components of the vectors in the vector space $\mathcal{V}$. $\mathcal{V}$ is the set of all possible sets of components; therefore, if $\mathcal{F}$ has an infinite number of elements as in $\mathbb{R}$, $\mathcal{V}$ is infinite even if its dimension is finite. Vector spaces are not quite fields; though a definition for the multiplication operation is given, it is not possible to use cancellation because the multiplicative inverse of a vector is not defined. For typical chemical physics research problems, one often deals with N particles and the

configuration space is the set $\{x_i, y_i, z_i \in R \mid i = 1, \ldots, N\}$ of $3N$ coordinates. The symol $\mathbb{R}^{3N}$ represents the space of all possible $3N$-tuples.

At the heart of the theory of vector spaces, there are two concepts, that of linear independence, and the basis set of a vector space.

Definition 3 (Linear independence). A set of vectors $\{v_1, v_2, \ldots, v_n\}$ is **linearly independent** over the field $\mathcal{F}$ if and only if, given $\{a_1, a_2, \ldots, a_n \in \mathcal{F}\}$,

$$a_1 v_1 + a_2 v_2 + \cdots + a_n v_n = 0$$

holds only when all elements of $\{a_1, a_2, \ldots, a_n \in \mathcal{F}\}$ are each separately equal to zero, $a_1 = 0, a_2 = 0, \ldots, a_n = 0$.

For example, in $\mathbb{R}^3$ the vectors $(1,0,0)$, $(0,1,0)$, and $(0,0,1)$ are clearly linearly independent since the only way that

$$a_1(1,0,0) + a_2(0,1,0) + a_3(0,0,1)$$

can equal the zero vector $(0,0,0)$ is if all three scalars a_1, a_2, a_3 are separately equal to zero.

Definition 4 (Basis set). Let $\mathcal{V}$ be a vector space over a field $\mathcal{F}$, and let $B = \{b_1, b_2, \ldots, b_n\}$ be a set of linearly independent vectors. $\mathcal{V}$ is said to span the set B if every element of $\mathcal{V}$ is a linear combination of the elements of B, and the set B is called a **basis set** of $\mathcal{V}$.

A basis set allows us to compute all elements of $\mathcal{V}$ as linear superpositions of the elements of B using variables of the type $\mathcal{F}$. A basis set B of a vector space $\mathcal{V}$ along with the components is a **representation** of an element of $\mathcal{V}$. The set $\{(1,0,0), (0,1,0), (0,0,1)\}$ is a basis set for $\mathbb{R}^3$ and the elements of this set can be represented using column or row vectors, or with the commonly used symbols **i**, **j**, **k**. All vectors in $\mathbb{R}^3$ can be represented as linear combinations of **i**, **j**, **k**. Note that basis sets are not unique, meaning a vector space $\mathcal{V}$ may have many basis sets, which determine it. Choosing a different basis set may create seemingly different objects. However, by carefully inspecting the behavior of the elements under the two operations will always reveal **isomorphism** between the representations of the same abstract object. For example, we can represent elements in $\mathbb{R}^3$ using

$$v = v_1\mathbf{i} + v_2\mathbf{j} + v_3\mathbf{k}$$

or we can use

$$v = v_1\begin{pmatrix}1\\0\\0\end{pmatrix} + v_2\begin{pmatrix}0\\1\\0\end{pmatrix} + v_3\begin{pmatrix}0\\0\\1\end{pmatrix} = \begin{pmatrix}v_1\\v_2\\v_3\end{pmatrix}.$$

We leave it to the reader as an exercise to show that these two representations are isomorphic. A vector space is often represented as the **span** of one of its basis set, for example,

$$\mathbb{R}^3 = \text{span}\left\{ \begin{pmatrix} 1 \\ 0 \\ 0 \end{pmatrix}, \begin{pmatrix} 0 \\ 1 \\ 0 \end{pmatrix}, \begin{pmatrix} 0 \\ 0 \\ 1 \end{pmatrix} \right\}.$$

1.2 Other examples of vector spaces

There are many sets that can be endowed with carefully defined sum and product by scalars. Therefore, there are many types of vector spaces one can build. For example, the set of all 2×2 skew symmetric matrices over $\mathbb{R}$ has these elements

$$\begin{pmatrix} a & b \\ -b & c \end{pmatrix} \quad a, b, c \in \mathbb{R}.$$

The sum in this vector space is defined as follows. Let $a_1, b_1, c_1, a_2, b_2, c_2 \in \mathbb{R}$,

$$\begin{pmatrix} a_1 & b_1 \\ -b_1 & c_1 \end{pmatrix} + \begin{pmatrix} a_2 & b_2 \\ -b_2 & c_2 \end{pmatrix} = \begin{pmatrix} a_1 + a_2 & b_1 + b_2 \\ -b_1 - b_2 & c_1 + c_2 \end{pmatrix}$$

and the multiplication by scalar is

$$d \begin{pmatrix} a & b \\ -b & c \end{pmatrix} = \begin{pmatrix} ad & bd \\ -bd & cd \end{pmatrix} \quad a, b, c, d \in \mathbb{R}.$$

A valid basis set for this vector space is

$$\left\{ \begin{pmatrix} 1 & 0 \\ 0 & 0 \end{pmatrix}, \begin{pmatrix} 0 & 1 \\ -1 & 0 \end{pmatrix}, \begin{pmatrix} 0 & 0 \\ 0 & 1 \end{pmatrix} \right\}.$$

When the components of $\mathbb{R}^n$ are elevated to functions of the three variables $x, y, z, \ldots \in \mathbb{R}$, the resulting object is a space of vector-valued functions. In physics, these objects are referred to as vector fields or force fields. Moreover, the symbol for a n-dimensional space of vector-valued functions is still $\mathbb{R}^n$, however, there is no ambiguity in such notation because the defining operations are not changed, and one can think of a constant component as a specific type of function. In spaces of vector-valued functions, one can perform calculus operations and the resulting analysis is very important in chemical physics. As an example, consider a particle in three-dimensional space subject to a potential energy $U(x, y, z)$. The **gradient** of U is an element of $\mathbb{R}^3$,

$$\nabla U = \frac{\partial}{\partial x} U\,\mathbf{i} + \frac{\partial}{\partial y} U\,\mathbf{j} + \frac{\partial}{\partial z} U\,\mathbf{k}.$$

The three partial derivatives in the gradient of U are the additive inverses of the respective components of the force vector acting on the particle located at position x, y, z. For N particles, the potential energy is a function of $3N$ variables and its gradient a member of $\mathbb{R}^{3N}$. Therefore, the component functions have a clear interpretation in this context and it is rare for them to be constants.

Another useful generalization of the vector space concept is achieved when basis sets are functions. For example, consider the following vector space over $\mathbb{R}$:

$$\mathcal{V} = \text{span}\{e^{-x}, xe^{-x}\} \quad x \in [0, \infty).$$

Here, $\mathcal{V}$ is the space of all functions that are linear combinations of the type $ae^{-x} + bxe^{-x}$, $a, b \in \mathbb{R}$. The vector sum is defined with the following operation. Let $a_1, a_2, b_1, b_2 \in \mathbb{R}$,

$$(a_1 e^{-x} + b_1 xe^{-x}) + (a_2 e^{-x} + b_2 xe^{-x}) = (a_1 + a_2)e^{-x} + (b_1 + b_2)xe^{-x}$$

whereas the multiplication by scalar is

$$c(ae^{-x} + bxe^{-x}) = cae^{-x} + cbxe^{-x}.$$

The two functions are linearly independent and this can be shown by constructing the **Wronskian** of the set. In general, for n distinct $\mathbb{R} \to \mathbb{R}$ functions $\{f_1, f_2, \ldots, f_n\}$ that are n-differentiable, the Wronskian matrix is constructed by the following procedure:

$$W\{f_1, f_2, \ldots, f_n\} = \det \begin{pmatrix} f_1 & f_2 & \cdots & f_n \\ f_1^1 & f_2^1 & \cdots & f_n^1 \\ \vdots & \vdots & & \vdots \\ f_1^n & f_2^n & \cdots & f_n^n \end{pmatrix}$$

where $f^n = d^n f / dx^n$ is the nth derivative of f. Let us briefly review how the determinant of a $n \times n$ is evaluated by mean of the cofactor expansion. The determinant of the $(n-1) \times (n-1)$ matrix formed by omitting row i and column j of the $n \times n$ matrix is denoted by M_{ij} and is called the **minor** of the a_{ij} element. The product $(-1)^{i+j} M_{ij}$ is called the **cofactor** of a_{ij}. Then expanding by row i, one computes

$$\det \mathbf{A} = \sum_{j=1}^{n} (-1)^{ij} \, a_{ij} M_{ij}$$

for a 1×1 matrix the determinant is simply its element, for a 2×2 expanding by row 1,

$$\det \begin{pmatrix} a_{11} & a_{12} \\ a_{21} & a_{22} \end{pmatrix} = a_{11} M_{11} + a_{12} M_{12} = a_{11} a_{22} - a_{12} a_{21}.$$

For our example,

$$W\{e^{-x}, xe^{-x}\} = \begin{pmatrix} 1 & x \\ -1 & 1-x \end{pmatrix} e^{-x}$$

and det $W\{e^{-x}, xe^{-x}\} = e^{-2x}$. Wrońki's theorem states that a set of functions are linearly independent in some interval of $\mathbb{R}$ if their Wronskian does not vanish in the interval. In our example, the determinant does not vanish for any value of $x \in [0, \infty)$. Therefore, the two functions are linearly independent over the interval. Vector spaces constructed using linearly independent functions as a basis set are of fundamental importance in the theory of differential equations.

1.3 Some key properties of the monomial set

Now we come to the first key result. The theorem we prove here is important for the construction and interpretation the nonrelativistic gauge theory. Specifically, we want to prove that the set of monomials $\{1, x, x^2, \ldots, x^n\}$ are linearly independent. This result is rarely mentioned in mathematics and mathematical physics books. Its proof is a great example of how the principle of **mathematical induction** is used. Before we can dive into the theorem itself, we need two intermediate results.

Theorem 1 (n^{th} derivative of x^n). *The following holds for the nth derivative of x^n:*

$$\frac{d^n}{dx^n} x^n = n!$$

Proof. The statement is obviously true for $n = 0$ and $n = 1$. The induction hypothesis is to assume the theorem holds for k. The induction step is to demonstrate that the statement is true for $k + 1$. Then the theorem is true for all natural numbers $\mathbb{N}$. So,

$$\frac{d^{k+1}}{dx^{k+1}} x^{k+1} = \frac{d^k}{dx^k} \left(\frac{d}{dx} x^{k+1} \right)$$

gives

$$\frac{d^{k+1}}{dx^{k+1}} x^{k+1} = (k+1) \frac{d^k}{dx^k} (x^k) = (k+1)k! = (k+1)!$$

where the second term follows from the derivative of x^{k+1}, the third by the induction step, and the fourth by the definition of the factorial operation. □

The second intermediate result has to do with the determinant of a specific type of matrix.

Theorem 2 (The determinant of an upper triangular $n \times n$ matrix). *The determinant of an upper triangular $n \times n$ matrix is the product of all its diagonal elements:*

$$\det\begin{pmatrix} a_{11} & a_{12} & \cdots & a_{1,n} \\ 0 & a_{22} & \cdots & a_{2,n} \\ \vdots & \vdots & & \vdots \\ 0 & 0 & 0 & a_{n,n} \end{pmatrix} = \prod_j^n a_{j,j}.$$

Proof. The $n = 1$ case is obvious. Let the statement hold for a $k \times k$ upper triangular matrix. Then consider a $(k+1) \times (k+1)$ matrix. The obvious choice is to expand the determinant by the last row because all its entries are zero except for $a_{k+1,k+1}$. The induction step is

$$\det\begin{pmatrix} a_{11} & a_{12} & \cdots & a_{1,k} & a_{1,k+1} \\ 0 & a_{22} & \cdots & a_{2,k} & a_{2,k+1} \\ \vdots & \vdots & \cdots & \vdots & \vdots \\ 0 & 0 & \cdots & a_{kk} & a_{k,k+1} \\ 0 & 0 & \cdots & 0 & a_{k+1,k+1} \end{pmatrix}$$
$$= (-1)^{2k+2}\, a_{k+1,k+1} \det\begin{pmatrix} a_{11} & a_{12} & \cdots & a_{1,k} \\ 0 & a_{22} & \cdots & a_{2,k} \\ \vdots & \vdots & \cdots & \vdots \\ 0 & 0 & \cdots & a_{k,k} \end{pmatrix},$$

and by the induction hypothesis, we have

$$\det\begin{pmatrix} a_{11} & a_{12} & \cdots & a_{1,k} & a_{1,k+1} \\ 0 & a_{22} & \cdots & a_{2,k} & a_{2,k+1} \\ \vdots & \vdots & \cdots & \vdots & \vdots \\ 0 & 0 & \cdots & a_{kk} & a_{k,k+1} \\ 0 & 0 & \cdots & 0 & a_{k+1,k+1} \end{pmatrix} = a_{k+1,k+1} \prod_j^k a_{j,j} = \prod_j^{k+1} a_{j,j}. \qquad \square$$

With the last two results, we can prove the following statement.

Theorem 3 (Linear independence of the monomials). *The set of monomials* $\{1, x, x^2, \ldots, x^n\}$ *with* $x \in \mathcal{F}$ *forms a linearly independent basis set for all polynomials up to degree* n *in* $\mathcal{F}$.

Proof. The Wronskian matrix $W\{1, x, x^2, \ldots, x^n\}$ is upper triangular since in the second row $a_{21} = 0$, $a_{22} = 1$, etc., and by Theorem 1 we get

$$W\{1, x, x^2, \ldots, x^n\} = \begin{pmatrix} 1 & x & x^2 & \cdots & x^n \\ 0 & 1 & 2x & \cdots & nx^{n-1} \\ 0 & 0 & 2 & \cdots & n(n-1)x^{n-2} \\ \vdots & \vdots & & \cdots & \vdots \\ 0 & 0 & 0 & \cdots & n! \end{pmatrix}.$$

By Theorem 2, it follows that

$$\det \begin{pmatrix} 1 & x & x^2 & \cdots & x^n \\ 0 & 1 & 2x & \cdots & nx^{n-1} \\ 0 & 0 & 2 & \cdots & n(n-1)x^{n-2} \\ \vdots & \vdots & & \cdots & \vdots \\ 0 & 0 & 0 & \cdots & n! \end{pmatrix} = \prod_j^n j!$$

The determinant is not zero for all values of x. Therefore, by Wroński's theorem the set of monomials is a linearly independent set. □

In light of this result, it makes sense to conjecture that the set

$$\{f(x), xf(x), x^2 f(x), \ldots, x^n f(x)\},$$

for some "well-behaved" function $f(x)$ is also a set of linearly independent vectors. That is indeed the case, but proving this statement by Wroński theorem is much more involved. A systematic way to develop viable vector spaces is to construct orthogonal vectors from an independent set. If the set is not linearly independent, the process fails. We will soon show one approach to orthogonalize sets of mutually independent vectors called the Gram–Schmidt procedure.

It is linear independence that allows one to form n equations when polynomial equalities, such as, e. g.,

$$ax^2 + bx + c = 15x^2 - 3$$

are solved for the coefficients a, b, and c. It is clear that for the example the solution is $a = 15$, $b = 0$, $c = -3$. In chemical physics applications, matrices can be quite large and the cofactor expansion becomes very inefficient. There are a number of ways to compute the determinant of a $n \times n$ matrix beyond $n = 3$. One uses similarity transformations to change the matrix into an upper or lower triangular, or diagonal, or turn it into a product of lower and upper triangular matrices such as, e. g., with the *LU* decomposition. Moreover, for many chemical physics applications matrices have a block diagonal structure. In that case, the determinant of the full matrix is the product of the determinants of its blocks.

1.4 The dot product

Before vector spaces can be implemented in the formulation of theories or algorithms, we have to build more structure. The dot product, also known as the inner product, or the scalar product is a binary map that takes two members of a vector space $a, b \in \mathcal{V}$ over $\mathcal{F}$, and produces a scalar $a \cdot b = c \in \mathcal{F}$. There are many ways the dot product is defined, depending on the field $\mathcal{F}$, the nature of the basis set, and on the choice of **metric**. Different choices of metric will produce different definitions of the dot product. The concept of metric and dot product are tightly linked and, therefore, they have to be discussed together. The best way to present them is by working through several examples of how the dot product has been defined. In $\mathbb{R}^3$, the dot product is the result of the following rules for the product over the basis set:

$$\begin{array}{cccc} \cdot & \mathbf{i} & \mathbf{j} & \mathbf{k} \\ \mathbf{i} & 1 & 0 & 0 \\ \mathbf{j} & 0 & 1 & 0 \\ \mathbf{k} & 0 & 0 & 1, \end{array}$$

which should be interpreted as $\mathbf{i} \cdot \mathbf{i} = 1$, $\mathbf{i} \cdot \mathbf{j} = \mathbf{i} \cdot \mathbf{k} = 0$, etc. With this table, it is simple to show that the dot product between two vectors a and b in $\mathbb{R}^3$ is

$$\begin{aligned} a \cdot b &= (a_x\mathbf{i} + a_y\mathbf{j} + a_z\mathbf{k}) \cdot (b_x\mathbf{i} + b_y\mathbf{j} + b_z\mathbf{k}) \\ &= a_x b_x + a_y b_y + a_z b_z. \end{aligned}$$

If the components of a and b are functions $\mathbb{R}^3 \to \mathbb{R}$, so will the dot product $a \cdot b$. There are geometric interpretations for the dot product in $\mathbb{R}^3$. For instance, the unit vectors **i**, **j**, **k** are mutually perpendicular. Moreover, any two vectors a and b whose dot product vanishes are mutually perpendicular, meaning the angle between the two straight lines through the origin and the points (a_x, a_y, a_z) and (b_x, b_y, b_z), respectively, is 90 degrees or $\pi/2$. The concept of two vectors being mutually perpendicular generalizes to the concept of orthogonality. Two vectors in a vector space $\mathcal{V}$ are mutually **orthogonal** if their dot product vanishes. With the dot product defined, one can compute the **norm** of a vector,

$$\|a\| = (a \cdot a)^{1/2}$$

and the angle between vectors a and b, θ_{ab} satisfies

$$\cos\theta_{ab} = \frac{a \cdot b}{\|a\|\,\|b\|}.$$

Vectors that are multiplied by the inverse of their norms are unit vectors (vectors of norm 1). By inspecting the dot product table for the set **i**, **j**, **k**, it is clear that **i**, **j**, **k** are unit vectors.

Next, let us investigate how the dot product over vector spaces is defined when the basis sets are functions. Using the previous example,

$$\mathcal{V} = \operatorname{span}\{e^{-x}, xe^{-x}\} \quad x \in [0, \infty),$$

over $\mathbb{R}$. The "dot product" between the two basis functions of this type $f_1(x), f_2(x)$ over an interval $I \in \mathbb{R}$ is typically defined as an integral

$$\langle f_1 \cdot f_2 \rangle = \int_I f_1 f_2 \, dx.$$

Then three separate cases need to be computed:

$$\langle e^{-x} \cdot e^{-x} \rangle = \int_0^\infty e^{-2x} \, dx = \frac{1}{2},$$

$$\langle e^{-x} \cdot xe^{-x} \rangle = \int_0^\infty xe^{-2x} \, dx = \frac{1}{4},$$

$$\langle xe^{-x} \cdot xe^{-x} \rangle = \int_0^\infty x^2 e^{-2x} \, dx = \frac{1}{4}.$$

It is clear from these results that the basis set for this example is not comprised of mutually orthogonal unit vectors.

Let us consider an example of importance in quantum theory, the vector space of mutually orthogonal polynomials in a finite interval. It is natural to choose the monomials as our basis set $\mathcal{V} = \text{span}\{1, x, x^2, \ldots, x^n\}$ $x \in [-1, 1]$, since we have shown that the elements are linearly independent. We now explore how the dot product works in this vector space. The dot product between two monomials is

$$\langle x^k \cdot x^l \rangle = \int_{-1}^{1} x^{k+l} \, dx = \frac{1}{k+l+1} [1 - (-1)^{k+l+1}]$$

where $1 - (-1)^{k+l+1}$ evaluates to zero whenever $k + l + 1$ is even and 2 when $k + l + 1$ is odd. A partial multiplication table is below:

$\cdot$	1	x	x^2	x^3
1	2	0	$\frac{2}{3}$	0
x	0	$\frac{2}{3}$	0	$\frac{2}{5}$
x^2	$\frac{2}{3}$	0	$\frac{2}{5}$	0
x^3	0	$\frac{2}{5}$	0	$\frac{2}{7}$.

The set of monomials is a nonorthogonal basis set, and in $[-1, 1]$ none are unit vectors, but we could create unit vectors by **normalizing** the set this way:

$$\{1, x, x^2, \ldots, x^n\} \to \left\{\frac{1}{2}, \frac{3}{2}x, \frac{5}{2}x^2, \ldots, \frac{2n+1}{2}x^n\right\}.$$

Normalization is, of course, important in quantum mechanics, but it is not critical that a basis is normalized at the outset. Frequently, the solutions of the Schrödinger equation are linear combinations of a given basis set that need to be renormalized even if it was normalized to begin with. A more critical property for a basis set is orthogonality. In fact, finding orthogonal basis sets is in some cases equivalent to finding the solutions of a differential equation. Therefore, it is now opportune to look at one approach to orthogonalization known as the Gram–Schmidt process.

1.5 The Gram–Schmidt process

We start by working through some examples. Consider the following three vectors in $\mathbb{R}^3$:

$$v_1 = \begin{pmatrix} -1 \\ 0 \\ 1 \end{pmatrix}, \quad v_2 = \begin{pmatrix} 2 \\ 1 \\ 4 \end{pmatrix}, \quad v_3 = \begin{pmatrix} 0 \\ 1 \\ 1 \end{pmatrix}$$

we want to build three unit vectors $u_1, u_2, u_3 \in \mathbb{R}^2$ that are mutually orthogonal from this set. The steps are as follows:

- Normalize v_1 to get u_1,

$$u_1 = \frac{v_1}{\|v_1\|} = \frac{1}{\sqrt{2}} \begin{pmatrix} -1 \\ 0 \\ 1 \end{pmatrix}.$$

- Find the dot product $u_1 \cdot v_2$ and use it to remove the projection of u_1 onto v_2 from v_2,

$$u_1 \cdot v_2 = \frac{2}{\sqrt{2}} \quad \widetilde{v}_2 = v_2 - (u_1 \cdot v_2)u_1$$

$$\widetilde{v}_2 = \begin{pmatrix} 2 \\ 1 \\ 4 \end{pmatrix} - \frac{2}{\sqrt{2}} \times \frac{1}{\sqrt{2}} \begin{pmatrix} -1 \\ 0 \\ 1 \end{pmatrix} = \begin{pmatrix} 3 \\ 1 \\ 3 \end{pmatrix}.$$

- Normalize $\widetilde{v}_2$ to find u_2,

$$u_2 = \frac{\widetilde{v}_2}{\|\widetilde{v}_2\|} = \frac{1}{\sqrt{19}} \begin{pmatrix} 3 \\ 1 \\ 3 \end{pmatrix}.$$

- Find the dot products $u_1 \cdot v_3$, $u_2 \cdot v_3$ and use them the remove from v_3 the projections of u_1 and u_2 on v_3,

$$u_1 \cdot v_3 = -\frac{2}{\sqrt{2}} \quad u_2 \cdot v_3 = -\frac{2}{\sqrt{19}},$$

$$\widetilde{v}_3 = v_3 - (u_1 \cdot v_3)u_1 - (u_2 \cdot v_3)u_2 = \frac{6}{19} \begin{pmatrix} 1 \\ -6 \\ 1 \end{pmatrix}.$$

Normalize $\tilde{v}_3$ to find u_3,

$$u_3 = \frac{1}{\sqrt{38}} \begin{pmatrix} 1 \\ -6 \\ 1 \end{pmatrix}.$$

In conclusion, starting with three vectors

$$v_1 = \begin{pmatrix} -1 \\ 0 \\ 1 \end{pmatrix}, \quad v_2 = \begin{pmatrix} 2 \\ 1 \\ 4 \end{pmatrix}, \quad v_3 = \begin{pmatrix} 0 \\ 1 \\ 1 \end{pmatrix}$$

we were able to construct from them three mutually orthogonal unit vectors,

$$u_1 = \frac{1}{\sqrt{2}} \begin{pmatrix} -1 \\ 0 \\ 1 \end{pmatrix}, \quad u_2 = \frac{1}{\sqrt{19}} \begin{pmatrix} 3 \\ 1 \\ 3 \end{pmatrix}, \quad u_3 = \frac{1}{\sqrt{38}} \begin{pmatrix} 1 \\ -6 \\ 1 \end{pmatrix}.$$

The procedure can be extended to any vector space for which a dot product has been defined.

When we apply the Gram–Schmidt procedure to sets of monomials over finite intervals in $\mathbb{R}$ or $\mathbb{C}$, we end up with orthogonal polynomials. In the next section, we redefine the dot product for vector spaces of functions over the $[0, \infty)$ and the $(-\infty, \infty)$ intervals.

1.6 Compact supported dot products

For many cases of interest, the interval I in $\mathbb{R}$ is not finite, and the integrals of a monomial of any power greater or equal to zero diverge. Therefore, the integral definition of the dot product has to be modified. This is done by introducing a weight function $\omega(x)$ into the definition that produces a convergent integral. The generic notation for the dot product of two monomials is $\langle x^k \cdot x^l \rangle_\omega$ and it is defined with the following integral:

$$\langle x^k \cdot x^l \rangle_\omega = \int_I \omega(x)\, x^{k+l}\, dx.$$

For the interval $[0, \infty)$, the typical weight function $\omega(x) = e^{-x}$ and the dot product between two monomials is

$$\langle x^k \cdot x^l \rangle_\omega = \int_0^\infty x^{k+l}\, e^{-x}\, dx = (k + l)!$$

whereas, for the $(-\infty, \infty)$ interval $\omega(x) = e^{-x^2/2}$ is used:

$$\langle x^k \cdot x^l \rangle_\omega = \int_{-\infty}^{\infty} x^{k+l} e^{-x^2/2}\, dx = \begin{cases} \sqrt{2\pi}(k+l-1)!! & \text{if } k+l \text{ is even} \\ 0 & \text{if } k+l \text{ is odd} \end{cases}$$

where $0!! = 1$, and $(2n-1)!! = 1\cdot 3\cdots(2n-1)$ is the product of all odd numbers from 1 up to $2n-1$. There are three main types of polynomials, that together with the specified weights find many applications in quantum mechanics. These are constructed by orthogonalizing the monomial set over the three intervals considered thus far with the weights as introduced. The Legendre polynomials, associated with angular momentum states are built by the Gram–Schmidt procedure in the interval $[-1, 1]$ with $\omega(x) = 1$, the Hermite polynomials associated with the harmonic oscillator are in the interval $(-\infty, \infty)$ with $\omega(x) = e^{-x^2/2}$, and the Laguerre polynomials when radial degrees of freedom are involved in $[0, \infty)$ with $\omega(x) = e^{-x}$. Here, we construct some Laguerre polynomials to demonstrate the procedure.

Starting with $\{1, x, x^2, \ldots, x^n\}$, the first Laguerre is $L_0(x) = 1$ since $0! = 1$ and the vector is already normalized:

$$\tilde{v}_1 = x - \langle x \cdot 1\rangle_\omega L_0 = x - 1$$

and its size becomes

$$\|\tilde{v}_1\| = [\langle (x-1)\cdot(x-1)\rangle_\omega = \langle x\cdot x\rangle_\omega - 2\langle 1\cdot x\rangle_\omega + \langle 1\cdot 1\rangle_\omega]^{1/2} = 1.$$

Therefore, $L_1 = x - 1$,

$$\begin{aligned}
\tilde{v}_2 &= x^2 - \langle x^2\cdot(x-1)\rangle_\omega L_1 - \langle x^2\cdot 1\rangle_\omega L_0 \\
\tilde{v}_2 &= x^2 - (\langle x^2\cdot x\rangle_\omega - \langle x^2\cdot 1\rangle_\omega)(x-1) - \langle x^2\cdot 1\rangle_\omega 1 \\
\tilde{v}_2 &= x^2 - 4x + 2 \\
\|\tilde{v}_2\| &= [\langle (x^2-4x+2)\cdot(x^2-4x+2)\rangle_\omega]^{1/2} \\
\|\tilde{v}_2\| &= [\langle x^2\cdot x^2\rangle_\omega - 8\langle x^2\cdot x\rangle_\omega + 4\langle x^2\cdot 1\rangle_\omega + 16\langle x\cdot x\rangle_\omega - 16\langle x\cdot 1\rangle_\omega + 4]^{1/2} = 2,
\end{aligned}$$

and, therefore, $L_2(x) = \frac{1}{2}(x^2 - 4x + 2)$. By hand, the procedure is quite tedious and error prone. To carry out the tasks in later chapters, we need to be able to compute orthogonal polynomials quickly and reliably. Later on, methods to compute the values of orthogonal polynomials using any weight and in any interval will be discussed.

1.7 Eigenanalysis

There are many other ways one can create orthogonal vectors. One of them is diagonalization. Before we begin a review of this subject, we need a definition.

Definition 5 (Transpose of a matrix). The transpose of a matrix A is denoted A^T and is obtained by changing the columns of A in to rows, starting from the leftmost column as row 1, etc. A matrix A is called symmetric if $A^T = A$ and antisymmetric if $A^T = -A$.

Linear algebra textbooks introduce the process to compute eigenvalues and eigenvectors for small sized $n \times n$ matrices by deriving their **characteristic polynomial** from the **secular determinant**, and then look for the corresponding **null space**. To see how this works, we present two examples. Consider the matrix

$$A = \begin{pmatrix} 3 & -1 & 0 \\ -1 & 2 & -1 \\ 0 & -1 & 3 \end{pmatrix}$$

and let λ be its eigenvalue. Then the secular determinant is

$$\det \begin{pmatrix} 3-\lambda & -1 & 0 \\ -1 & 2-\lambda & -1 \\ 0 & -1 & 3-\lambda \end{pmatrix} = 0.$$

The characteristic polynomial is $-\lambda^3 + 8\lambda^2 - 19\lambda + 12$ and setting it to zero gives three distinct roots, $\lambda = 1, 3, 4$. To find the eigenvector associated with $\lambda_1 = 1$, one seeks the solution of

$$\begin{pmatrix} 3-\lambda_1 & -1 & 0 \\ -1 & 2-\lambda_1 & -1 \\ 0 & -1 & 3-\lambda_1 \end{pmatrix} \begin{pmatrix} a \\ b \\ c \end{pmatrix} = 0.$$

Because the determinant of the above matrix is zero, only two equations are linearly independent, which means that one of the three variables in the column vector cannot be determined. That becomes a degree of freedom of the null space. The equations read

$$\begin{aligned} 2a - b &= 0 \\ -a + b - c &= 0 \\ -b + 2c &= 0. \end{aligned}$$

Choosing a as the degree of freedom, we immediately get $b = 2a$ and $c = a$ from the top two equations. Then we have the pair

$$\lambda_1 = 1 \quad a \begin{pmatrix} 1 \\ 2 \\ 1 \end{pmatrix}.$$

The other two pairs are

$$\lambda_2 = 3 \quad b\begin{pmatrix} 1 \\ 0 \\ -1 \end{pmatrix},$$

$$\lambda_3 = 4 \quad c\begin{pmatrix} 1 \\ -1 \\ 1 \end{pmatrix}.$$

The three vectors are typically normalized and arranged into the matrix O,

$$O = \begin{pmatrix} \frac{1}{\sqrt{6}} & \frac{1}{\sqrt{2}} & \frac{1}{\sqrt{3}} \\ \frac{2}{\sqrt{6}} & 0 & -\frac{1}{\sqrt{3}} \\ \frac{1}{\sqrt{6}} & -\frac{1}{\sqrt{2}} & \frac{1}{\sqrt{3}} \end{pmatrix}.$$

A set of mutually orthogonal unit vector is often called an **orthonormal** set. One can show that $O^T AO = \Lambda$. Moreover, the transforming matrix O is orthogonal, meaning $O^T O = I$ the unit matrix. The columns of O are the eigenvectors of A and the corresponding elements of Λ the eigenvalues. The procedure of transforming a matrix by $P^{-1}AP$ is called a **similarity transformation.**

The detailed proofs of the following list of important theorems can be found in many linear algebra textbooks. Let A be a $n \times n$ matrix.

1. If A is symmetric with real entries $a_{ij} \in \mathbb{R}$, then each member of the set of its eigenvalues $\{\lambda_i\}$ is real.
2. If A is not symmetric with real entries, then the eigenvalues are either real or complex conjugate in pairs.
3. There exists at least one eigenvector associated with each distinct eigenvalue λ_i.
4. If $\{\lambda_1, \lambda_2, \dots, \lambda_n\}$ is a set of distinct eigenvalues, the associated eigenvectors are linearly independent and mutually orthogonal.

1.8 Exercises

1. Show that the set of complex numbers $\{a + ib \mid a, b \in \mathbb{R}, i = \sqrt{-1}\}$ satisfy the axioms for the addition and multiplication operations and, therefore, the set of all complex numbers is a field.
2. A vector space combined with a binary vector product that satisfies associativity, closure, existence of a multiplicative identity, and existence of inverse for all elements other than zero is known as an **algebra.**
   (a) Consider the set of all 2×2 matrices of the form

   $$\begin{pmatrix} a & b \\ -b & a \end{pmatrix} \quad a, b \in \mathbb{R}$$

   show that it is a vector space $\mathcal{V}$, and propose a basis set.

(b) Define the binary operation $*$ as the matrix product between two elements of the vector space $\mathcal{V}$,

$$\begin{pmatrix} a_1 & b_1 \\ -b_1 & a_1 \end{pmatrix}, \quad \begin{pmatrix} a_2 & b_2 \\ -b_2 & a_2 \end{pmatrix}.$$

Carry out the product and show that $\mathcal{V}$ endowed with the product $*$ is indeed an algebra. Recall that the i, j element of a product of two $n \times n$ matrices $C = AB$ is the dot product of row i of A and column j of B,

$$c_{ij} = \sum_{k=1}^{n} a_{ik} b_{kj}.$$

Also, recall than the inverse of a 2×2 matrix is

$$A^{-1} = \begin{pmatrix} a_{11} & a_{12} \\ a_{21} & a_{22} \end{pmatrix}^{-1} = \frac{1}{\det A} \begin{pmatrix} a_{22} & -a_{12} \\ -a_{21} & a_{11} \end{pmatrix}.$$

(c) Show that the algebra constructed over this particular set of skew symmetric matrices is isomorphic to $\mathbb{C}$ by consulting the first exercise.

3. Consider again the vector space of all 2×2 skew symmetric matrices over $\mathbb{R}$ with elements

$$\begin{pmatrix} a & b \\ -b & c \end{pmatrix} \quad a, b, c \in \mathbb{R}.$$

(a) Show that the three vectors

$$\begin{pmatrix} 1 & 0 \\ 0 & 0 \end{pmatrix}, \quad \begin{pmatrix} 0 & 1 \\ -1 & 0 \end{pmatrix}, \quad \begin{pmatrix} 0 & 0 \\ 0 & 1 \end{pmatrix}$$

are linearly independent.

(b) Does the matrix product satisfy all the requirements to build an algebra over this vector space? Show that closure is not satisfied in this case.

4. Show that the three vectors

$$\begin{pmatrix} 1 \\ 2 \\ 1 \end{pmatrix}, \quad \begin{pmatrix} 1 \\ 0 \\ -1 \end{pmatrix}, \quad \begin{pmatrix} 1 \\ -2 \\ 1 \end{pmatrix}$$

are linearly independent using two methods:

(a) Show that the only way that

$$a \begin{pmatrix} 1 \\ 2 \\ 1 \end{pmatrix} + b \begin{pmatrix} 1 \\ 0 \\ -1 \end{pmatrix} + c \begin{pmatrix} 1 \\ -2 \\ 1 \end{pmatrix} = \begin{pmatrix} 0 \\ 0 \\ 0 \end{pmatrix}$$

can be true is if $a = 0$, $b = 0$, $c = 0$ by setting up and solving a system of equations.

(b) Produce a 3×3 matrix that has as columns the given vectors. Then compute its determinant using any method you like. In Python, the numpy function that evaluates determinants is `linalg.det(a)`. Here is a suggested minimal Python code. Note the spacing when entering elements of array is part of the syntax and as such very important. To make the output more human readable, we suggest you add some `print('string')` statements before the print statements in the lines below, such as, e. g., `print('The a matrix')` right above the `print(a)` statement.

```
import numpy as np
a = np.array([[1, 1, 1], [2, 0, -2], [1, -1, 1]])
print(a)
print(a.shape)
print(np.linalg.det(a))
```

5. Form an orthonormal set from the following three vectors:

$$\begin{pmatrix} 1 \\ 1 \\ 1 \\ -1 \end{pmatrix}, \begin{pmatrix} 2 \\ -1 \\ -1 \\ -1 \end{pmatrix}, \begin{pmatrix} -1 \\ 2 \\ 2 \\ 1 \end{pmatrix}.$$

6. Form an orthogonal set of vectors from the set of monomials $\{v_1, v_2, v_3, v_4\} = \{1, x, x^2, x^3\}\ x \in [-1, 1]$.

7. Show that the three vectors

$$\frac{1}{\sqrt{2}}\begin{pmatrix} -1 \\ 0 \\ 1 \end{pmatrix}, \frac{1}{\sqrt{19}}\begin{pmatrix} 3 \\ 1 \\ 3 \end{pmatrix}, \frac{1}{\sqrt{38}}\begin{pmatrix} 1 \\ -6 \\ 1 \end{pmatrix}$$

are orthonormal in two ways:

(a) Compute all 6 dot products and verify that the three dot products among pairs of these vectors are zero, and the three dot products of the vectors with themselves are 1.

(b) Organize the three vectors as columns of a 3×3 matrix A, find its transpose A^T, and show that $A^T A = AA^T = I$ where I is a 3×3 matrix with ones along the main diagonals and zeros otherwise. Here is a suggested minimal Python code for this exercise.

```
import numpy as np
s2 = np.sqrt(2)
s19 = np.sqrt(19)
s38 = np.sqrt(38)
a = np.array([[-1/s2, 3/s19, 1/s38], [0, 1/s19, -6/s38],
```

```
    [1/s2, 3/s19, 1/s38]])
b = np.transpose(a)
print(np.matmul(a,b))
print(np.matmul(b,a))
```

8. Verify that the three vectors of the previous exercise are linearly independent by adding one or more lines to your code.
9. Does the Gram–Schmidt procedure work if not all vectors are linearly independent of each other? To answer this question, consider two vectors v_1 and v_2. Set $v_2 = cv_1$, where c is a scalar in $\mathcal{F}$. Follow the orthogonalization procedure to find expressions for u_1 and u_2, then answer the question.
10. For the following function in $\mathbb{R}^3$,

$$\phi = 2xy - e^z - xz^2$$

(a) Find its gradient $\nabla\phi$

(b) Given a vector field

$$\mathbf{F}(x,y,z) = F_x(x,y,z)\,\mathbf{i} + F_y(x,y,z)\,\mathbf{j} + F_z(x,y,z)\,\mathbf{k}$$

the divergence of $\mathbf{F}$ is

$$\frac{\partial F_x}{\partial x} + \frac{\partial F_y}{\partial y} + \frac{\partial F_z}{\partial z}$$

a scalar quantity. Find the divergence of $\nabla\phi$. The divergence of a gradient of a function ϕ is known as the Laplacian of ϕ.

(c) Given a vector field $\mathbf{F}(x,y,z)$ as in part (b), the curl of $\mathbf{F}$ is

$$\nabla \times \mathbf{F}$$

where ∇ is the “nabla” operator

$$\nabla = \mathbf{i}\frac{\partial}{\partial x} + \mathbf{j}\frac{\partial}{\partial y} + \mathbf{k}\frac{\partial}{\partial z}$$

and $\times$ is the cross-product of two vectors, a vector whose components are the cofactors of the first row of this determinant, specialized for the case at hand

$$\nabla \times \mathbf{F} = \det\begin{vmatrix} \mathbf{i} & \mathbf{j} & \mathbf{k} \\ \frac{\partial}{\partial x} & \frac{\partial}{\partial y} & \frac{\partial}{\partial z} \\ F_x & F_y & F_z \end{vmatrix},$$

and find the curl of $\nabla\phi$ and show that it vanishes.

(d) Prove in general that for "well-behaved" functions, the $\nabla \times (\nabla\phi) = 0$. By well behaved, it suffices to show that ϕ and its first and second derivatives are continuous. Recall that for continuous and differentiable functions in more than one dimension, the mixed partials can be evaluated in any order.

11. The Laguerre polynomials can be derived using Rodrigues' formula

$$L_n(x) = (-1)^n e^x \frac{d^n}{dx^n}(x^n e^{-x}).$$

Use this to find expressions for $L_0(x)$, $L_1(x)$, and $L_2(x)$. Compare your result with the expressions in the text, but note that Rodrigues' formula does not give normalized polynomials.

12. The Legendre polynomials can be generated with a similar Rodrigues' formula,

$$P_n(x) = \frac{1}{2^n n!}\frac{d^n}{dx^n}[(x^2-1)^n].$$

Use this to compute the first three polynomials and compare your answers with those of exercise 6.

13. It is possible to derive a generic three-term recursion relation for all orthogonal polynomials constructed with the Gram–Schmidt procedure. For the Legendre polynomials, it reads

$$(n+1)P_{n+1}(x) = (2n+1)xP_n(x) - nP_{n-1}(x).$$

Starting with $P_0(x) = 1$ and $P_1(x) = x$, find P_n for $n = 2$ through 10.

14. Fill in the details of the derivations of the eigenvectors of

$$A = \begin{pmatrix} 3 & -1 & 0 \\ -1 & 2 & -1 \\ 0 & -1 & 3 \end{pmatrix}$$

associated with the two eigenvalues $\lambda_2 = 3$ and $\lambda_3 = 4$.

15. For matrix A in exercise 13, the similarity matrix that diagonalizes it

$$O = \begin{pmatrix} \frac{1}{\sqrt{6}} & \frac{1}{\sqrt{2}} & \frac{1}{\sqrt{3}} \\ \frac{2}{\sqrt{6}} & 0 & -\frac{1}{\sqrt{3}} \\ \frac{1}{\sqrt{6}} & -\frac{1}{\sqrt{2}} & \frac{1}{\sqrt{3}} \end{pmatrix}.$$

(a) Show that $O^T O = I$ a three-by-three unit matrix.

(b) Show that $O^T A O = \Lambda$, a diagonal matrix with 1, 3, 4 as the entries.

16. The numpy function that diagonalizes $n \times n$, real symmetric matrices is `linalg.eig(a)`. Write a Python code that diagonalizes the matrix A in exercise 14, and performs the end-to-end checks in exercise 15. The output of `linalg.eig(a)`

is a stacked array containing both eigenvalues and eigenvectors. To unpack them, you will need this line:

```
eigenvalues, eigenvectors =  np.linalg.eig(a)
```

assuming you have `import numpy as np` earlier in your code.

17. Let A be a $n \times n$ matrix with eigenvalues λ_i. Use the code from the previous exercise to numerically verify the following facts:
    (a) The transpose of A has the same eigenvalues of A.
    (b) The matrix cA, where $c \in \mathbb{R}$ has eigenvalues $c\lambda_i$.
    (c) The matrix $A + cI$ has eigenvalues $\lambda_i + c$.
18. Prove the three statements you verified in the previous exercise.
19. Similar transformations are powerful tools for the analysis that we carry out in later chapters. Suppose we want to know the eigenvalues of some power of a real matrix A given its eigenvalues $\{\lambda_i\}$, and eigenvectors arranged into the columns of the matrix O. Show that the eigenvalues of A^p are $\{\lambda_i^p\}$ by following the steps below:
    (a) Let begin with a specific case: Given

$$A = \begin{pmatrix} 3 & -5 \\ -5 & 3 \end{pmatrix}$$

    and that

$$O = \frac{1}{\sqrt{2}} \begin{pmatrix} 1 & 1 \\ -1 & 1 \end{pmatrix}$$

    diagonalizes A. Find the eigenvalues of A.
    (b) Referring to part (a), verify that $O^T A^2 O$ is also diagonal.
    (c) Prove that if $O^T AO$ transforms A into a diagonal matrix Λ, it will do so for any power of A. Moreover, if $\Lambda = \text{diag}\{\lambda_1, \ldots\}$ contains the eigenvalues of A, $\Lambda^n = \text{diag}\{\lambda_1^n, \ldots\}$ contains those of A^n. Start with the equation

$$A = O \Lambda O^T,$$

    multiply it by A from the right on both sides, then insert the unit matrix OO^T in a strategic place to simplify the expression. Complete the proof by induction.

2 Applications of vector spaces

I am no friend of probability theory, I have hated it from the first moment when our dear friend Max Born gave it birth.

Erwin Schrödinger

2.1 Eigenstates of the Hamiltonian

The solutions of the Schrödinger equation are represented by vector spaces. Assuming that we know all the solutions of

$$-\frac{\hbar^2}{2m}\frac{d^2}{dx^2}\psi_n + U\,\psi_n = E_n\psi_n \tag{2.1}$$

over $\mathbb{R}$ with a generic potential U, the set $\{\psi_0, \psi_1, \ldots, \psi_n\}$ can be used as the basis set for a vector space that contains all possible states of the quantum system. In that case, $\{\psi_0, \psi_1, \ldots, \psi_n\}$ is called the set of *eigenfunctions* and $\{E_0, E_1, \ldots, E_n\}$ the set of *eigenvalues* of the *Hamiltonian operator* H:

$$H = -\frac{\hbar^2}{2m}\frac{d^2}{dx^2} + U.$$

The potential energy U, for realistic applications is finite, meaning $\lim_{x\to\pm\infty} U(x)$ approaches a constant U_∞. Then there are two types of solutions: those that are bound with eigenvalues below U_∞, and scattering states, where energy values are in a continuous interval $[U_\infty, \infty)$. The number of eigenvalues and eigenfunctions is infinite in both sets. Scattering theory and the computation of scattering states are topics outside of the scope of this textbook. A "state of the system," in this book should be understood as a bound state. Any bound state can be determined by any one of the eigenfunctions eigenvector pair, or as a linear superposition of them,

$$\psi(x) = A\sum_{j=0}^{n} c_j\psi_j(x),$$

where A is the normalization constant and we assume that the eigenfunctions ψ_j are normalized. A state described by a single eigenvalue and eigenvector is called a "pure state" or an *eigenstate*. A linear superposition of eigenstates is called a *wavepacket*. In theory, the linear combination of states could have an infinite amount of terms. In practical applications, the linear superposition is incremented with eigenstates of higher and higher energies until some convergence criteria is met.

In general, the coefficients and the wave functions are complex, meaning $c_i \in \mathbb{C}$, and $\psi_i : \mathbb{R} \to \mathbb{C}$. Because of this fact, the dot product needs to be redefined with a small modification of the definition in Chapter 1,

https://doi.org/10.1515/9783111610207-002

$$\langle \psi_j | \psi_k \rangle = \int_R \psi_j^* \psi_k \, dx,$$

where ψ_j^* is the complex conjugate of ψ_j. In many quantum theory textbooks, the symbol $\langle \psi_j | \psi_k \rangle$ is abbreviated using Dirac's notation $\langle j | k \rangle$. The Dirac notation is indeed quite convenient, especially when vector quantities are manipulated. If we assume that ψ_j is normalized, then $\langle j | k \rangle = \delta_{jk}$, where the Kronecker delta δ_{jk} is either zero if $j \neq k$ or 1 if $j = k$. We leave it to the reader as an exercise to find an expression for A in terms of $\{c_j\}$ and the energy E in terms of $\{c_j, E_j\}$.

Now we turn to an example of how vector spaces can be used to find the eigenvalues and eigenvectors of H, when these are not known a priori. Because we have many uses for vector spaces, let us call this approach the direct vector space approach. One uses a vector space $\text{span}\{\chi_0, \chi_1, \ldots, \chi_n\}$, perhaps constructed with orthogonal polynomials, computes the elements of the Hamiltonian matrix,

$$\left\langle j \left| -\frac{\hbar^2}{2m} \frac{d^2}{dx^2} + U \right| k \right\rangle = \int_R \chi_j^* \left(-\frac{\hbar^2}{2m} \frac{d^2}{dx^2} \chi_k + U \chi_k \right) dx$$

and then diagonalizes it, using a numerical procedure like the numpy Python class `linalg.eig(a)`. The procedure returns eigenvalues, namely the energies associated with each eigenstate and eigenvectors as the columns of the transformation matrix. The wave function for eigenstate ψ_n is reconstructed with

$$\psi_k = \sum_{j=1}^{n} c_{jk} \chi_j \qquad c_{ik} \in \mathbb{C}$$

and ψ_k is normalized if the functions in $\{\chi_j\}$ are also normalized. While this tool is powerful and can be used to solve the Schrödinger equations in many dimensions, problems involving multiple particles quickly become computationally intractable. The reason for the difficulties with the direct vector space approach is the exponential growth of the basis set with respect to the number of dimensions. Even a two-dimensional problem requires specialized vector spaces, sparse matrix technology, and could take several hours on a desktop computer. Nonetheless, having multiple ways of finding the answers we seek helps to build confidence in the numerical strategies we develop throughout.

2.1.1 A singular Sturm–Liouville theorem

In this section, we prove an important result about the Schrödinger equation. We begin with an important definition.

Definition 6 (Degeneracy). Two distinct states i and j are said to be *degenerate* if $E_i = E_j$.

Degenerate states have different wave functions but the same energy. The degenerate states of a particle in a cube are a great example. It is possible for two states to be degenerate even in one dimension, if the potential energy U has symmetric wells. The wave functions of degenerate states may or may not be linearly independent and may not be mutually orthogonal. The following theorem deals with states that are not degenerate.

Theorem 4. *States that are not degenerate have linearly independent orthogonal eigenfunctions. Moreover, the excited states are mutually orthogonal functions under the weight φ_0^2.*

Proof. Consider the generic Schrödinger equation in (2.1) and let φ_0 represent the unnormalized ground state with energy E_0. Let $H_n(x)$ represent such polynomials with $H_0 = 1$. The transformation

$$\psi_n = H_n \varphi_0$$

changes equation (2.1) into

$$\varphi_0 \frac{d^2 H_n}{dx^2} + 2\left(\frac{dH_n}{dx}\right)\left(\frac{d\varphi_0}{dx}\right) + H_n \frac{d^2\varphi_0}{dx^2} + \left[\frac{2m}{\hbar^2}(E_n - U)\right] H_n \varphi_0 = 0.$$

This can be rearranged to cancel U, by using equation (2.1) and

$$\begin{aligned}\left(\frac{d\varphi_0}{dx}\right) &= \left(\frac{d\ln\varphi_0}{dx}\right)\varphi_0\\ \frac{d^2\varphi_0}{dx^2} &= \left[\frac{2m}{\hbar^2}(U - E_0)\right]\varphi_0.\end{aligned}$$

Then, canceling U and dropping φ_0 on both sides yield

$$\frac{d^2 H_n}{dx^2} + 2\left(\frac{dH_n}{dx}\right)\left(\frac{d\ln\varphi_0}{dx}\right) + \frac{2m}{\hbar^2}(E_n - E_0)H_n = 0. \tag{2.2}$$

The last equation is a specific case of the singular Sturm–Liouville equation,

$$\frac{d}{dx}\left(r\frac{dy}{dx}\right) + (q + \lambda p)y = 0, \tag{2.3}$$

as shown by identifying: $y = H_n$,

$$r(x) = \exp\left(2\int\left(\frac{d\ln\varphi_0}{dx}\right)dx\right) = \varphi_0^2 = p(x), \tag{2.4}$$

$q(x) = -2mE_0\varphi_0^2/\hbar^2$ and $\lambda = 2mE_n/\hbar^2$. It follows from (2.3) that

$$\varphi_0^2 \frac{d^2 H_n}{dx^2} + 2\varphi_0^2 \left(\frac{dH_n}{dx}\right)\left(\frac{d\ln\varphi_0}{dx}\right) + \frac{2m}{\hbar^2}(-E_0 + E_n)\varphi_0^2 H_n = 0.$$

Equation (2.2) follows by dropping φ_0^2. Note that $r(x)$ is continuous and that it approaches zero at both ends of $(-\infty + \infty)$, making this a singular Sturm–Liouville problem [103] of the third kind. The theorem states that

$$\int_{-\infty}^{\infty} H_n H_k \varphi_0^2 \, dx = \delta_{kn}$$

and its proof proceeds as follows. Writing for k and for $n \neq k$,

$$\frac{d}{dx}\left(\varphi_0^2 \frac{dH_n}{dx}\right) + \frac{2m}{\hbar^2}(-E_0 + E_n)\varphi_0^2 H_n = 0,$$
$$\frac{d}{dx}\left(\varphi_0^2 \frac{dH_k}{dx}\right) + \frac{2m}{\hbar^2}(-E_0 + E_k)\varphi_0^2 H_k = 0,$$

then multiplying the first by H_k and the second by H_n, we arrive at two expressions:

$$H_k \frac{d}{dx}\left(\varphi_0^2 \frac{dH_n}{dx}\right) + \frac{2m}{\hbar^2}(-E_0 + E_n)\varphi_0^2 H_n H_k = 0,$$
$$H_n \frac{d}{dx}\left(\varphi_0^2 \frac{dH_k}{dx}\right) + \frac{2m}{\hbar^2}(-E_0 + E_k)\varphi_0^2 H_k H_n = 0.$$

Subtracting the second from the first, and integrating over $(-\infty + \infty)$ on both sides gives

$$\int_{-\infty}^{\infty} H_k \frac{d}{dx}\left(\varphi_0^2 \frac{dH_n}{dx}\right) dx - \int_{-\infty}^{\infty} H_n \frac{d}{dx}\left(\varphi_0^2 \frac{dH_k}{dx}\right) dx$$
$$= \frac{2m}{\hbar^2}(E_n - E_k) \int_{-\infty}^{\infty} \varphi_0^2 H_n H_k.$$

The left-hand side can be integrated by parts,

$$H_k \varphi_0^2 \frac{dH_n}{dx}\bigg|_{-\infty}^{\infty} - \int_{-\infty}^{\infty} \left(\frac{dH_k}{dx}\right)\left(\frac{dH_n}{dx}\right)\varphi_0^2 \, dx$$
$$-H_n \varphi_0^2 \frac{dH_k}{dx}\bigg|_{-\infty}^{\infty} + \int_{-\infty}^{\infty} \left(\frac{dH_n}{dx}\right)\left(\frac{dH_k}{dx}\right)\varphi_0^2 \, dx$$
$$= \frac{2m}{\hbar^2}(E_n - E_k) \int_{-\infty}^{\infty} \varphi_0^2 H_n H_k = 0$$

since the boundary terms vanish as the property of φ_0, and the second term on the left is the opposite of the fourth. Therefore, if the eigenvalues of the one-dimensional Schrödinger equation are distinct, $E_n - E_k$ cannot be zero and the theorem follows. □

It is important to note at this point that the functions H_n of this theorem can always be approximated with polynomials, however, there are only a handful of cases where the maximum order of the polynomial that approximates H_n is n. Two popular sets of functions where the maximum order of the polynomial [90] that computes H_n exactly is n, are the Hermite, and the associate Legendre polynomials. These, when multiplied by the ground state wave function, are the exact solutions of the harmonic oscillator and the angular momentum Hamiltonian, respectively. We make use of Hermite polynomials in this book. Therefore, in rest of this chapter, we explore methods to automate their computation.

2.2 Building excited states from the ground up

The statement proved in the previous section is a weaker version of the Sturm–Liouville theorem, but it is quite valuable because it gives us a way to define the weight function for the dot product in the vector space we wish to construct numerically. Namely, if by some means we are able to find the ground state of the Schrödinger equation, we can set $\omega(x) = \varphi_0^2$ and build good approximations for the rest of the states by the Gram–Schmidt procedure of Chapter 1. There are several useful examples where the ground state of the Hamiltonian and the integrals of the type,

$$\mathcal{I}_n = \int_R x^n \, \omega(x) \, dx \tag{2.5}$$

are available analytically. These integrals are the moments of the positive definite density $\omega(x)$, but we also call these the monomial integrals. Therefore, we can create a number of vector spaces that can be deployed in advanced computational methods. At this point, we need to clarify the notation that we use throughout regarding the weight function $\omega(x)$. In what follows, it is set to φ_0^2, however, in our notation and in our codes, φ_0 *is not* normalized. The *normalized* ground state wave function is the product of the constant polynomial P_0, which we derive below, with φ_0. Similarly, the normalized wave function for an arbitrary state j is $\psi_j = P_j \varphi_0$. This convention is used throughout the book. With regard to the computation of the wave functions, this is strictly not necessary. One can define the weight function using a normalized ψ_0, in which case $\mathcal{I}_0$ should evaluate to one. The codes can be easily modified and in some cases such modifications could be advantageous.

2.2.1 The polynomial coefficients

Here, we derive equations for the set $\{\mathscr{I}_n\}$ and compute the coefficients of the orthogonal set of polynomials $\{P_n(x)\}$. Specifically, introducing some notation to define the set $\{w_k^{(n)}\}$,

$$P_n(x) = \sum_{k=0}^{n} w_k^{(n)} x^k, \tag{2.6}$$

we prove that one can in principle compute the values of $\{w_k^{(n)}\}$ given the values of the set $\{\mathscr{I}_n\}$. Starting with

$$P_0(x) = w_0^{(0)},$$

the size of this vector is

$$\|P_0\|_\omega = w_0^{(0)} \left(\int_{-\infty}^{\infty} \varphi_0^2 \, dx \right)^{1/2} = w_0^{(0)} \mathscr{I}_0^{1/2}$$

and since we choose the norm to be one, we obtain the first coefficient:

$$w_0^{(0)} = \mathscr{I}_0^{-1/2}. \tag{2.7}$$

Using the Gram–Schmidt procedure, we construct $P_1(x)$ with

$$P_1(x) = \frac{xP_0(x) - \langle P_0, xP_0 \rangle_\omega P_0(x)}{\|xP_0(x) - \langle P_0, xP_0 \rangle_\omega P_0(x)\|_\omega^2},$$

where

$$\langle P_0, xP_0 \rangle_\omega = (w_0^{(0)})^2 \int_{-\infty}^{\infty} x \, \varphi_0^2 \, dx = (w_0^{(0)})^2 \mathscr{I}_1 = \mathscr{I}_0^{-1} \mathscr{I}_1.$$

Therefore,

$$P_1(x) = \frac{\mathscr{I}_0^{-1/2} x - \mathscr{I}_1 \mathscr{I}_0^{-3/2}}{\|\mathscr{I}_0^{-1/2} x - \mathscr{I}_1 \mathscr{I}_0^{-3/2}\|_\omega^2},$$

and the norm in the denominator is

$$\|\mathscr{I}_0^{-1/2} x - \mathscr{I}_1 \mathscr{I}_0^{-3/2}\|_\omega^2 = \left\{ \mathscr{I}_0^{-1} \int_{-\infty}^{\infty} x^2 \varphi_0^2 \, dx \right.$$
$$\left. + \mathscr{I}_1^2 \mathscr{I}_0^{-3} \int_{-\infty}^{\infty} \varphi_0^2 \, dx - 2 \mathscr{I}_1 \mathscr{I}_0^{-2} \int_{-\infty}^{\infty} x \, \varphi_0^2 \, dx \right\}^{1/2}$$

$$\|\mathscr{I}_0^{-1/2}x - \mathscr{I}_1\mathscr{I}_0^{-3/2}\|_\omega^2 = (\mathscr{I}_0^{-1}\mathscr{I}_2 - \mathscr{I}_1^2\mathscr{I}_0^{-2})^{1/2}.$$

Putting this back into the expression for P_1, simplifying

$$P_1(x) = \frac{x - \mathscr{I}_1\mathscr{I}_0^{-1}}{(\mathscr{I}_2 - \mathscr{I}_1^2\mathscr{I}_0^{-1})^{1/2}},$$

and comparing with $P_1(x) = w_0^{(1)} + w_1^{(1)}x$ gives

$$w_1^{(1)} = \frac{1}{(\mathscr{I}_2 - \mathscr{I}_1^2\mathscr{I}_0^{-1})^{1/2}} \tag{2.8}$$

$$w_0^{(1)} = -\frac{\mathscr{I}_1\mathscr{I}_0^{-1}}{(\mathscr{I}_2 - \mathscr{I}_1^2\mathscr{I}_0^{-1})^{1/2}}. \tag{2.9}$$

Thus, we have found the coefficients of the first excited state,

$$\psi_1 = (w_1^{(1)}x + w_0^{(1)})\varphi_0. \tag{2.10}$$

Expressing $w_k^{(n)}$ in terms of $\mathscr{I}_n$ grows enormously in complexity as the equations for the coefficients in terms of $\mathscr{I}_n$ become formidable. We clearly have to abandon the idea of finding all the equations for $\{w_k^{(n)}\}$ this way. However, explicit expressions of the polynomial coefficients in terms of the $\mathscr{I}_n$ integral are not necessary to automate their computation. There are at least three strategies that allow us to systematically compute the polynomial coefficients for $n \geq 2$ once values of $w_0^{(0)}$, $w_0^{(1)}$, and $w_1^{(1)}$ are known. The following theorem provides one such way.

Theorem 5. *Let P_i be defined as in equation (2.6). The coefficient of x^{n+1} in the expression for P_{n+1} is*

$$w_{n+1}^{(n+1)} = \frac{w_n^{(n)}}{d_{n+1}} \tag{2.11}$$

where

$$d_{n+1} = \left(\langle xP_n, xP_n\rangle_\omega - \sum_{i=0}^{n} \langle P_i, xP_n\rangle_\omega^2 \right)^{1/2}, \tag{2.12}$$

is the norm of the $n+1$ polynomial. The constant coefficient is computed with

$$w_0^{(n+1)} = -\sum_{i=0}^{n} \frac{\langle P_i, xP_n\rangle_\omega}{d_{n+1}} w_0^{(i)} \tag{2.13}$$

where as for $1 \leq j \leq n$, we get

$$w_j^{(n+1)} = \frac{w_{j-1}^{(n)}}{d_{n+1}} - \sum_{i=0}^{n} \frac{\langle P_i, xP_n\rangle_\omega}{d_{n+1}} w_j^{(i)}. \tag{2.14}$$

Proof. The Gram–Schmidt process for polynomials can be expressed with a single equation,

$$P_{n+1}(x) = \frac{xP_n(x) - \sum_{i=0}^{n}\langle P_i, xP_n\rangle_\omega P_i(x)}{\|xP_n(x) - \sum_{i=0}^{n}\langle P_i, xP_n\rangle_\omega P_i(x)\|_\omega^2}. \tag{2.15}$$

Here, we assume that the coefficients of $P_n(x)$, $n = 0, 1, \ldots n$ are already known, and show that those of $P_{n+1}(x)$ can be obtained in terms of $\{w_k^{(n)}\}$ and $\mathscr{I}_n$. First, we expand the norm in the denominator,

$$\begin{aligned} d_{n+1} &= \left\| xP_n(x) - \sum_{i=0}^{n}\langle P_i, xP_n\rangle_\omega P_i(x)\right\|_\omega^2 \\ &= \Bigg(\langle xP_n, xP_n\rangle_\omega + \sum_{i,j=0}^{n}\langle P_i, xP_n\rangle_\omega \langle P_j, xP_n\rangle_\omega \langle P_j, P_i\rangle_\omega \\ &\quad - 2\sum_{i=0}^{n}\langle P_i, xP_n\rangle_\omega \langle xP_n, P_i\rangle_\omega \Bigg)^{1/2}. \end{aligned}$$

Noting that $\langle P_j, P_i\rangle_\omega = \delta_{ij}$ by construction and that $\langle xP_n, P_i\rangle_\omega = \langle P_i, xP_n\rangle_\omega$ allows for some simplifications from which equation (2.12) follows. We next note that all the inner products can be expressed in terms of known quantities:

$$\begin{aligned} \langle P_i, xP_n\rangle_\omega &= \int_{-\infty}^{\infty} \sum_{m=0}^{i}\sum_{k=0}^{n} w_m^{(i)} w_k^{(n)} x^{m+k+1}\, \omega(x)\, dx \\ &= \sum_{m=0}^{i}\sum_{k=0}^{n} w_m^{(i)} w_k^{(n)}\, \mathscr{I}_{m+k+1}. \end{aligned} \tag{2.16}$$

A similar expansion also proves

$$\langle xP_n, xP_n\rangle_\omega = \sum_{m=0}^{n}\sum_{k=0}^{n} w_m^{(n)} w_k^{(n)}\, \mathscr{I}_{m+k+2}. \tag{2.17}$$

Therefore, by expanding both sides of (2.15) and using (2.12), one arrives at

$$\sum_{j=0}^{n+1} w_j^{(n+1)} x^j = \sum_{k=0}^{n} \frac{w_j^{(n)}}{d_{n+1}} x^{k+1} - \sum_{i=0}^{n}\sum_{m=0}^{i} \frac{\langle P_i, xP_n\rangle_\omega}{d_{n+1}} w_m^{(i)} x^m.$$

The theorem follows by equating the coefficients for equal powers of x on both sides of the last equation. For the $n = 2$ case, the highest power of x is 3, and there is only one term that matches that power on the righ-hand side. Moreover, the coefficient of the highest power of x is the expression in equation (2.11) since the $j = n + 1$ case on the left matches with the last term of the first sum on the right-hand side. □

In pseudocode, the procedure to compute all the desired coefficients is:
1. Compute $w_0^{(0)} = \mathscr{I}_0^{-1/2}$ and set n to zero.
2. Compute $\langle P_i, xP_n\rangle_\omega$ with equation (2.16) for $i = 0$ to n.
3. Compute $\langle xP_n, xP_n\rangle_\omega$ with equation (2.17).
4. Compute d_{n+1} with equation (2.12).
5. Compute the set $\{w_j^{(n+1)}\}$ by means of (2.11), (2.13), and (2.14) for the respective cases.
6. Increment n by one and repeat steps 2 through 5 until n reaches the maximum desired polynomial order.

Note that the process not only produces the coefficients of the polynomials up to order $n_{\max} + 1$, but it also yields the elements of the position matrix $\langle P_i, xP_n\rangle_\omega$ for $0 \le i, n \le n_{\max}$. A good test to verify that the code is working properly is to check that $\langle P_i, xP_n\rangle_\omega$ is tridiagonal and symmetric. A more stringent test is to verify that the polynomials satisfy a three term recursion we now propose and prove.

Theorem 6. *The coefficients of P_{n+1} satisfy a three term recursion. For $1 \le j \le n$,*

$$d_{n+1}w_j^{(n+1)} = w_{j-1}^{(n)} - \langle P_n, xP_n\rangle_\omega w_j^{(n)} - \langle P_{n-1}, xP_n\rangle_\omega w_j^{(n-1)}.$$

The $j = 0$ terms are

$$d_{n+1}w_0^{(n+1)} = -\langle P_n, xP_n\rangle_\omega w_0^{(n)} - \langle P_{n-1}, xP_n\rangle_\omega w_0^{(n-1)},$$

and the $j = n + 1$ term is

$$d_{n+1}w_{n+1}^{(n+1)} = w_n^{(n)}.$$

Proof. Starting with equation (2.15) rewritten slightly,

$$\begin{aligned} xP_n(x) &= d_{n+1}P_{n+1}(x) + \langle P_n, xP_n\rangle_\omega P_n(x) \\ &\quad + \langle P_{n-1}, xP_n\rangle_\omega P_{n-1}(x) + \cdots + \langle P_0, xP_n\rangle_\omega P_0(x) \end{aligned}$$

and changing n to $n - 2$ we get

$$\begin{aligned} xP_{n-2}(x) &= d_{n-1}P_{n-1}(x) + \langle P_{n-2}, xP_{n-2}\rangle_\omega P_{n-2}(x) \\ &\quad + \langle P_{n-3}, xP_{n-2}\rangle_\omega P_{n-3}(x) + \cdots + \langle P_0, xP_{n-2}\rangle_\omega P_0(x). \end{aligned}$$

Now let us use the last equation to create an expression for the inner product $\langle P_n, xP_{n-2}\rangle_\omega$,

$$\langle P_n, xP_{n-2}\rangle_\omega = d_{n-1}\langle P_n, P_{n-1}\rangle_\omega + \langle P_{n-2}, xP_{n-2}\rangle_\omega\langle P_n, P_{n-2}\rangle_\omega + \langle P_{n-3}, xP_{n-2}\rangle_\omega\langle P_n, P_{n-3}\rangle_\omega + \cdots + \langle P_0, xP_{n-2}\rangle_\omega\langle P_n, P_0\rangle_\omega$$

but since the polynomials up to order n are orthogonal by construction, we get

$$\langle P_n, xP_{n-2}\rangle_\omega = \cdots = \langle P_n, xP_0\rangle_\omega = 0,$$

which proves that the position matrix is tridiagonal in general. Armed with this knowledge, we can rewrite equation (2.15) once more

$$d_{n+1}P_{n+1}(x) = xP_n(x) - \langle P_n, xP_n\rangle_\omega P_n(x) - \langle P_{n-1}, xP_n\rangle_\omega P_{n-1}(x). \tag{2.18}$$

In terms of the coefficients, (2.18) reads

$$d_{n+1}\sum_{j=0}^{n+1} w_j^{(n+1)}x^j = \sum_{j=0}^{n} w_j^{(n)}x^{j+1} - \langle P_n, xP_n\rangle_\omega \sum_{j=0}^{n} w_j^{(n)}x^j - \langle P_{n-1}, xP_n\rangle_\omega \sum_{j=0}^{n-1} w_j^{(n-1)}x^j.$$

Reindexing the first sum on the right with $j+1 \to j'$ gives

$$\sum_{j=0}^{n} w_j^{(n)}x^{j+1} = \sum_{j'=1}^{n} w_{j-1}^{(n)}x^{j'}.$$

When j is n, j' is $n+1$ but $w_{n+1}^{(n)}$ is zero. The recursion becomes

$$d_{n+1}\sum_{j=0}^{n+1} w_j^{(n+1)}x^j = \sum_{j=1}^{n} w_{j-1}^{(n)}x^j - \langle P_n, xP_n\rangle_\omega \sum_{j=0}^{n} w_j^{(n)}x^j - \langle P_{n-1}, xP_n\rangle_\omega \sum_{j=0}^{n-1} w_j^{(n-1)}x^j.$$

Note that the $j = 0$ terms gives an expression equivalent to equation (2.13) written without the inner products that vanish. The statement of the theorem can be obtained from (2.14) by suppressing the inner products that vanish as well. Finally, for $j = n+1$, there are only two terms that survive. The result is just equation (2.11). □

Theorem 6 can be used to compute coefficients directly, or can be used to check the code for the algorithm given earlier. The two algorithms suggested by Theorems 5 and 6 are sufficient to compute the first few excited states, but become unstable when n increases beyond a dozen states. The source of numerical instability is the difference in both the numerator and denominator of equation (2.15) of two increasingly larger and approximately equal terms. These differences lead to loss of precision that accumulates rapidly during the recursion. There are alternative strategies to compute polynomials systematically. One frequently used is based on evaluating the cofactors of the Hankel determinant representation. Most applications of vector spaces we consider here need fewer than ten states and the Hankel determinant method is not necessary in those cases. Moreover, it too becomes unstable when more than a dozen excited state coefficients are computed. Nevertheless, we present the approach later in the chapter because it paves the way of computing orthogonal polynomials in many dimensions as discussed in Chapter 17.

2.2.2 The energy levels

Now that we have constructed a vector space, we proceed by projecting the Hamiltonian operator. Here, we assume that φ_0 is the ground state of the same operator and that the integrals

$$\mathscr{I}_n = \int_{-\infty}^{\infty} x^n \, \varphi_0^2 \, dx$$

$$\mathscr{Q}_n = \int_{-\infty}^{\infty} x^n \left(\frac{d \ln \varphi_0}{dx} \right) \varphi_0^2 \, dx$$

$$\mathscr{P}_n = \int_{-\infty}^{\infty} x^n \left(\varphi_0^{-1} \frac{d^2 \varphi_0}{dx^2} \right) \varphi_0^2 \, dx$$

and

$$\mathscr{U}_n = \int_{-\infty}^{\infty} x^n \, U \, \varphi_0^2 \, dx$$

can be computed by some means. Let us begin with the derivatives of $\psi_n = P_n(x)\varphi_0$,

$$\frac{d}{dx} P_n(x)\varphi_0 = P_n(x) \frac{d}{dx}\varphi_0 + \left(\frac{d}{dx} P_n(x) \right) \varphi_0$$

$$\frac{d^2}{dx^2} P_n(x)\varphi_0 = 2\left(\frac{d}{dx} P_n(x) \right) \frac{d}{dx}\varphi_0 + P_n(x) \frac{d^2}{dx^2}\varphi_0 + \left(\frac{d^2}{dx^2} P_n(x) \right) \varphi_0.$$

Inserting the last result into the Schrödinger equation,

$$-\frac{\hbar^2}{2m} \frac{d^2}{dx^2} P_n(x)\varphi_0 + U P_n(x)\varphi_0 = E_n P_n(x)\varphi_0$$

produces

$$-\frac{\hbar^2}{m}\left(\frac{d}{dx}P_n(x)\right)\frac{d}{dx}\varphi_0 - \frac{\hbar^2}{2m}P_n(x)\frac{d^2}{dx^2}\varphi_0 - \frac{\hbar^2}{2m}\left(\frac{d^2}{dx^2}P_n(x)\right)\varphi_0 + UP_n(x)\varphi_0$$
$$= E_nP_n(x)\varphi_0.$$

By construction, $P_n(x)\varphi_0$ is the eigenfunction of the Hamiltonian; therefore, we simply multiply both sides by $P_n(x)\varphi_0$ and integrate. In terms of the polynomial coefficients defined with

$$P_n(x) = \sum_{k=0}^{n} w_k^{(n)} x^k,$$

this becomes

$$\frac{dP_n(x)}{dx} = \sum_{k=1}^{n} k\, w_k^{(n)} x^{k-1}$$
$$\frac{d^2P_n(x)}{dx^2} = \sum_{k=2}^{n} k(k-1)\, w_k^{(n)} x^{k-2}.$$

Then

$$E_n = -\frac{\hbar^2}{m}\sum_{l=0}^{n}\sum_{k=1}^{n} k\, w_l^{(n)}\, w_k^{(n)} \int_{-\infty}^{\infty} x^{k+l-1}\left(\frac{d\ln\varphi_0}{dx}\right)\varphi_0^2 dx$$
$$- \frac{\hbar^2}{2m}\sum_{k,l=0}^{n} w_k^{(n)} w_l^{(n)} \int_{-\infty}^{\infty} x^{k+l}\left(\varphi_0^{-1}\frac{d^2}{dx^2}\varphi_0\right)\varphi_0^2 dx$$
$$- \frac{\hbar^2}{2m}\sum_{l=0}^{n}\sum_{k=2}^{n} k(k-1)\, w_l^{(n)}\, w_k^{(n)} \int_{-\infty}^{\infty} x^{k+l-2}\varphi_0^2 dx$$
$$+ \sum_{k,l=0}^{n} w_k^{(n)} w_l^{(n)} \int_{-\infty}^{\infty} x^{k+l}\, U\, \varphi_0^2\, dx$$

or using the notation introduced earlier,

$$E_n = -\frac{\hbar^2}{m}\sum_{l=0}^{n}\sum_{k=1}^{n} k\, w_l^{(n)}\, w_k^{(n)}\, \mathscr{Q}_{k+l-1}$$
$$- \frac{\hbar^2}{2m}\sum_{k,l=0}^{n} w_k^{(n)} w_l^{(n)}\, \mathscr{P}_{k+l}$$
$$- \frac{\hbar^2}{2m}\sum_{l=0}^{n}\sum_{k=2}^{n} k(k-1)\, w_l^{(n)}\, w_k^{(n)}\, \mathscr{I}_{k+l-2}$$
$$+ \sum_{k,l=0}^{n} w_k^{(n)} w_l^{(n)}\, \mathscr{U}_{k+l}\,. \tag{2.19}$$

2.2.3 A test case: the harmonic oscillator

For the ground state of the harmonic oscillator with mass m and force constant k, the differential equation is

$$-\frac{\hbar^2}{2m}\frac{d^2}{dx^2}\varphi_0 + \frac{1}{2}kx^2\varphi_0 = E\varphi_0. \tag{2.20}$$

Choosing $\varphi_0(x) = \exp(-ax^2/2)$ gives

$$-\frac{\hbar^2}{2m}(a^2x^2 - a)\varphi_0 + \frac{1}{2}kx^2\varphi_0 = E\varphi_0.$$

Dropping φ_0 and rearranging the terms yield

$$-\frac{\hbar^2}{m}a^2x^2 + \frac{\hbar^2}{2m}a + \frac{1}{2}kx^2 = E.$$

The last expression holds for every value of x if and only if the coefficient of every power of x is equal on both sides. Since there are two powers of x present (2 and 0), two equations need to be satisfied. For the constant terms, we get an expression for the energy,

$$E_0 = \frac{\hbar^2}{2m}a,$$

and for the quadratic terms, we get a value for a,

$$a = \frac{\sqrt{km}}{\hbar}. \tag{2.21}$$

Note that the analysis yields the ground state energy of the harmonic oscillator with natural frequency $\omega = \sqrt{k/m}$,

$$E_0 = \frac{\hbar}{2}\omega,$$

as well as its (unnormalized) wave function $\varphi_0(x) = \exp(-ax^2/2)$. We are now going to use the strategies developed in this chapter to compute wave functions and energies of the excited states in terms of ω. The integrals we need for the task are

$$\mathscr{I}_{2n} = \int_{-\infty}^{\infty} x^{2n}\exp(-ax^2)\,dx = \frac{(2n-1)!!}{2^n a^n}\sqrt{\frac{\pi}{a}} \tag{2.22}$$

$$\mathscr{Q}_n = -a\mathscr{I}_{n+1} \tag{2.23}$$

$$\mathscr{P}_n = a^2\mathscr{I}_{n+2} - a\mathscr{I}_n \tag{2.24}$$

and

$$\mathcal{U}_n = \frac{1}{2} k \mathcal{I}_{n+2}. \tag{2.25}$$

Moreover, $\mathcal{I}_n = 0$ if n is odd.

The coefficients for the ground and first excited state, derived from (2.9) and (2.8) are

$$w_0^{(0)} = \mathcal{I}_0^{-1/2} = \left(\frac{a}{\pi}\right)^{1/4},$$
$$w_0^{(1)} = 0,$$
$$w_1^{(1)} = \left(\frac{a}{\pi}\right)^{1/4} (2a)^{1/2}.$$

Therefore, $\psi_0 = (a/\pi)^{1/4} \exp(-ax^2/2)$, $\psi_1 = (a/\pi)^{1/4}(2a)^{1/2} x \exp(-ax^2/2)$ with a node at $x = 0$ as expected. The first excited energy is

$$E_1 = -\frac{\hbar^2}{m}[w_1^{(1)}]^2(-a\mathcal{I}_2) - \frac{\hbar^2}{2m}[w_1^{(1)}]^2(a^2\mathcal{I}_4 - a\mathcal{I}_2) + \frac{1}{2}k[w_1^{(1)}]^2\mathcal{I}_4, \tag{2.26}$$

derived from equation (2.19). After the expressions for a, $\mathcal{I}_2$ and $\mathcal{I}_4$ are substituted in, and some simplifications are performed, we get the expected result $E_1 = (3/2)\hbar\omega$.

2.2.4 Some numerical results for the test case

The following Python code makes use of the results of Theorem 5 to compute some of the polynomial coefficients of the states of a particle in a harmonic oscillator with a spring constant k and the mass m equal to one. Note that atomic units are used, meaning $\hbar = 1$, the mass of an electron is 1 and the ratio $e/(4\pi\epsilon_0)$ is one. The unit of energy is the hartree and the unit of time is hartree^{-1}. The code assumes that the set of integrals $\{\mathcal{I}_n\}$ is available. The elements of this set are stored in the array `inx[i]`. The code to compute them is left as an exercise:

```
w = np.zeros((maxn+1,maxn+1),np.float64)
xm = np.zeros((maxn+1,maxn+1),np.float64)
w[0,0] = 1/np.sqrt(inx[0])  # Step 1 compute the first coefficient
for n in range(maxn):
    for i in range(n+1):
        sum1 = 0
        for m in range(i+1): # Step 2 compute < P_i | xP_n >
            for k in range(n+1):
                sum1 += w[i,m]*w[n,k]*inx[m+k+1]
```

```
        xm[i,n] = sum1
    sum1 = 0
    for m in range (n+1): # Step 3 compute < xP_n | xP_n >
        for k in range(n+1):
            sum1 += w[n,m]*w[n,k]*inx[m+k+2]
    for i in range(n+1): # Step 4 compute d_{n+1}
        sum1 -= xm[i,n]*xm[i,n]
    dnp1 = np.sqrt(sum1)
    sum1 = 0
    for i in range(n+1): # Step 5 compute w[n+1,j]
        sum1 += xm[i,n]*xm[i,0]
    w[n+1,0] = -sum1/dnp1   # constant coefficient
    w[n+1,n+1] = w[n,n]/dnp1 # coefficient of x^{n+1} in P_{n+1}(x)
    for j in range(n+1):
        sum1 = 0
        for i in range(n+1):
            sum1 += xm[i,n]*w[i,j]/dnp1
        w[n+1,j] = w[n,j-1]/dnp1 - sum1
```

The code produces values for the polynomial coefficients of the lowest six states. These are listed in Table 2.1. Only the upper triangular part of the position matrix is needed for the computation of the coefficients. The position matrix is symmetric, but this property is not necessary in the implementation of the algorithm to compute the polynomial coefficients. Here are some values as a check: $\langle P_0, xP_1 \rangle = 0.0701$, $\langle P_1, xP_2 \rangle = 1.000$, $\langle P_2, xP_3 \rangle = 1.2247$.

Table 2.1: Coefficients of the first six normalized Hermite polynomials for a mass and spring constant equal to one.

$P_j(x)$	w_0^j	w_1^j	w_2^j	w_3^j	w_4^j	w_5^j
P_0	0.7511	0	0	0	0	0
P_1	0	1.0623	0	0	0	0
P_2	−0.5311	0	1.0623	0	0	0
P_3	0	−1.3010	0	0.8673	0	0
P_4	0.4600	0	−1.8399	0	0.6133	0
P_5	0	1.4545	0	−1.9394	0	0.3879

The next snippet of Python code computes the first few energy levels,

```
for n in range(maxn-1):
    sum1 = 0.0
    sum2 = 0.0
```

```
sum3 = 0.0
sum4 = 0.0
for m in range(n+1):
    for k in range(n+1):
        if k > 0:
            sum1 += -k*w[n,m]*w[n,k]*qnx[k+m-1]/mass
        sum2 += -w[n,m]*w[n,k]*pnx[k+m]/(2*mass)
        if k > 1:
            sum3 +=  -(k*(k-1))*w[n,m]*w[n,k]*inx[k+m-2]/(2*mass)
        sum4 +=  w[n,m]*w[n,k]*unx[k+m]
en = sum1+sum2+sum3+sum4
```

Of course, these should agree with the known energy spectrum of the harmonic oscillator Hamiltonian,

$$E_n = \left(n + \frac{1}{2}\right)\hbar\omega,$$

and the check is left as an exercise as well.

At this point, we have seen a few examples where excited states can be constructed by multiplying a linear polynomial to φ_0 for the first excitation, quadratic for the second, etc. However, it is important to state that the version of the Sturm–Liouville theorem we prove here does not guarantee that such process works in general. Instead, the theory only tells us that we can expand H_1 in $\psi_1 = H_1\varphi_0$ into a polynomial since polynomials are dense in compact spaces (or with compact support as we do here) by the Weierstrass approximation theorem. It does not tell us the order of such polynomial, however. A simple case in point is the particle in a box for $x \in (0, 1)$, where the ground and excited states are sinusoidal functions, a polynomial of high order is required to fit the ground state alone, and the ratio $\psi_1/\psi_0 = \sin(2\pi x)/\sin(\pi x)$ is clearly not a linear polynomial.

2.3 The Hankel determinant representation

We conclude this chapter with an alternative approach for the construction of orthogonal polynomials in general. The method uses matrix decomposition (typically LU) to compute a set of determinants. Consider the following function:

$$\mathcal{D}_n(x) = \det \begin{vmatrix} \mathcal{I}_0 & \mathcal{I}_1 & \mathcal{I}_2 & \cdots & \mathcal{I}_{n-1} & \mathcal{I}_n \\ \mathcal{I}_1 & \mathcal{I}_2 & \mathcal{I}_3 & \cdots & \mathcal{I}_n & \mathcal{I}_{n+1} \\ \vdots & & & & & \vdots \\ \mathcal{I}_{n-1} & \mathcal{I}_n & \mathcal{I}_{n+1} & \cdots & \mathcal{I}_{2n-2} & \mathcal{I}_{2n-1} \\ 1 & x & x^2 & \cdots & x^{n-1} & x^n \end{vmatrix},$$

where $\mathscr{I}_n$ are the nth moments of the measure $\omega(x)$ in equation (2.5) and $\mathcal{D}_0(x) = 1$. Clearly, $\mathcal{D}_n(x)$ is a polynomial of degree n, as can be expressed by using the cofactor expansion about the last row,

$$\mathcal{D}_n(x) = \sum_{k=0}^{n} C_k^{(n)} x^k$$

where, e. g.,

$$C_2^{(n)} = (-1)^{n+2} \det \begin{vmatrix} \mathscr{I}_0 & \mathscr{I}_1 & \mathscr{I}_3 & \cdots & \mathscr{I}_{n-1} & \mathscr{I}_n \\ \mathscr{I}_1 & \mathscr{I}_2 & \mathscr{I}_4 & \cdots & \mathscr{I}_n & \mathscr{I}_{n+1} \\ \vdots & & & & & \vdots \\ \mathscr{I}_{n-1} & \mathscr{I}_n & \mathscr{I}_{n+2} & \cdots & \mathscr{I}_{2n-2} & \mathscr{I}_{2n-1} \end{vmatrix},$$

$$C_n^{(n)} = \det \begin{vmatrix} \mathscr{I}_0 & \mathscr{I}_1 & \mathscr{I}_2 & \cdots & \mathscr{I}_{n-1} \\ \mathscr{I}_1 & \mathscr{I}_2 & \mathscr{I}_3 & \cdots & \mathscr{I}_n \\ \vdots & & & & \\ \mathscr{I}_{n-1} & \mathscr{I}_n & \mathscr{I}_{n+1} & \cdots & \mathscr{I}_{2n-2} \end{vmatrix}.$$

The expression $\mathcal{D}_n(x)$ is known as the Hankel determinant representation of a polynomial of degree n.

For example,

$$\mathcal{D}_1(x) = \det \begin{vmatrix} \mathscr{I}_0 & \mathscr{I}_1 \\ 1 & x \end{vmatrix} = \mathscr{I}_0 x - \mathscr{I}_1.$$

Note that $\mathcal{D}_1(x)$ is not normalized. The following theorem gives us the normalization constants and the polynomial coefficients in terms of the cofactors of the last row of $\mathcal{D}_n(x)$.

Theorem 7. *The polynomial determinants $\mathcal{D}_n(x)$ and $\mathcal{D}_m(x)$ are orthogonal, the polynomials*

$$p_n(x) = \frac{\mathcal{D}_n(x)}{\sqrt{C_{n+1}^{(n+1)} C_n^{(n)}}}$$

are normalized, and

$$w_k^{(n)} = \frac{C_k^{(n)}}{\sqrt{C_{n+1}^{(n+1)} C_n^{(n)}}}.$$

Proof. Consider the following inner product:

$$\langle \mathcal{D}_n(x), x^m \rangle_\omega = \sum_{k=0}^{n} C_k^{(n)} \int_R x^{k+m}\, \omega(x)\, dx,$$

for $m \le n$. This can be expressed as a determinant

$$\langle \mathcal{D}_n(x), x^m \rangle_\omega = \det \begin{vmatrix} \mathcal{I}_0 & \mathcal{I}_1 & \mathcal{I}_2 & \cdots & \mathcal{I}_n \\ \mathcal{I}_1 & \mathcal{I}_2 & \mathcal{I}_3 & \cdots & \mathcal{I}_{n+1} \\ \vdots & & & & \vdots \\ \mathcal{I}_{n-1} & \mathcal{I}_n & \mathcal{I}_{n+1} & \cdots & \mathcal{I}_{2n-1} \\ \mathcal{I}_m & \mathcal{I}_{m+1} & \mathcal{I}_{m+2} & \cdots & \mathcal{I}_{m+n} \end{vmatrix} = \delta_{m,n} C_{n+1}^{(n+1)}$$

where

$$C_{n+1}^{(n+1)} = \det \begin{vmatrix} \mathcal{I}_0 & \mathcal{I}_1 & \mathcal{I}_2 & \cdots & \mathcal{I}_n \\ \mathcal{I}_1 & \mathcal{I}_2 & \mathcal{I}_3 & \cdots & \mathcal{I}_{n+1} \\ \vdots & & & & \vdots \\ \mathcal{I}_{n-1} & \mathcal{I}_n & \mathcal{I}_{n+1} & \cdots & \mathcal{I}_{2n-1} \\ \mathcal{I}_n & \mathcal{I}_{n+1} & \mathcal{I}_{n+2} & \cdots & \mathcal{I}_{2n} \end{vmatrix},$$

for if $m \neq n$, say, $m = n - 1$, then

$$\langle \mathcal{D}_n(x), x^{n-1} \rangle_\omega = \det \begin{vmatrix} \mathcal{I}_0 & \mathcal{I}_1 & \mathcal{I}_2 & \cdots & \mathcal{I}_n \\ \mathcal{I}_1 & \mathcal{I}_2 & \mathcal{I}_3 & \cdots & \mathcal{I}_{n+1} \\ \vdots & & & & \vdots \\ \mathcal{I}_{n-1} & \mathcal{I}_n & \mathcal{I}_{n+1} & \cdots & \mathcal{I}_{2n-1} \\ \mathcal{I}_{n-1} & \mathcal{I}_n & \mathcal{I}_{n+1} & \cdots & \mathcal{I}_{2n-1} \end{vmatrix}.$$

The last row always equals another row of the matrix and the determinant evaluates to zero unless $m = n$. Any degree $m < n$ will cause two rows in the resulting determinant to become equal. Consequently every term

$$\langle \mathcal{D}_n(x), \mathcal{D}_m(x) \rangle_\omega = \sum_{\ell=0}^{m} C_\ell^{(m)} \sum_{k=0}^{n} C_k^{(n)} \int_R x^{k+l}\, \omega(x)\, dx$$

will yield a zero determinant $\sum_{k=0}^{n} C_k^{(n)} \int_R x^{k+l}\, \omega(x)\, dx$ with two identical rows. Moreover,

$$\langle \mathcal{D}_n(x), \mathcal{D}_n(x) \rangle_\omega = C_n^{(n)} \langle \mathcal{D}_n(x), x^n \rangle_\omega = C_n^{(n)} C_{n+1}^{(n+1)}$$

since the determinant $C_n^{(n)}$ is the coefficient of x^n. This last expression is used to normalize $\mathcal{D}_n(x)$. □

The last is an example of a "constructive proof," which not only proves the theorem, but indicates the strategy to mathematically construct the determinant function to be orthogonal to determinant functions of lower degree. Moreover, Theorem 6 gives us a way of computing the coefficients by evaluating a number of determinants. Most determinants are of nonsymmetric matrices and are best evaluated using the LU decomposition or the Singular Value Decomposition (SVD).

2.4 Exercises

1. Prove that ψ_1 in equation (2.10) is normalized.
2. Fill in the details of the derivation of equation (2.19).
3. For the generic linear superposition of eigenfunctions

$$\psi(x) = A\sum_{j=0}^{n} c_j\psi_j(x),$$

find an expression for the normalization constant A and the energy E in terms of the coefficients and the eigenenergies.
4. Check that substitution of $y = H_n$, r in equation (2.4), $q(x) = -2mE_0\varphi_0^2/\hbar^2$ and $\lambda = 2mE_n/\hbar^2$ back into equation (2.3) yields equation (2.2) after ψ_0^2 is canceled on both sides.
5. Consider the Harmonic oscillator DE:

$$-\frac{\hbar^2}{2m}\frac{d^2}{dx^2}\psi_n + \frac{1}{2}kx^2\psi_n = E_n\psi_n.$$

(a) Transform the independent variable x as follows:

$$y = x\left(\frac{mk}{\hbar^2}\right)^{1/4}$$

and derive the following form:

$$\frac{d^2}{dy^2}\psi_n - y^2\psi_n + \epsilon_n\psi_n = 0,$$

with $\epsilon_n = 2E_n/(\hbar\omega)$.

(b) Rescale the dependent variable as follows:

$$\psi_n = H_n e^{-y^2/2},$$

and derive Hermite's differential equation,

$$\frac{d^2}{dy^2}H_n - 2y\frac{d}{dy}H_n + (\epsilon_n - 1)H_n = 0.$$

(c) Let $H_0(y) = 1$, $H_1(y) = y$, $H_2(y) = y^2 - 1/2$, and $H_3(y) = y^3 - 3y/2$. Show that these satisfy the differential equation for special values of $\epsilon_n = 1, 3, 5$, and 7, respectively.

(d) Show that

$$\int_{-\infty}^{\infty} H_2(y)H_0(y)\, e^{-y^2}\, dy, \quad \text{and} \quad \int_{-\infty}^{\infty} H_3(y)H_1(y)\, e^{-y^2}\, dy = 0.$$

(e) The polynomials in part (c) have the coefficient of the highest power of y equal to one. Such polynomials are called *monic* $\Pi_n(y) = x^n + a_{n-1}^{(n)}x^{n-1} + \cdots + a_0^{(n)}$. Monic Hermite polynomials can be generated with the following Rodrigues' formula:

$$\Pi_n(y) = -e^{y^2}\frac{d^n}{dy^2}e^{-y^2}$$

derive expressions for $\Pi_n(y)$ for $n = 0$ through 3, and show that you get exactly $H_0(y)$, $H_1(y)$, $H_2(y)$, and $H_3(y)$ of part (c).

(f) The monic polynomials can be obtained from the Hankel determinant representation,

$$\Pi_n(y) = \frac{\mathcal{D}_n(y)}{C_{(n)}^n}.$$

Prove this statement in general.

(g) Use the Hankel determinant representation for Hermite polynomials $\mathcal{D}_1(y)$, $\mathcal{D}_2(y)$, and $\mathcal{D}_3(y)$. Then use the result of the last section to derive the same expressions for $H_0(y)$, $H_1(y)$, $H_2(y)$, and $H_3(y)$ of part (c). Begin by showing that

$$\mathcal{I}_{2n} = \frac{(2n-1)!!}{2^n}\sqrt{\pi},$$

and zero otherwise, as a special case of equation (2.22). Then

$$\mathcal{I}_0 = \sqrt{\pi}, \quad \mathcal{I}_2 = \frac{1}{2}\sqrt{\pi}, \quad \mathcal{I}_4 = \frac{3}{4}\sqrt{\pi}, \quad \mathcal{I}_6 = \frac{15}{8}\sqrt{\pi} \quad \text{etc.}$$

6. The **Discrete Variable Representation** (DVR) is a vector space built on a lattice of points in configuration space. Points are typically evenly spaced, and projecting derivatives in this vector space is relatively simple. For $x \in [a, b]$ with $a, b \in \mathbb{R}$, $b > a$, an evenly space grid with $n+1$ points, $x_0, x_1, \ldots, x_n$ has a spacing $h = (b-a)/n$. The derivative of a function $f(x)$ on such lattice can be approximated on this vector space using the central definition of the derivative

$$\frac{d}{dx}f(x_i) \approx \frac{f(x_{i+1}) - f(x_{i-1})}{2h}$$

for $i > 0$, $i < n$. When $i = 0$,

$$\frac{d}{dx}f(x_0) \approx f(x_0)/(2h),$$

and when $i = n$,

$$\frac{d}{dx}f(x_n) \approx -f(x_{n-1})/(2h).$$

(a) For simplicity, consider the $n = 3$ case. Let D be the matrix representation of the derivative operator d/dx on this vector space. Then, by the central difference approximation,

$$D\begin{pmatrix} f(x_0) \\ f(x_1) \\ f(x_2) \\ f(x_3) \end{pmatrix} \approx \frac{1}{2h}\begin{pmatrix} f(x_1) \\ f(x_2) - f(x_0) \\ f(x_3) - f(x_0) \\ -f(x_2) \end{pmatrix}.$$

Show by explicitly expanding the matrix vector product that

$$D = \frac{1}{2h}\begin{pmatrix} 0 & 1 & 0 & 0 \\ -1 & 0 & 1 & 0 \\ 0 & -1 & 0 & 1 \\ 0 & 0 & -1 & 0 \end{pmatrix}.$$

(b) Use the same procedure in part (a) together with the definition of the second derivative on a lattice,

$$\frac{d^2 f(x_i)}{dx^2} \approx \frac{f(x_{i+1}) - 2f(x_i) + f(x_{i-1})}{h^2}$$

to find the representation of D^2, the Laplacian operator for the $n = 3$ case.

(c) The nice property of the DVR is that there are no integrals involved, the representation of the potential energy is a diagonal $(n+1) \times (n+1)$ matrix with $U(x_i)$ as the U_{ii} element. Run the following Python code to calculate the DVR Hamiltonian for the harmonic oscillator with mass and spring constant equal to 1838 and 0.01 atomic unit, respectively. Print out the eigenvalues and eigenvectors:

```
import numpy as np
mass = 1838.0          # atomic units are used throughout
spring_constant = 0.010
maxn = 40
d = 5.0  # The length of [-a,a]
```

```
a = -d/2.0
h = d/maxn
h2 = h*h
mh = mass*h2
hm = np.zeros((maxn+1,maxn+1),np.float64)
for i in range(maxn+1):
    x = a + i*h
    for j in range(maxn+1):
        if i == j:
            hm[i,j] = 1.0/mh + spring_constant*x*x/2.0
        if i == (j-1):
            hm[i,j] = -1.0/(2.0*mh)
        if i == (j+1):
            hm[i,j] = -1.0/(2.0*mh)
eigenvalues, eigenvectors =  np.linalg.eig(hm)
idx = eigenvalues.argsort()[::+1]
for n in range(10):
    print(n,eigenvalues[idx[n]])
```

Note that indentation here is very important. All the commands inside the `for i in range(maxn):` loop must begin in column 5. Inside the first nested control sequence, the commands begin in column 9, in the second nested sequence in column 13, and so on. Command that begin four spaces to the left of the previous line indicate the end of the control sequence block. The line `idx = eigenvalues.argsort()[::+1]` creates an array of indexes needed to sort the `eigenvalues` array in ascending order.

(d) Write code that computes the frequency ω, and the expected values of the energies. Then compute the differences between the eigenvalues and the expected energies.

(e) There are two sources of errors with the direct vector space approach and these can be appreciated in the present DVR example. The sources are the basis set truncation error and the energy cutoff error. Both of these can be controlled systematically and this exercise demonstrates how. Increase the value of `maxn` by 10 until the relative error of the first four DVR energies is below 1×10^{-5} hartree to monitor the basis set size convergence.

(f) The cutoff error affects excited states that approach from below the cutoff energy. For $U = kx^2/2$, the cutoff energy is $kd^2/8$. Compute the cutoff energy when $d = 5.0$. Set `maxn` back to 40, increase the value of `d` so that ten states are below the cutoff, and gradually increment the basis set size until the first ten states converge.

(g) Colbert and Miller [86] developed a n-point representation of the second derivative that can accelerate the basis set convergence of the DVR. The Hamiltonian matrix is computed with

$$\langle i|H|j\rangle = \begin{cases} \frac{\hbar^2\pi^2}{6m\epsilon^2} + U(x_j) & i = j \\ (-1)^{(i-j)} \frac{\hbar^2\pi^2}{m\epsilon^2(i-j)^2} & i \neq j \end{cases}$$

Write code that computes this representation and repeat the convergence analysis. To compute $(-1)^{(i-j)}$, one approach is to use modular arithmetics with the lines of code below:

```
sign = -1.0
if (i-j) % 2 == 0:
   sign = 1.0
hm[i,j] = sign/(mh*((i-j)**2))
```

These will need to be placed inside the two loops and in the control sequence that handles the $i \neq j$ case. In Python, the modulo operator $a\%b$ divides integer a by integer b and returns the integer remainder. If the remainder of $(i - 1)$ divided by 2 is zero, then $(i - 1)$ is even and (-1) raised to an even number is 1.

7. Fill in the details of the derivation of equation (2.26), then show that when $w_1^{(1)}$, a, $\mathcal{I}_2$ and $\mathcal{I}_4$ are substituted in, E_1 is the expected result $E_1 = 3\hbar\omega/2$.
8. Prove the result in equation (2.22) by starting with

$$\int_{-\infty}^{\infty} e^{-ax^2}\,dx = \sqrt{\frac{\pi}{a}},$$

and using the strategy suggested by R. P. Feynman, i. e., by evaluating the derivative with respect to the scaling factor a on both sided of the last equation,

$$-\frac{d}{da}\int_{-\infty}^{\infty} e^{-ax^2}\,dx = \frac{1}{2a}\sqrt{\frac{\pi}{a}}$$

where the derivative is evaluated on the right-hand side, and the derivative operation can be permuted with the integration on the left since the integrand is continuous and bound in $(-\infty, \infty)$,

$$\int_{-\infty}^{\infty} -\frac{d}{da} e^{-ax^2}\,dx = \frac{1}{2a}\sqrt{\frac{\pi}{a}}.$$

Next, the evaluation of the derivative in the integrand yields

$$\int_{-\infty}^{\infty} x^2 e^{-ax^2}\,dx = \frac{1}{2a}\sqrt{\frac{\pi}{a}}.$$

Now note that the expression you seek to prove works for the $n = 0$ and 1 case. Use induction to complete the proof.

9. Integrals of the type

$$\int_{-a}^{a} f(x)dx$$

where $f(x)$ is an odd function of x vanish by symmetry. Odd functions integrated over symmetric domains vanish as can be shown by breaking the integrals into two terms. The first integrates from $-a$ to 0, and the second from 0 to a. If $f(-x) = -f(x)$, a change of variable $y = -x$ can prove the integral to vanish. Show that it is true in general, then use the same steps to show that

$$\int_{-\infty}^{\infty} x^{2n-1} e^{-ax^2} dx = 0 \quad n \geq 1.$$

10. A minimal Python code to compute values of $\mathscr{I}_n$ for the harmonic oscillator is below:

```
import numpy as np
mass = 1          # atomic units are used throughout
spring_constant = 1
frequency = np.sqrt(spring_constant/mass)
a = np.sqrt(spring_constant*mass)
maxn = 10  # maximum number of states
inx  = np.zeros(2*maxn+1)  # first 21 moments are stored here
factorial_factorial = 1
spoa = np.sqrt(np.pi/a)
tan = 1
for n in range(maxn+1):  # Computes the moments
    inx[2*n] = factorial_factorial*spoa/tan
    factorial_factorial *= (2*(n+1) - 1)
    tan = tan*(2*a)
```

Print values of `n`, `factorial_factorial`, `inx[2*n]`, and compare them with n, $(2n-1)!!$, and $\mathscr{I}_{2n}$. Compare the value of $\mathscr{I}_{20}$ with 1133278.38....

11. Use the code from the previous exercise to compute values for $\mathscr{Q}_n$, $\mathscr{P}_n$, and $\mathscr{U}_n$ by mean of equations (2.23) through (2.25).

12. After completing the two previous exercises perform the following tasks:
    (a) Append the Python code for the computation of the polynomial coefficients, run the script and check with the result in Table 2.1. The command `print(w)` will create a hard to read output. A better strategy, albeit not very elegant, is to use, e. g.,

```
for i in range(3):
    print('{0:8f} {1:8f} {2:8f}'.format(w[i,0],w[i,1],w[i,2]))
```

    to print a three by three table of coefficients.

(b) Append the snippet of code that computes the first few energies and compare with the expected energies, 0.5, 1.5, 2.5, etc.
(c) When the code is working, mute the output except for the energy values. Increase the number of states (`max`) to 50 and compute the deviations between the expected and computed values of the energies. The deviations are caused by the accumulated rounding errors. After $n = 36$, you should see lots of `nan` the dreaded "not a number" warning, which means erroneous operations are carried out.
(d) The likely culprit for those erroneous operation messages is the line
`dnp1 = np.sqrt(sum1)`
where the square root of a negative number is attempted. Verify that statement by adding print statements to the code.

13. This set of exercises is designed to gain some experience with the Hankel determinant expression of orthogonal polynomials in one variable.
(a) Beginning with

$$\mathcal{D}_0 = 1,$$
$$\mathcal{D}_1(x) = \det\begin{vmatrix} \mathcal{I}_0 & \mathcal{I}_1 \\ 1 & x \end{vmatrix}$$

evaluate the determinant.
(b) Now normalize by expressing

$$C_2^2 = \det\begin{vmatrix} \mathcal{I}_0 & \mathcal{I}_1 \\ \mathcal{I}_1 & \mathcal{I}_2 \end{vmatrix}$$

and $C_1^1 = \mathcal{I}_0$. Show that

$$p_1(x) = \frac{\mathcal{I}_0 x - \mathcal{I}_1}{(\mathcal{I}_0^2 \mathcal{I}_2 - \mathcal{I}_0 \mathcal{I}_1^2)^{1/2}}.$$

(c) Compare the expression in part (b) with the one derived in the chapter using the Gram–Schmidt procedure

$$p_1(x) = \frac{x - \mathcal{I}_1 \mathcal{I}_0^{-1}}{(\mathcal{I}_2 - \mathcal{I}_0^{-1} \mathcal{I}_1^2)^{1/2}}$$

and show that they are the same after some trivial algebra.
(d) Expand the determinant function

$$\mathcal{D}_2(x) = \det\begin{vmatrix} \mathcal{I}_0 & \mathcal{I}_1 & \mathcal{I}_2 \\ \mathcal{I}_1 & \mathcal{I}_2 & \mathcal{I}_3 \\ 1 & x & x^2 \end{vmatrix}$$

and use it to show that

$$\langle \mathcal{D}_2(x), 1\rangle_\omega = \det\begin{vmatrix} \mathscr{I}_0 & \mathscr{I}_1 & \mathscr{I}_2 \\ \mathscr{I}_1 & \mathscr{I}_2 & \mathscr{I}_3 \\ \mathscr{I}_0 & \mathscr{I}_1 & \mathscr{I}_2 \end{vmatrix} = 0$$

$$\langle \mathcal{D}_2(x), x\rangle_\omega = \det\begin{vmatrix} \mathscr{I}_0 & \mathscr{I}_1 & \mathscr{I}_2 \\ \mathscr{I}_1 & \mathscr{I}_2 & \mathscr{I}_3 \\ \mathscr{I}_1 & \mathscr{I}_2 & \mathscr{I}_3 \end{vmatrix} = 0.$$

(e) Use the results in part (d) to show that

$$\langle \mathcal{D}_2(x), \mathcal{D}_1(x)\rangle_\omega = 0.$$

14. Use the Hankel determinant expansion to show that

$$p_2(x) = w_0^{(2)} + w_1^{(2)}x + w_2^{(2)}x^2$$

where

$$\begin{aligned} w_2^{(2)} &= \frac{\mathscr{I}_1^2 - \mathscr{I}_0\mathscr{I}_2}{d_2}, \\ w_1^{(2)} &= \frac{\mathscr{I}_0\mathscr{I}_3 - \mathscr{I}_1\mathscr{I}_2}{d_2} \\ w_0^{(2)} &= \frac{\mathscr{I}_2^2 - \mathscr{I}_1\mathscr{I}_3}{d_2} \end{aligned}$$

$$d_2 = \left[\mathscr{I}_0(\mathscr{I}_2\mathscr{I}_4 - \mathscr{I}_3^2) - \mathscr{I}_1(\mathscr{I}_1\mathscr{I}_4 - \mathscr{I}_2\mathscr{I}_3) + \mathscr{I}_2(\mathscr{I}_1\mathscr{I}_3 - \mathscr{I}_2^2)\right]^{1/2} \times \left[\mathscr{I}_0\mathscr{I}_2 - \mathscr{I}_1^2\right]^{1/2}.$$

15. It is possible to produce the coefficients of a few polynomials from a list of known moments using your favorite spreadsheet application if the values of the moments, what we have been calling the monomial integrals $\mathscr{I}_n$, are known.
(a) Recall the example in Chapter 1,

$$\mathscr{I}_n = \int_{-1}^{1} x^n dx = \frac{1}{n+1}\left[1 - (-1)^{n+1}\right].$$

Use this equation to create a list of integrals for values of n between 0 and 8.
(b) Use these values to construct the matrix needed for the evaluations of $\{w_k^{(4)}\}_{k=0}^4$. Follow the steps below.

(i) Construct the 5×5 matrix whose determinant is C_5^5, then use the `MDETERM()` function or its equivalent, to compute the numerical value of C_5^5. See Table 2.2 for reference.

Table 2.2: Exercise 15 reference data.

2	0	2/3	0	2/5
0	2/3	0	2/5	0
2/3	0	2/5	0	2/7
0	2/5	0	2/7	0
2/5	0	2/7	0	2/9

(ii) Construct another matrix by eliminating the last row of the matrix in part (i) as well as the rightmost column. Its determinant is C_4^4.

(iii) Use the results in parts (i) and (ii) to find the normalization constant.

(iv) Construct four other matrices by eliminating the last row of the matrix in part (i) and column $j \in \{1, 2, 3, 4\}$, and use them to construct the proper cofactors of $\mathcal{D}_4(x)$, and the coefficients of the normalized polynomial $p_4(x)$.

(c) Prove that C_{n+1}^{n+1}, and C_n^n should always be greater than zero.

3 Elements of group theory

The actual solution of quantum mechanical equations is, in general, so difficult that one obtains by direct calculations only crude approximations to the real solutions. It is gratifying, therefore, that a large part of the relevant results can be deduced by considering the fundamental symmetry operations.

Eugene P. Wigner

In this chapter, we introduce finite and continuous groups. The most practical aspect of groups explored in this chapter is the transformation properties of vectors carried out by members of both finite and continuous groups. The group test performed by verifying closure is demonstrated using several examples. Subgroups, classes, cyclic groups, the permutation group, the order of groups, and the order of elements of a group are all defined in Section 3.1. Additionally, in the same section, the Cayley table is computed for a couple of examples, and its role in identifying subgroups and isomorphisms is explained. In Section 3.2, we define and produce examples of the most important classical Lie groups of continuous transformations. Among the examples provided are rotations and reflections in $\mathbb{R}^2$ and $\mathbb{R}^3$, similarity transformations, norm preserving transformations, and finally the trace invariance theorem is proved.

3.1 Basic definitions

An algebraic group is defined as follows.

Definition 7 (Group). A group is a set of objects $\mathcal{G} = \{a, b, c, \ldots\}$ together with a binary group operation, $*$ that satisfies the following three properties:

1. The group operation satisfies closure, given $a, b \in \mathcal{G}$. Then $a * b = c \in \mathcal{G}$.
2. The group operation is associative, for $a, b, c \in \mathcal{G}$, $(a * b) * c = a * (b * c)$.
3. There exists an identity element $e \in \mathcal{G}$ under $*$ such that, for $a \in \mathcal{G}$, $a * e = e * a = a$.
4. Every element $a \in \mathcal{G}$ has a unique inverse $a^{-1} \in \mathcal{G}$ such that $a^{-1} * a = a * a^{-1} = e$.

We have already seen in Chapter 1 three types of binary operations that satisfy closure: multiplication of scalars, multiplication between two matrices when allowed, and summation. The operation $*$ can represent any of these, however, other group operations exist. A set is usually called a group under the particular operation that defines it. Groups can have a simpler algebraic structure than fields and vector spaces because only one binary operation is required to define them. Vector spaces are examples of groups under the vector addition with an additional operation, the multiplication by a scalar.

Example 1 ($\mathrm{Gl}(2, \mathbb{R})$). As a first example, the group of 2×2 matrices with nonzero determinant $\mathrm{Gl}(2, \mathbb{R})$ is defined as

https://doi.org/10.1515/9783111610207-003

$$\mathrm{Gl}(2,\mathbb{R}) = \left\{ A = \begin{pmatrix} a & b \\ c & d \end{pmatrix} \,\middle|\, a,b,c,d \in \mathbb{R},\ ad - bc \neq 0 \right\}.$$

As defined, $\mathrm{Gl}(2,\mathbb{R})$ is a group under matrix multiplication. To see how closure is satisfied under matrix multiplication, consider

$$A_1A_2 = \begin{pmatrix} a_1 & b_1 \\ c_1 & d_1 \end{pmatrix}\begin{pmatrix} a_2 & b_2 \\ c_2 & d_2 \end{pmatrix} = \begin{pmatrix} a_1a_2 + b_1c_2 & a_1b_2 + b_1d_2 \\ c_1a_2 + d_1c_2 & c_1b_2 + d_1d_2 \end{pmatrix}.$$

Then

$$\begin{aligned} &\det(A_1A_2) \\ &= \det\begin{pmatrix} a_1a_2 + b_1c_2 & a_1b_2 + b_1d_2 \\ c_1a_2 + d_1c_2 & c_1b_2 + d_1d_2 \end{pmatrix} \\ &= (a_1a_2 + b_1c_2)(c_1b_2 + d_1d_2) - (a_1b_2 + b_1d_2)(c_1a_2 + d_1c_2). \end{aligned}$$

After expanding, permuting the components in each term since in $\mathbb{R}$ the product is invariant under permutation, and canceling, we end up with four terms

$$\begin{aligned} &\det(A_1A_2) \\ &\quad = a_1d_1a_2d_2 + b_1c_1b_2c_2 - a_1d_1b_2c_2 - b_1d_1c_2d_2. \end{aligned}$$

This is shown by directly expanding the product of the binomials to be equivalent to

$$\det A_1 \det A_2 = (a_1d_1 - b_1c_1)(a_2d_2 - b_2c_2).$$

Therefore, $\det(A_1A_2) = \det A_1 \det A_2 \neq 0$ since neither $\det A_1$, nor $\det A_2$ can be zero by assumption. The proof that $\det AB = \det A \det B$ for $n \times n$ matrices in general requires the development of elementary row reduction operations to transform them into row-echelon form. The reader is guided through the details of the proof in two exercises at the end of the chapter.

Groups can have *subgroups* and these are defined as follows.

Definition 8 (Subgroup). A subgroup H of a group $\mathcal{G}$ is a set of elements of $\mathcal{G}$ that behaves like its own group under the operation of $\mathcal{G}$. If $\mathcal{H}$ is a subgroup of $\mathcal{G}$, we write $\mathcal{H} \leq \mathcal{G}$. If not all elements of $\mathcal{G}$ are in $\mathcal{H}$, then we call $\mathcal{H} < \mathcal{G}$ a *proper subgroup* of $\mathcal{G}$.

Example 2 (The group of 2×2 matrices with unit determinant $\mathrm{Sl}(2,\mathbb{R})$). The set of all 2×2 matrices that have determinants equal to one under multiplication is called the special linear group,

$$\mathrm{Sl}(2,\mathbb{R}) = \left\{ A_i = \begin{pmatrix} a_i & b_i \\ c_i & d_i \end{pmatrix} \,\middle|\, a_i, b_i, c_i, d_i \in \mathbb{R},\ a_id_i - b_ic_i = 1 \right\},$$

and is a proper subgroup of $\mathrm{Gl}(2,\mathbb{R})$. The group of $n \times n$ matrices with nonzero determinant under matrix multiplication is $\mathrm{Gl}(n,\mathbb{R})$, and $\mathrm{Sl}(n,\mathbb{R}) < \mathrm{Gl}(n,\mathbb{R})$ is the subgroup of $n\times n$ matrices with unit determinant. The properties of the group operation $*$ listed in the definition is used to prove the cancellation properties and uniqueness of the identity and inverse. The following is another important result regarding multiplicative inverses.

Theorem 8 (The sock and shoe relationship). *For $a, b \in \mathcal{G}$ $(ab)^{-1} = b^{-1}a^{-1}$.*

Proof. Since $a^{-1}a = e$, $b^{-1}b = e$ by definition and $c = ab \in \mathcal{G}$ by definition, $\exists\, c^{-1} \in \mathcal{G}$ such that $c^{-1}c = e$ or

$$e = b^{-1}a^{-1}ab = b^{-1}eb = b^{-1}be = ee = e.$$

The next definition deals with the number of elements in a group, and these can be finite or infinite. □

Definition 9 (Order of a group). The number of elements in a group $\mathcal{G}$, denoted $|\mathcal{G}|$ is called the *order* of the group.

The examples we have looked at so far are groups with infinite order since there are an infinite number of a, b, c, d combinations of real numbers where $ad - bc \neq 0$ and $ad - bc = 1$. We shall be interested in finite groups as well as infinite ones.

Example 3 (The dihedral group D_3). Consider the group composed of the following six matrices. $\{E, A, B, C, D, F\}$ where

$$E = \begin{pmatrix} 1 & 0 \\ 0 & 1 \end{pmatrix}, \quad A = \begin{pmatrix} 1 & 0 \\ 0 & -1 \end{pmatrix}, \quad B = \begin{pmatrix} -\frac{1}{2} & \frac{\sqrt{3}}{2} \\ \frac{\sqrt{3}}{2} & \frac{1}{2} \end{pmatrix}, \tag{3.1}$$

$$C = \begin{pmatrix} -\frac{1}{2} & -\frac{\sqrt{3}}{2} \\ -\frac{\sqrt{3}}{2} & \frac{1}{2} \end{pmatrix}, \quad D = \begin{pmatrix} -\frac{1}{2} & \frac{\sqrt{3}}{2} \\ -\frac{\sqrt{3}}{2} & -\frac{1}{2} \end{pmatrix}, \quad F = \begin{pmatrix} -\frac{1}{2} & -\frac{\sqrt{3}}{2} \\ \frac{\sqrt{3}}{2} & -\frac{1}{2} \end{pmatrix}. \tag{3.2}$$

By multiplying each element with all the other elements, including itself, we can construct a so-called *Cayley table*. For example,

$$CD = \frac{1}{4}\begin{pmatrix} -1 & -\sqrt{3} \\ -\sqrt{3} & 1 \end{pmatrix}\begin{pmatrix} -1 & \sqrt{3} \\ -\sqrt{3} & -1 \end{pmatrix} = \begin{pmatrix} 1 & 0 \\ 0 & -1 \end{pmatrix} = A$$

*	E	A	B	C	D	F
E	E	A	B	C	D	F
A	A	E	D	F	B	C
B	B	F	E	D	C	A
C	C	D	F	E	A	B
D	D	C	A	B	F	E
F	F	B	C	A	E	D

(3.3)

where the element on the left of $*$ comes from the first column, and the element to the right from the first row. Using the Cayley table in equation (3.3), we can verify that the group has 5 proper subgroups, one of order one, one of order three, and three of order two: $\{E\}$, $\{E, A\}$, $\{E, B\}$, $\{EC\}$, $\{E, D, F\}$. Note that from the table $DC = B \neq CD = A$, therefore, the order of operation is important. Groups under a commutative operation are called *Abelian* groups. The Cayley table of an Abelian group is symmetric about the main diagonal. The composition of the Cayley table is proof by exhaustion that $*$ satisfies closure. Therefore, D_3 is a group and its order is 6 because there are six elements in it. Note that every element has an inverse; E, A, B, C are their own inverses, e. g., $BB = E$, and the inverse of D is F since $DF = FD = E$. An important class of groups is constructed by evaluating successive group operations of a single element and continue until the identity is reached. Such groups are called cyclic.

Definition 10 (Cyclic groups). Let a be an element of a group. The cyclic set of a, $\langle a\rangle = \{e, a, a * a, a * a * a, \ldots\}$ is a group under the operation $*$.

Note that all the subgroups of D_3 in example 3 are cyclic groups. Cyclic groups are powerful tools in the identification of subgroups. The order of an element of a group is defined using the cyclic group it generates.

Definition 11 (Order of an element). In a group $\mathcal{G}$, the order of $a \in \mathcal{G}$ is the smallest integer power n of a, such that $a^n = e$ where e is the identity. The order of a is denoted $|a|$.

To find the order of a from the Cayley table, one simply forms the set $a, a^2, \ldots$ until the identity is found. Inspecting the Cayley table from the previous example, we immediately note that $|A| = 2$ since $AA = E$, and the same goes for B and C; $|B| = 2$, $|C| = 2$. However, $|D| = 3$ because $D^2 = F$ and $DF = D^3 = E$. Here is another example of a cyclic group.

Example 4 (C_8). Consider the following element:

$$A = \frac{1}{\sqrt{2}}\begin{pmatrix} 1 & -1 \\ 1 & 1 \end{pmatrix}.$$

By direct matrix multiplication, we arrive at the following representations:

$$A^2 = \begin{pmatrix} 0 & -1 \\ 1 & 0 \end{pmatrix}, \quad A^3 = \frac{1}{\sqrt{2}}\begin{pmatrix} -1 & -1 \\ 1 & -1 \end{pmatrix},$$
$$A^4 = \begin{pmatrix} -1 & 0 \\ 0 & -1 \end{pmatrix}, \quad A^5 = \frac{1}{\sqrt{2}}\begin{pmatrix} -1 & 1 \\ -1 & -1 \end{pmatrix},$$
$$A^6 = \begin{pmatrix} 0 & 1 \\ -1 & 0 \end{pmatrix}, \quad A^7 = \frac{1}{\sqrt{2}}\begin{pmatrix} 1 & 1 \\ -1 & 1 \end{pmatrix},$$

and finally $A^8 = I$, the 2×2 identity matrix. It should be clear that this is a group since $A^{n+m} = A^k$ where $n + m \equiv k$ mod(8). The mod(8) operation divides the integer $m + n$ by 8 and returns the integer remainder k. One says $n + m$ is *congruent* to k *modulo* 8. For example, $A^3A^6 = A^{3+6} = A$ since $9 \equiv 1$ mod(8). Using modular arithmetic, one can construct the Cayley table without performing any matrix multiplication. Moreover, it is simple to show that cyclic groups C_n are Abelian since $A^lA^m = A^mA^l = A^k$ with $k \in 0, 1, 2, \ldots, n-1$.

Definition 12 (Permutation groups). The permutation of n objects is performed by a group of operations S_n.

Example 5 (S_3). Consider three balls with different colors, r for red, b for blue, and g for green. The arrangement of the three can be rbg but it can also be brg. To get to the second arrangement, we exchange the first two objects. The following matrix permutes the first two rows of a column vector in the same way:

$$p_{213} = \begin{pmatrix} 0 & 1 & 0 \\ 1 & 0 & 0 \\ 0 & 0 & 1 \end{pmatrix} \begin{pmatrix} r \\ b \\ g \end{pmatrix} = \begin{pmatrix} b \\ r \\ g \end{pmatrix}. \tag{3.4}$$

Now note that $|p_{213}| = 2$. To build the rest of the possible permutation operations of three distinct objects, we proceed systematically. To permute the second and third element of the vector, and to permute the first and third, respectively, we have p_{132} and p_{321},

$$p_{132} = \begin{pmatrix} 1 & 0 & 0 \\ 0 & 0 & 1 \\ 0 & 1 & 0 \end{pmatrix}, \quad p_{321} = \begin{pmatrix} 0 & 0 & 1 \\ 0 & 1 & 0 \\ 1 & 0 & 0 \end{pmatrix}, \tag{3.5}$$

and we again note that $|p_{132}| = |p_{321}| = 2$. Therefore, taking successive powers of p_{213}, p_{132}, and p_{321} does not provide any additional elements. But two new elements can be found when pairs of the three operators are multiplied together. Therefore,

$$p_{213}p_{132} = \begin{pmatrix} 0 & 1 & 0 \\ 1 & 0 & 0 \\ 0 & 0 & 1 \end{pmatrix} \begin{pmatrix} 1 & 0 & 0 \\ 0 & 0 & 1 \\ 0 & 1 & 0 \end{pmatrix} = \begin{pmatrix} 0 & 0 & 1 \\ 1 & 0 & 0 \\ 0 & 1 & 0 \end{pmatrix} = p_{312} \tag{3.6}$$

$$p_{213}p_{321} = \begin{pmatrix} 0 & 1 & 0 \\ 1 & 0 & 0 \\ 0 & 0 & 1 \end{pmatrix} \begin{pmatrix} 0 & 0 & 1 \\ 0 & 1 & 0 \\ 1 & 0 & 0 \end{pmatrix} = \begin{pmatrix} 0 & 1 & 0 \\ 0 & 0 & 1 \\ 1 & 0 & 0 \end{pmatrix} = p_{231}. \tag{3.7}$$

Of course, we need the identity matrix e. At this point, we have some information to start building the Cayley table for S_3. The rest of the calculations are left as an exercise:

S_3	e	p_{132}	p_{213}	p_{321}	p_{231}	p_{312}
e	e	p_{132}	p_{213}	p_{321}	p_{231}	p_{312}
p_{132}	p_{132}	e	p_{231}	p_{321}	p_{213}	p_{321}
p_{213}	p_{213}	p_{312}	e	p_{231}	p_{321}	p_{132}
p_{321}	p_{321}	p_{231}	p_{312}	e	p_{132}	p_{213}
p_{231}	p_{231}	p_{321}	p_{132}	p_{213}	p_{312}	e
p_{312}	p_{312}	p_{213}	p_{321}	p_{132}	e	p_{231}

(3.8)

The Cayley table reveals the structure of the group and is the best way to define the group in the abstract. Abstraction is always a powerful tool. In this case, after close inspection of the tables in equation (3.3) and in equation (3.8), we realize that the two have exactly the same structure if we make the substitution $A = p_{132}$, $B = p_{213}$, etc. We have just discovered a *group isomorphism*. Mathematicians would say that there are two isomorphisms, one that takes the abstract object defined by either one of the Cayley tables into D_3 and the another to S_3.

Definition 13 (Group isomorphism). Two groups $\mathcal{G}$ and $\mathcal{G}'$ are said to be isomorphic if a one to one map $f : \mathcal{G} \to \mathcal{G}'$ that preserves the group operation can be found. The map f preserves the group operation $*$, if $f(a * b) = f(a) * f(b)$. If group $\mathcal{G}$ is isomorphic to group $\mathcal{G}'$, we write $\mathcal{G} \equiv \mathcal{G}'$. The one- to-one map that preserves the group operation is called an isomorphism.

Note that the operation in $\mathcal{G}'$ can be different from the one in $\mathcal{G}$ and the group can be isomorphic nonetheless. Additionally, it is a corollary that an isomorphism of $\mathcal{G}$ preserves the subgroups of $\mathcal{G}$. A practical way to discover isomorphisms among two groups is to identify their respective subgroups.

3.2 The classic continuous groups

In the previous section, we encountered examples of two types of groups: the finite groups in examples 3 and 5, and those groups in examples 1 and 2, which are of infinite order as the result of having matrix representation with entries from a field like $\mathbb{R}$ or $\mathbb{C}$. The next two definitions introduce a generalization of the group in examples 1 and 2.

Definition 14 (The general linear groups $\mathrm{Gl}(n, \mathbb{R})$). The set of $n \times n$ matrices with nonzero determinant are a group under the matrix multiplication.

The closure requirement is satisfied by extending the result obtained in example 1, $\det AB = \det A \det B$ to $n \times n$ matrices with arbitrary n. Note that $\mathrm{Gl}(n, \mathbb{R})$ is also a vector space under matrix addition, and the size of the basis set is n^2. For example, the general linear groups $\mathrm{Gl}(2, \mathbb{R})$ has a vector space that spans 4 linearly independent basis,

$$\mathbf{e}_1 = \begin{pmatrix} 1 & 0 \\ 0 & 0 \end{pmatrix} \quad \mathbf{e}_2 = \begin{pmatrix} 0 & 0 \\ 0 & 1 \end{pmatrix} \quad \mathbf{e}_3 = \begin{pmatrix} 0 & 0 \\ 1 & 0 \end{pmatrix} \quad \mathbf{e}_4 = \begin{pmatrix} 0 & 1 \\ 0 & 0 \end{pmatrix}.$$

Equivalently, one can say that a member of $\mathrm{Gl}(2, \mathbb{R})$ requires 4 independent degrees of freedom.

Definition 15 (Dimensionality of a continuous group). The dimensionality of a continuous group is the dimension of the associated vector space.

It was shown that $\mathrm{Sl}(2, \mathbb{R})$ is a subgroup of $\mathrm{Gl}(2, \mathbb{R})$. This is an example of a general set of special orthogonal groups.

Definition 16 (The special linear group $\mathrm{Sl}(n, \mathbb{R})$). The set of $n \times n$ matrices with determinant equal to $+1$ are a group under the matrix multiplication.

Another important subgroup of $\mathrm{Gl}(2, \mathbb{R})$ is the orthogonal group.

Definition 17 ($O(n, \mathbb{R})$). The orthogonal group of $n \times n$ matrices over the field of real numbers is denoted $O(n, \mathbb{R})$, and $A \in O(n, \mathbb{R})$ if and only if the column (or rows) of A are orthogonal.

The requirement of orthogonality reduces the dimensionality of $O(n, \mathbb{R})$ to $n(n+1)/2$ since the $n(n-1)/2$ defining equations (orthogonality requirements) introduce dependencies among the n^2 degrees of freedom of $\mathrm{Gl}(n, \mathbb{R})$.

Example 6 (The representation of $O(2, \mathbb{R})$). Let

$$A = \begin{pmatrix} a & b \\ c & d \end{pmatrix}.$$

If the first column to the left is to be orthogonal to the second, then $ab + cd = 0$, solving this for d yields

$$A = \begin{pmatrix} a & b \\ c & -ab/c \end{pmatrix}.$$

With $n = 2$, we have $n(n+1)/2 = 3$ degrees of freedom, a, b, and c. If we also restrict the $O(n, \mathbb{R})$ set to those that have a determinant equal to $+1$, then we have a member of the special orthogonal group, a subgroup of $O(n, \mathbb{R})$.

Definition 18 ($\mathrm{SO}(n, \mathbb{R})$). The special orthogonal group of $n \times n$ matrices over the field of real numbers is denoted $\mathrm{SO}(n, \mathbb{R})$ and $A \in \mathrm{SO}(n, \mathbb{R})$ if and only if

$$A^T A = AA^T = I$$

where I is the $n \times n$ unit matrix.

The dimensionality of $\mathrm{SO}(n, \mathbb{R})$ is $n(n-1)/2$ because from the orthogonality conditions we have $n(n-1)/2$ equations of constraint, and for the normalization of each vector we have another n such relationships. Subtracting from n^2 these dependent degrees of freedom yields

$$n^2 - \frac{n(n-1)}{2} - n = \frac{n(n-1)}{2}.$$

Next, let us show that the two definitions $\mathrm{SO}(n, \mathbb{R})$ are equivalent.

Theorem 9. *Let A be a $n \times n$ matrix that satisfies $A^T A = I$, where I is the $n \times n$ unit matrix. Then* $\det A = 1$.

Proof. Using the fact that $\det AB = \det A \det B$, we write

$$\det I = \det AA^T = \det A \det A^T.$$

Now $\det A^T = \det A$ as can be seen by expanding the determinant by a column instead of a row. Since $\det I = 1$, the theorem follows. □

Example 7 (The representation of $\mathrm{SO}(2, \mathbb{R})$). Let us return to the previous example:

$$A = \begin{pmatrix} a & b \\ c & d \end{pmatrix}.$$

If the first column to the left is to be orthogonal to the second, then $ab + cd = 0$. Moreover, we want both vectors to be normalized, meaning $a^2 + c^2 = 1$ and $b^2 + d^2 = 1$. Eliminating b using the first equation of constraint, c from the second, and d from the third, we arrive at

$$A = \begin{pmatrix} a & -\sqrt{1-a^2} \\ \sqrt{1-a^2} & a \end{pmatrix}. \tag{3.9}$$

The dimensionality of $\mathrm{SO}(2, \mathbb{R})$ is 1 in agreement with $n(n-1)/2$. At this point, with a little trigonometry one can recognize that all members of $\mathrm{SO}(2, \mathbb{R})$ are *rotation matrices* $R_z(\theta)$. In fact, $\mathrm{SO}(n, \mathbb{R})$ is also known as the $n \times n$ rotation group. Using $a = \cos\theta$, it follows that $\sqrt{1-a^2} = \sin\theta$, and

$$R_z(\theta) = \begin{pmatrix} \cos\theta & -\sin\theta \\ \sin\theta & \cos\theta \end{pmatrix} \tag{3.10}$$

represents a counterclockwise rotation of a vector $a = a_x \mathbf{i} + a_y \mathbf{j}$ in the xy plane. Then

$$a' = R_z(\theta) a = (a_x \cos\theta - a_y \sin\theta)\mathbf{i} + (a_x \sin\theta + a_y \cos\theta)\mathbf{j}.$$

With this equation, it is possible to show that $\|a'\| = \|a\|$. That is, rotations are norm preserving. A vector starting on the positive x axis rotated by $2\pi/4$ is along the positive y axis, etc. Note that rotations preserve the dot product of two vectors,

$$a' \cdot b' = (a^T R_z(\theta)^T) \cdot (R_z(\theta) b) = a \cdot b.$$

The group properties (namely closure under multiplication) follows from

$$\begin{aligned} R_z(\theta) R_z(\phi) &= \begin{pmatrix} \cos\theta & -\sin\theta \\ \sin\theta & \cos\theta \end{pmatrix} \begin{pmatrix} \cos\phi & -\sin\phi \\ \sin\phi & \cos\phi \end{pmatrix} \\ &= \begin{pmatrix} \cos(\theta+\phi) & -\sin(\theta+\phi) \\ \sin(\theta+\phi) & \cos(\theta+\phi) \end{pmatrix} \end{aligned} \tag{3.11}$$

by using appropriate trigonometry identities.

For vectors in $\mathbb{R}^3$, we can build a representation of the SO(3, $\mathbb{R}$) in a number of ways. One approach is to rotate once about the three axis in turn. The dimensionality is 3, so three angles are needed. The standard approach is to first rotate about the z axis by an angle ψ, followed by a rotation about the x' axis by θ (by multiplying from the left), followed by another rotation about z'' by ϕ. We have for the first rotation about the z axis,

$$R_z(\psi) = \begin{pmatrix} \cos\psi & -\sin\psi & 0 \\ \sin\psi & \cos\psi & 0 \\ 0 & 0 & 1 \end{pmatrix},$$

whereas a rotation about the x axis is

$$R_x(\theta) = \begin{pmatrix} 1 & 0 & 0 \\ 0 & \cos\theta & -\sin\theta \\ 0 & \sin\theta & \cos\theta \end{pmatrix}.$$

The three rotations combined become

$$\begin{aligned} &R_z(\phi) R_x(\theta) R_z(\psi) \\ &= \begin{pmatrix} \cos\psi\cos\phi - \cos\theta\sin\psi\sin\phi & -\sin\psi\cos\phi - \cos\theta\sin\psi\cos\phi & \sin\phi\sin\theta \\ \sin\psi\cos\phi + \cos\theta\sin\psi\cos\phi & -\sin\psi\sin\phi + \cos\theta\cos\psi\cos\phi & \cos\phi\sin\theta \\ \sin\psi\sin\theta & \cos\psi\sin\theta & \cos\theta \end{pmatrix}. \end{aligned} \tag{3.12}$$

There remains two other subgroups of Gl(n, $\mathbb{R}$) to be defined.

Definition 19 ($SO_2(n, \mathbb{R})$). The special orthogonal group of $n \times n$ matrices over the field of real numbers is denoted $SO_2(n, \mathbb{R})$ and $A \in SO(n, \mathbb{R})$ if and only if

$$\det A = -1.$$

Example 8. Let us work through the algebraic construction of the $SO_2(2, \mathbb{R})$ members. Note that the dimensionality is one, since the number of equations of constraints is the same as for the $SO(n, \mathbb{R})$ group. A matrix $A \in SO_2(2s, \mathbb{R})$ is

$$A = \begin{pmatrix} a & b \\ c & d \end{pmatrix}, \quad ad - cb = -1$$

and $a^2 + c^2 = 1$, $b^2 + d^2 = 1$. In terms of a, we derive

$$A = \begin{pmatrix} a & \sqrt{1-a^2} \\ \sqrt{1-a^2} & -a \end{pmatrix}.$$

As in the $SO(2, \mathbb{R})$ case, we can represent the group using trigonometric functions

$$A = \begin{pmatrix} \cos\theta & \sin\theta \\ \sin\theta & -\cos\theta \end{pmatrix}.$$

However, these are not rotations, they are reflections. The best way to see this is to set θ to $\pi/2$, and operate on a generic two vector,

$$A\begin{pmatrix} x \\ y \end{pmatrix} = \begin{pmatrix} 0 & 1 \\ 1 & 0 \end{pmatrix}\begin{pmatrix} x \\ y \end{pmatrix} = \begin{pmatrix} y \\ x \end{pmatrix}.$$

The outcome is a reflection about the $y = x$ plane.

Before defining the last important class of groups, let us observe that members of continuous groups are used to transform other objects aside from vectors. For example, a matrix can be transformed by the product,

$$A' = B^{-1}AB$$

known as a *similarity transformation*. In Chapter 1, we have already encountered a special type of similarity transformation, one that diagonalizes a symmetric matrix,

$$\Lambda = R^T AR$$

where $R^T R = I$. It is possible to show that, if $A^T = A$, then R is either a reflection or a rotation matrix. With the knowledge of groups, we can explore this idea a little further.

Example 9. Let

$$A = \begin{pmatrix} a & b \\ b & c \end{pmatrix}$$

and R as in (3.10). The idea is to find an equation for the off diagonal element $(R^T AR)_{12}$ and solve for the value of θ that causes it to vanish. After some algebra, we get

$$0 = (c - a)\cos\theta\sin\theta + b(\cos^2\theta - \sin^2\theta),$$

and with two trigonometric identities,

$$0 = (c - a)\sin(2\theta) + 2b\cos(2\theta)$$

one arrives at

$$\theta = \frac{1}{2}\tan^{-1}\left(\frac{2b}{a - c}\right).$$

The last important subgroup of $\mathrm{Gl}(n, \mathbb{R})$ is comprised of those matrices that preserve the *symplectic norm.*

Definition 20. The $2n$-dimensional symplectic norm J is a $2n\times 2n$ matrix with ones along the first n minor diagonal entries and minus ones for the remaining n entries. The minor diagonal of a $n \times n$ matrix runs from the top right to the bottom left corner:

$$J = \begin{pmatrix} 0 & 0 & \cdots & 0 & 1 \\ 0 & 0 & \cdots & 1 & 0 \\ \vdots & & \vdots & & \vdots \\ 0 & -1 & \cdots & 0 & 0 \\ -1 & 0 & \cdots & 0 & 0 \end{pmatrix}.$$

Definition 21 ($\mathrm{Sp}(2n, \mathbb{R})$). A member of the symplectic group is a $2n \times 2n$ matrix that preserves the symplectic metric. That is, $A \in S_p(2n, \mathbb{R})$ if and only if,

$$A^T JA = J.$$

Example 10. Let us derive the general form of a 2×2 member of $\mathrm{Sp}(2, \mathbb{R})$, i. e., the $n = 1$ case. Let

$$A = \begin{pmatrix} a & b \\ c & d \end{pmatrix}$$

be such that $\det A \neq 0$, and $A^T JA = J$, in matrix form we translate the statement to

$$\begin{pmatrix} a & c \\ b & d \end{pmatrix} \begin{pmatrix} 0 & 1 \\ -1 & 0 \end{pmatrix} \begin{pmatrix} a & b \\ c & d \end{pmatrix} = \begin{pmatrix} 0 & 1 \\ -1 & 0 \end{pmatrix}.$$

After multiplication of the three matrices on the left, we derive

$$\begin{pmatrix} o & ad - cb \\ cb - ad & 0 \end{pmatrix} = \begin{pmatrix} 0 & 1 \\ -1 & 0 \end{pmatrix},$$

from which only one equation emerges. Using $ad - cb = 1$, to eliminate d we end up with the following representation of $\mathrm{Sp}(2, \mathbb{R})$:

$$A = \begin{pmatrix} a & b \\ c & (1 + cb)/a \end{pmatrix}.$$

There are complex field counterparts to $\mathrm{Gl}(n, \mathbb{R})$ and all its subgroups. $\mathrm{Gl}(n, \mathbb{C})$ is simply the set of $n \times n$ matrices with nonzero determinant and entries from the complex field. The operation of transposition has to be replaced with the *Hermitian conjugate* version, namely $A^\dagger = (A^*)^T$ where A^* is A with all $n \times n$ of its entries being the complex conjugates of those in A, as shown in the next example.

Example 11 (Hermitian conjugate). Let

$$A = \begin{pmatrix} 1 & 3 - 5i \\ 4 + 3i & 7 - 2i \end{pmatrix},$$

then

$$A^* = \begin{pmatrix} 1 & 3 + 5i \\ 4 - 3i & 7 + 2i \end{pmatrix}, \quad A^\dagger = \begin{pmatrix} 1 & 4 - 3i \\ 3 + 5i & 7 + 2i \end{pmatrix}.$$

Matrices that satisfy $A^\dagger = A$ are called *Hermitian.* Matrices that satisfy $A^\dagger = -A$ are called *anti-Hermitian.* The complex equivalent of the orthogonal group is called the *unitary group* $U(n, \mathbb{C})$, the space of all $n \times n$ matrices that satisfies $A^\dagger A = I$. Members of $U(n, \mathbb{C})$ that have unit determinant are a group denoted $\mathrm{SU}(n, \mathbb{C})$. Members of $U(n, \mathbb{C})$ with unimodular complex determinant ($e^i a$) are a group denoted $\mathrm{SU}_2(n, \mathbb{C})$.

Definition 22 (Trace). The *trace* of an $n \times n$ matrix is the sum of its elements along the main diagonal

$$\operatorname{trace}(A) = \sum_{i=1}^{n} a_{ii}.$$

The next theorem establishes the trace invariance of a matrix under an arbitrary similarity transformation.

Theorem 10 (Trace invariance). *Let*

$$A' = B^{-1}AB$$

where $B \in \mathrm{Gl}(n, F)$, *then* $\mathrm{trace}(A') = \mathrm{trace}(A)$.

Proof. By the definition of multiplicative inverse, we have $B^{-1}B = I$. In terms of components, the equation reads

$$\delta_{ij} = \sum_{k=1}^{n} b_{ik}^{-1} b_{kj}.$$

Therefore,

$$\mathrm{trace}(A') = \mathrm{trace}(B^{-1}AB) = \sum_{i=1}^{n}\sum_{k=1}^{n}\sum_{k'=1}^{n} b_{ik}^{-1} a_{kk'} b_{k'i}.$$

Changing the order of summation,

$$\mathrm{trace}(A') = \sum_{k=1}^{n}\sum_{k'=1}^{n}\sum_{i=1}^{n} b_{ik}^{-1} a_{kk'} b_{k'i}$$

and moving $a_{kk'}$ outside of the innermost sum,

$$\begin{aligned}\mathrm{trace}(A') &= \sum_{k=1}^{n}\sum_{k'=1}^{n} a_{kk'} \sum_{i=1}^{n} b_{ik}^{-1} b_{k'i} = \sum_{k=1}^{n}\sum_{k'=1}^{n} a_{kk'} \delta_{kk'} \\ &= \sum_{k=1}^{n} a_{kk} = \mathrm{trace}(A)\end{aligned}$$

where the definition of multiplicative inverse has been used. □

This result is of great importance because it tells us that no matter what representation we choose, the trace of a matrix remains unchanged. When A is a member of a group of transformation, its trace is called a *character*. For finite groups in particular, the characters are used to classify symmetry operations and create *character tables*. Therefore, the easiest way to find the character of a member of a group is to diagonalize its matrix representation if one can be found [95].

3.3 Exercises

1. Use the matrix product definition for $n \times n$ matrices,

$$(AB)_{ij} = \sum a_{ik} b_{jk},$$

to prove the associativity of the matrix product $A(BC) = (AB)C$.

2. Consider two matrices in Gl(2),

$$A = \begin{pmatrix} a_{11} & a_{12} \\ a_{21} & a_{22} \end{pmatrix}, \quad \text{and} \quad B = \begin{pmatrix} b_{11} & b_{12} \\ b_{21} & b_{22} \end{pmatrix}$$

by direct calculation prove that
   (a) $\det AB = \det A \det B$
   (b) $(AB)^{-1} = B^{-1}A^{-1}$.
3. Is the set of 2×2 symmetric matrices,

$$\left\{ A = \begin{pmatrix} a & b \\ b & c \end{pmatrix} \middle| \; a, b, c \in R \; ac - b^2 \neq 0 \right\}$$

a group under matrix multiplication?
4. Is the set of 2×2 symmetric matrices,

$$\left\{ A = \begin{pmatrix} a & b \\ b & a \end{pmatrix} \middle| \; a, b \in R \; a^2 - b^2 \neq 0 \right\}$$

a group under matrix multiplication?
5. Prove that the set of all three by three matrices of the form

$$\begin{pmatrix} 1 & x & y \\ 0 & 1 & z \\ 0 & 0 & 1 \end{pmatrix}$$

is a group under matrix multiplication.
6. Confirm that the operators in equations (3.4) through (3.7) produce the proper permutations of the *rgb* column vector.
7. Show that $\{E\}, \{E, A\}, \{E, B\}, \{EC\}, \{E, D, F\}$ are proper subgroups of the group defined in equations (3.1) and (3.2).
8. Carry out the rest of the calculations to fill the Cayley table in equation (3.8). Use of Python `numpy.matmul` is recommended.
9. Use modular arithmetic to construct the Cayley table for C_8 in example 4.
10. Use the map

$$f : \{E, A, B, C, D, F\} \rightarrow \{e, p_{132}, p_{213}, p_{321}, p_{231}, p_{312}\}$$

to prove that the group defined in equations (3.1) and (3.2) and the group defined in equations (3.4) through (3.7) are isomorphic.
11. The following exercise is a demonstration that every nonsingular $n \times n$ matrix has a reduced row-echelon form equal to the $n \times n$ unit matrix. Consider the following matrix:

$$A = \begin{pmatrix} 2 & 1 & -2 \\ 4 & -1 & 2 \\ 2 & -1 & 1 \end{pmatrix}.$$

(a) Show that multiplying row 1 by −1 and adding the result to row 3 produces the following changes. Only row 3 is changed by the operation, rows 1 and 2 remain unchanged:

$$\begin{pmatrix} 2 & 1 & -2 \\ 4 & -1 & 2 \\ 2 & -1 & 1 \end{pmatrix} \xrightarrow{r_3+(-2)r_2 \to r_3} \begin{pmatrix} 2 & 1 & -2 \\ 4 & -1 & 2 \\ 0 & -2 & 3 \end{pmatrix}.$$

(b) Show that the same row operation can be expressed as a matrix product,

$$\begin{pmatrix} 2 & 1 & -2 \\ 4 & -1 & 2 \\ 0 & -2 & 3 \end{pmatrix} = \begin{pmatrix} 1 & 0 & 0 \\ 0 & 1 & 0 \\ -1 & 0 & 1 \end{pmatrix} \begin{pmatrix} 2 & 1 & -2 \\ 4 & -1 & 2 \\ 2 & -1 & 1 \end{pmatrix}.$$

(c) The matrix to the left of A is called an elementary row reduction matrix. Calculate its determinant.

(d) There are seven additional operations to reduce A to its reduced row-echelon form, a 3×3 unit matrix. Perform them in this order:
   i. $r_2 + (-2)r_1 \to r_2$
   ii. $r_3 + (-2/3)r_2 \to r_3$
   iii. $(1/2)r_1 \to r_1$, $(-1/3)r_2 \to r_2$, $(-1)r_3 \to r_3$
   iv. $r_1 + (-1/2)r_2 \to r_1$
   v. $r_2 + (2)r_3 \to r_2$ where, e. g., the elementary matrix for the operation in step iii is

$$E_3 = \begin{pmatrix} \frac{1}{2} & 0 & 0 \\ 0 & -\frac{1}{3} & 0 \\ 0 & 0 & -1 \end{pmatrix},$$

   the product of three elementary operations.

12. Consider the following lemmas:
   - All elementary row reduction matrices for operations of the type $r_k + ar_j \to r_k$ have 1 as the determinant.
   - All elementary row reduction matrices that multiply a row by c have a determinant equal to c.
   - If all elements of a row or column of a matrix are multiplied by c, then the determinant is multiplied by c, as can be shown by expanding the determinant about the effected row or column.

Use these to show that in all cases

$$\det EA = \det E \, \det A$$

where E is an elementary row reduction matrix.

13. Use the results of the previous two exercises to prove the following theorem:

$$\det AB = \det A \det B$$

for $A, B \in \mathrm{Gl}(n, \mathbb{R})$.

14. Use the result from the previous exercise to prove the following facts:
    (a) If $P \in \mathrm{SO}(n, \mathbb{R})$, then $\det P = 1$.
    (b) If A is similar to B, i. e., if $A = P^{-1}BP$, then $\det(A) = \det(B)$.
15. The group of symmetric transformations of regular n-gones (e. g., triangle, the square, ...) centered at the origin and on the xy plane is called the dihedral group D_n. The group contains several types of operations on the vectors from the origin to the corners of the planar figure, such as rotation by $2\pi/n$, and reflection about symmetry planes perpendicular to the xy plane. D_3 can be represented by the following set of matrices:

$$D_3 = \left\{ E = \begin{pmatrix} 1 & 0 \\ 0 & 1 \end{pmatrix}, R_{2\pi/3} = \begin{pmatrix} -\frac{1}{2} & -\frac{\sqrt{3}}{2} \\ \frac{\sqrt{3}}{2} & -\frac{1}{2} \end{pmatrix}, \right.$$
$$R_{4\pi/3} = \begin{pmatrix} -\frac{1}{2} & \frac{\sqrt{3}}{2} \\ -\frac{\sqrt{3}}{2} & -\frac{1}{2} \end{pmatrix}, \sigma = \begin{pmatrix} \frac{1}{2} & -\frac{\sqrt{3}}{2} \\ -\frac{\sqrt{3}}{2} & -\frac{1}{2} \end{pmatrix},$$
$$\left. \sigma' = \begin{pmatrix} \frac{1}{2} & \frac{\sqrt{3}}{2} \\ \frac{\sqrt{3}}{2} & -\frac{1}{2} \end{pmatrix}, \sigma'' = \begin{pmatrix} -1 & 0 \\ 0 & 1 \end{pmatrix} \right\}.$$

Each of these operations applied to the three vectors from the origin to the corners of the equilateral triangle,

$$v_1 = \begin{pmatrix} \frac{\sqrt{3}}{2} \\ -\frac{1}{2} \end{pmatrix}, \quad v_2 = \begin{pmatrix} 0 \\ 1 \end{pmatrix}, \quad v_3 = \begin{pmatrix} -\frac{\sqrt{3}}{2} \\ -\frac{1}{2} \end{pmatrix},$$

transforms the geometric figure into another one indistinguishable from the original. See, e. g., the effect of $R_{2\pi/3}$ in Figure 3.1. The two rotations are clearly about the z axis and the notation should be self- explanatory. The element σ is a reflection about the $y = -x$ plane, σ' a reflection about the $y = x$ plane, and finally σ'' a reflection about the $x = 0$ plane. The geometric features, points, axis (e. g., the z axis), planes (e. g., the $y = x$) are the symmetry elements of the equilateral triangle, whereas the elements of D_3 are the symmetry operators. Confirm that D_3 is a group by building a Cayley table using the following two approaches:

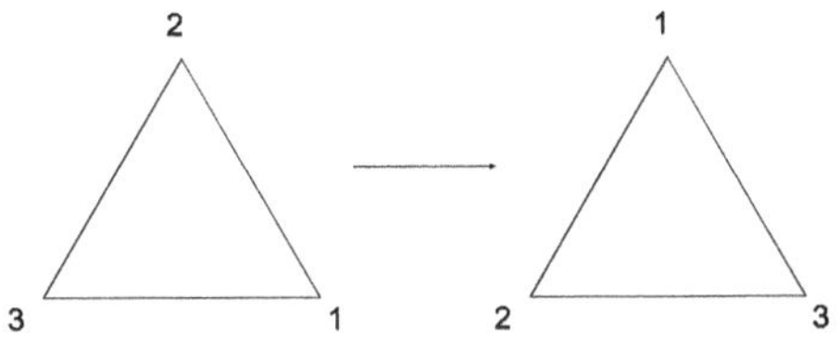

Figure 3.1: An equilateral triangle centered at the origin transformed by a $R_{2\pi/3}$ counterclockwise rotation.

(a) Build the diagrams as in Figure 3.1 to find all the products of each operation on the vectors and the product of two operations.

(b) Show that the Cayley table becomes

D_3	E	$R_{2\pi/3}$	$R_{4\pi/3}$	σ	σ'	σ''
E	E	$R_{2\pi/3}$	$R_{4\pi/3}$	σ	σ'	σ''
$R_{2\pi/3}$	$R_{2\pi/3}$	$R_{4\pi/3}$	E	σ'	σ''	σ
$R_{4\pi/3}$	$R_{4\pi/3}$	E	$R_{2\pi/3}$	σ''	σ	σ'
σ	σ	σ''	σ'	E	$R_{4\pi/3}$	$R_{2\pi/3}$
σ'	σ'	σ	σ''	$R_{2\pi/3}$	E	$R_{4\pi/3}$
σ''	σ''	σ'	σ	$R_{4\pi/3}$	$R_{2\pi/3}$	E

by sketching the outcome of two successive operations on the same equilateral triangle as in part (a).

16. Find the isomorphism that maps the elements of D_3 listed in the previous exercise to the elements of the permutation group S_3.
17. Consider the symmetry operations for a square:

$$\{R_{2\pi/4}, R^2_{2\pi/4}, R^3_{2\pi/4}, \sigma, \sigma', \sigma'', \sigma'''\}$$

where σ is the reflection about the $y = -x$ plane, σ' about the $y = 0$ plane, σ'' about the $y = x$ plane, and σ''' about the $x = 0$ plane.

(a) Construct a 2×2 matrix representation of the group.

(b) An element a of a group along with all its powers up to an including $|a|$ is a group called a *cyclic* group denoted by $\langle a \rangle$. Show that $\langle R_{2\pi/4} \rangle$ is a group and a proper subgroup of D_4.

18. Show that $R_z(-\theta) = R_z(\theta)^T = R_z(\theta)^{-1}$.
19. Use the requirement $A^T A = I$ where I is the identity matrix to derive a general representation of SO(2, ℝ), the special orthogonal group of 2×2 matrices over the field of reals by filling in the details of the derivation of equation (3.9).
20. Let

$$A = \begin{pmatrix} a & b \\ b & c \end{pmatrix}.$$

Show that a reflection

$$\begin{pmatrix} \cos\theta & \sin\theta \\ \sin\theta & -\cos\theta \end{pmatrix}$$

can diagonalize this matrix as well, and that the angle θ is exactly the same.

21. Use the requirement that $A^{\dagger}A = I$ where I is the identity matrix to derive a general representation of $SU(2, \mathbb{C})$, the special unitary group of 2×2 matrices over the field of complex numbers.
22. Derive equations (3.11) and (3.12).
23. Prove by induction on $n \in \mathbb{N}$ that

$$\begin{pmatrix} \cos\theta & -\sin\theta \\ \sin\theta & \cos\theta \end{pmatrix}^n = \begin{pmatrix} \cos n\theta & -\sin n\theta \\ \sin n\theta & \cos n\theta \end{pmatrix}.$$

4 An introduction to Lie algebras

> Mathematics must therefore remain an essential element of the knowledge and abilities which we have to teach, of the culture we have to transmit, to the next generation.
>
> Hermann Weyl

This chapter contains a brief introduction of the theory of Lie algebras, along with their connections to symmetry and Lie groups of continuous transformations introduced in the previous chapter. After the basic definitions, we introduce some simple examples of Taylor's theorem in conjunction with some practical methods to find the exponential of selected classes of matrices. We then show how the cross-product is a Lie algebra, and in the last section, we demonstrate with some simple examples how coupled sets of differential equations are solved by Lie groups of continuous transformations, and how the generators of the corresponding Lie algebras can be deduced from a given group.

4.1 Definitions

Definition 23 (Algebra). An algebra $\mathcal{A}$ is a vector space $\mathcal{V} = \{v_1, v_2, v_3, \ldots\}$ over a field $\mathcal{F}$ endowed with Vector addition $+$, scalar multiplication $a \in \mathcal{F}$, and an additional binary operation, a vector product $\square$ that satisfies the following axioms:

- The vector product $\square$ satisfies closure. If $v_1, v_2 \in \mathcal{V}$, then

$$v_1 \square v_2 \in \mathcal{V}.$$

- The vector product $\square$ is distributive. Let $v_1, v_2, v_3 \in \mathcal{V}$, then

$$\begin{aligned} v_1 \square (v_2 + v_3) &= v_1 \square v_2 + v_1 \square v_3, \\ (v_1 + v_2) \square v_3 &= v_1 \square v_3 + v_2 \square v_3. \end{aligned}$$

Note that in general the vector product $\square$ is not commutative. Beyond these two basic requirements, there exist other restrictions that create numerous flavors of algebras. An Abelian algebra is one where the vector product commutes $v_1 \square v_2 = v_2 \square v_1$. An associative algebra is endowed with an associative vector product such that given $v_1, v_2, v_3 \in \mathcal{V}$, then $v_1 \square (v_2 \square v_3) = (v_1 \square v_2) \square v_3 = v_1 \square v_2 \square v_3$. The existence of identity, namely $\exists \mathbf{1} \in \mathcal{A}$ for $\square$ such that if $v_1 \in \mathcal{V}$, then $\mathbf{1} \square v_1 = v_1 \square \mathbf{1} = v_1$, and $\exists\, v_1^{-1}$ such that $\mathbf{1} = v_1^{-1} \square v_1 = v_1 \square v_1^{-1}$ create an algebra with the cancellation property.

Example 12. The space of all real $n \times n$ matrices is a vector space under matrix addition, and is also a distributive algebra. To prove the distributive property, we use the definition of the matrix product,

https://doi.org/10.1515/9783111610207-004

$$[A(B+C)]_{ij} = \sum_{k=1}^{n} a_{ik}(b_{kj} + c_{kj}).$$

The components are real numbers and multiplication of real numbers is distributive, so

$$[A(B+C)]_{ij} = \sum_{k=1}^{n} a_{ik}b_{kj} + \sum_{k=1}^{n} a_{ik}c_{kj} = AB + AC.$$

A similar argument can be used to show that $(A+B)C = AC + BC$. The set of nonsingular real $n \times n$ matrices is a vector space under matrix addition; it is a group under the matrix product [$\mathrm{Gl}(n, \mathbb{R})$] and, therefore, it is also a distributive algebra under the regular matrix product endowed with the cancellation property. However, the cancellation property is not crucial for our purposes.

Example 13. Consider the vector space of 2×2 symmetric matrices over the real numbers,

$$\mathcal{V} = \left\{ \begin{pmatrix} a & b \\ b & c \end{pmatrix} \middle| \; a, b, c \in \mathbb{R} \right\}.$$

If we define $\square$ to be the symmetrized matrix (anticommutation) product, namely let $A, B \in \mathcal{V}$, and

$$A \square B = \{A, B\} = AB + BA,$$

then the resulting algebra is associative. The closure property can be shown for this example by direct multiplication,

$$\begin{aligned}
&\left\{ \begin{pmatrix} a_1 & b_1 \\ b_1 & c_1 \end{pmatrix}, \begin{pmatrix} a_2 & b_2 \\ b_2 & c_2 \end{pmatrix} \right\} \\
&= \begin{pmatrix} a_1 & b_1 \\ b_1 & c_1 \end{pmatrix}\begin{pmatrix} a_2 & b_2 \\ b_2 & c_2 \end{pmatrix} + \begin{pmatrix} a_2 & b_2 \\ b_2 & c_2 \end{pmatrix}\begin{pmatrix} a_1 & b_1 \\ b_1 & c_1 \end{pmatrix} \\
&= \begin{pmatrix} a_1a_2 + b_1b_2 & a_1b_2 + b_1c_2 \\ b_1a_2 + c_1b_2 & b_1b_2 + c_1c_2 \end{pmatrix} + \begin{pmatrix} a_1a_2 + b_1b_2 & b_1a_2 + c_1b_2 \\ a_1b_2 + b_1c_2 & b_1b_2 + c_1c_2 \end{pmatrix}.
\end{aligned}$$

It now suffices to confirm that after the summation the 1,2 element

$$a_1b_2 + b_1c_2 + b_1a_2 + c_1b_2$$

is the same as the 2,1 element,

$$b_1a_2 + c_1b_2 + a_1b_2 + b_1c_2.$$

Note, however, that the regular matrix product does not satisfy closure for 2×2 symmetric matrices. The $\{,\}$ operator is called the *anticommutator*. If we define the vector product □ as the symmetrized matrix product, then the set of $n \times n$ real symmetric matrices forms an algebra. The details are left as an exercise.

Example 14. Consider the vector space of 3×3 antisymmetric matrices over the real numbers

$$\mathcal{V} = \left\{ \begin{pmatrix} 0 & a & b \\ -a & 0 & c \\ -b & -c & 0 \end{pmatrix} \middle| \, a, b, c \in R \right\}$$

such that $A^T = -A$. If we define □ to be the antisymmetrized (commuted) matrix product, namely let $A, B \in \mathcal{V}$, and

$$A \,\square\, B = [A,\, B] = AB - BA,$$

then closure is satisfied as can be shown by direct multiplication,

$$\begin{aligned} &\left[\begin{pmatrix} 0 & a_1 & b_1 \\ -a_1 & 0 & c_1 \\ -b_1 & -c_1 & 0 \end{pmatrix}, \begin{pmatrix} 0 & a_2 & b_2 \\ -a_2 & 0 & c_2 \\ -b_2 & -c_2 & 0 \end{pmatrix} \right] \\ &= \begin{pmatrix} 0 & -b_1 c_2 + c_1 b_2 & a_1 c_2 - a_2 c_1 \\ -c_1 b_2 + b_1 c_2 & 0 & -a_1 b_2 + b_1 a_2 \\ c_1 a_2 - a_1 c_2 & -b_1 a_2 + a_1 b_2 & 0 \end{pmatrix}. \end{aligned} \tag{4.1}$$

Note that the regular matrix product does not satisfy closure for this vector space either. The [,] operator is called the *commutator*. If we define the vector product □ as the antisymmetrized matrix product, then the vector space $\mathcal{V}$ of all $n \times n$ real antisymmetric matrices $A^T = -A$ forms an algebra. The general proof for the closure property goes as follows. Let $A, B \in \mathcal{V}$. Using the matrix product,

$$[A,\, B]_{ij} = \sum_{k=1}^{n} a_{ik} b_{kj} - b_{ik} a_{kj}.$$

First, we need to show that

$$[A,\, B]_{ji} = -[A,\, B]_{ij},$$

and then that

$$[A,\, B]_{ii} = 0.$$

To get the first result, note that

$$[A,\, B]_{ji} = \sum_{k=1}^{n} a_{jk} b_{ki} - b_{jk} a_{ki}$$

$$= -\sum_{k=1}^{n} b_{jk}a_{ki} - a_{jk}b_{ki}$$
$$= -\sum_{k=1}^{n} a_{ki}b_{jk} - b_{ki}a_{jk}$$
$$= -\sum_{k=1}^{n} a_{ik}b_{kj} - b_{ik}a_{kj} = -[A,B]_{ij},$$

where the second line just rearranges the two terms inside the sum, the third makes use of the commutative properties of the product of real numbers, and the fourth uses $a_{ik} = -a_{ki}$ if $k \neq i$, etc. To show that the diagonal elements are zero, we use the matrix product again

$$[A, B]_{ii} = \sum_{k=1}^{n} a_{ik}b_{ki} - b_{ik}a_{ki},$$

when we apply the antisymmetric property of the components on the last term on the right, and we get

$$[A, B]_{ii} = \sum_{k=1}^{n} a_{ik}b_{ki} - a_{ik}b_{ki} = 0.$$

Of course, when $k = i$, the two terms in the sum vanish. To prove the distributive property for the commutator, consider

$$\begin{aligned}[A,(B+C)]_{ij} &= \{A(B+C) - (B+C)A\}_{ij}\\ &= \sum_{k=1}^{n} a_{ik}(b_{kj}+c_{kj}) - (b_{ik}+c_{ij})a_{kj}\\ &= \sum_{k=1}^{n} a_{ik}b_{kj} + a_{ik}c_{kj} - b_{ik}a_{kj} - c_{ij}a_{kj}\\ &= \sum_{k=1}^{n} a_{ik}b_{kj} - b_{ik}a_{kj} + a_{ik}c_{kj} - c_{ij}a_{kj} = [A,B]_{ij} + [A,C]_{ij}.\end{aligned}$$

Finally, to conclude this example, we want to show that this algebra is not associative. That is, given $A, B, C \in \mathcal{V}$ $A\Box(B\Box C)$ is not equal to $(A\Box B)\Box C$. To see this, consider the first term,

$$A\Box(B\Box C) = [A,[B,C]] = [A,(BC - CB)] = ABC - ACB - BCA + CBA,$$

whereas from the second term

$$(A\Box B)\Box C = [[A,B],C] = [(AB - BA),C] = ABC - BAC - CAB + CBA.$$

Clearly, $A\square(B\square C) \neq (A\square B)\square C$. However, the reader is urged to demonstrate that

$$[A,[B,C]] + [C,[A,B]] + [B,[C,A]] = 0. \tag{4.2}$$

This relationship is called the Jacobi identity.

Definition 24 (Lie algebra). A Lie algebra $\mathcal{A}$, named after the Norwegian mathematician Sophus Lie, is an algebra endowed with a vector product that satisfies the Jacobian identity,

$$A\square(B\square C) + C\square(A\square B) + B\square(C\square A).$$

Examples 15, 16, and 17 are examples of Lie algebras.

Example 15 (so(2, $\mathbb{R}$)). The set of 2×2 antisymmetric matrices over the real numbers is an example of a Lie algebra. The basis set for the underlying vector space contains a single element,

$$\begin{pmatrix} 0 & 1 \\ -1 & 0 \end{pmatrix}.$$

Example 16 (su(2, $\mathbb{C}$)). The vector space of 2×2 anti-Hermitian matrices over the complex numbers is a Lie algebra. A matrix H is Hermitian if it satisfies $H^\dagger = H$ and it is anti-Hermitian if $H^\dagger = -H$. Let

$$H = \begin{pmatrix} a & b \\ c & d \end{pmatrix}, \quad a, b, c, d \in \mathbb{C}.$$

Then

$$H^\dagger = \begin{pmatrix} a^* & c^* \\ b^* & d^* \end{pmatrix}$$

where a^*, b^*, c^*, d^* are the respective complex conjugates of a, b, c, d. If A is Hermitian, then

$$a^* = a, \quad d^* = d$$
$$b^* = c \quad c^* = b.$$

These conditions yield eight equations when the real and imaginary parts of each equations are considered. For instance, let $a = a_r + ia_i$, $a_r, a_i \in R$. From $a^* = a$, we obtain $(a_r + ia_i)^* = a_r + ia_i$ or from the imaginary portion $a_i = -a_i$ leads to $a_i = 0$, and a_r undetermined. Proceeding along the same lines, $d_i = 0$ and d_r is undetermined, $b_r = c_r$, and finally $b_i = -c_i$. Therefore, a total of four degrees of freedom are needed to fully define a 2×2 Hermitian matrix,

$$H = \begin{pmatrix} a_r & b_r - ib_i \\ b_r + ib_i & d_r \end{pmatrix}.$$

Four basis vectors are required:

$$\begin{pmatrix} 1 & 0 \\ 0 & 0 \end{pmatrix}, \quad \begin{pmatrix} 0 & 0 \\ 0 & 1 \end{pmatrix}, \quad \begin{pmatrix} 0 & 1 \\ 1 & 0 \end{pmatrix}, \quad \begin{pmatrix} 0 & -i \\ i & 0 \end{pmatrix}.$$

If we combine linearly the first two as the sum and the difference, and include the remaining two, we obtain the *Pauli spin matrices* in their standard form,

$$\sigma_0 = \begin{pmatrix} 1 & 0 \\ 0 & 1 \end{pmatrix}, \quad \sigma_1 = \begin{pmatrix} 0 & 1 \\ 1 & 0 \end{pmatrix},$$
$$\sigma_2 = \begin{pmatrix} 0 & -i \\ i & 0 \end{pmatrix}, \quad \sigma_3 = \begin{pmatrix} 1 & 0 \\ 0 & -1 \end{pmatrix}. \tag{4.3}$$

Finally, to convert a Hermitian matrix H into an anti-Hermitian one, it suffices to multiply H by $i = \sqrt{-1}$, for then

$$A^\dagger = (iH)^\dagger = -iH^\dagger = -A.$$

Therefore, the generators of $\mathrm{su}(2, \mathbb{C})$ are $x_k = i\sigma_k/2$.

Example 17 ($\mathrm{so}(3, \mathbb{R})$). The 3×3 antisymmetric matrices over the real numbers is another example of a Lie algebra. The basis set for the underlying vector space contains three elements:

$$J_x = \begin{pmatrix} 0 & 0 & 0 \\ 0 & 0 & 1 \\ 0 & -1 & 0 \end{pmatrix}, \quad J_y = \begin{pmatrix} 0 & 0 & -1 \\ 0 & 0 & 0 \\ 1 & 0 & 0 \end{pmatrix}, \quad J_z = \begin{pmatrix} 0 & 1 & 0 \\ -1 & 0 & 0 \\ 0 & 0 & 0 \end{pmatrix}.$$

Definition 25 (Structure constants). The structure constants C_{ij}^k are defined by the following equation. Let $\mathcal{A} = \mathrm{span}\{x_1, x_2, \ldots, x_n\}$ be the basis set of an algebra

$$x_i \square x_j = C_{ij}^k x_k = C_{ij}^1 x_1 + C_{ij}^2 x_2 + \cdots + C_{ij}^n x_n$$

where k in the C_{ij}^k symbol is not a power; it is a tensor index. Moreover, whenever a tensor index is repeated in a subscript and superscript in a term a sum over all degrees of freedom is implied. This is an example of *Einstein's sum notation.*

Since algebras are vector spaces, we can study their structure by inspecting how the product behaves with the basis set.

Example 18. A valid basis set the algebra of 2×2 matrices over the real numbers under the regular matrix product is

$$X_1 = \begin{pmatrix} 1 & 0 \\ 0 & 0 \end{pmatrix}, \quad X_2 = \begin{pmatrix} 0 & 1 \\ 0 & 0 \end{pmatrix}, \quad X_3 = \begin{pmatrix} 0 & 0 \\ 1 & 0 \end{pmatrix}, \quad X_4 = \begin{pmatrix} 0 & 0 \\ 0 & 1 \end{pmatrix}.$$

Using the regular matrix product $\times$, the following results are obtained:

$$X_1X_1 = X_1 \quad X_1X_2 = X_2, \quad X_1X_3 = 0$$

etc. The following product table results:

$\times$	X_1	X_2	X_3	X_4
X_1	X_1	X_2	0	0
X_2	0	0	X_1	X_2
X_3	X_3	X_4	0	0
X_4	0	0	X_3	X_4

The surviving structure constants are all equal to 1,

$$C_{11}^{1} = C_{12}^{2} = C_{23}^{1} = C_{24}^{2} = C_{31}^{3} = C_{32}^{4} = C_{43}^{3} = C_{44}^{4} = 1.$$

Note that this is not a Lie algebra, but we can define and compute structure constants nonetheless.

Example 19. Let us compute the structure constants for the Lie algebra so(3, ℝ) of example 17. Using the commutator, it is straightforward, albeit somewhat tedious, to produce the following table:

[,]	J_x	J_y	J_z
J_x	0	J_z	$-J_y$
J_y	$-J_z$	0	J_x
J_z	J_y	$-J_x$	0

The two nonzero values from the top row of the table are

$$C_3^{12} = 1 \quad C_2^{13} = -1,$$

and from the other two rows, we get

$$C_3^{21} = -1 \quad C_1^{23} = 1$$
$$C_2^{31} = 1 \quad C_1^{32} = -1.$$

Notably, the structure constants for this Lie algebra is the Levi–Civita tensor. Every odd permutation of the indices a, b, c in C_c^{ab} gives a -1, every even permutation a $+1$ and no two indices can take the same numerical value. Structure constants are useful for the classification of Lie algebras, to identify subalgebras, to discover isomorphisms among algebras, to construct physically meaningful operators, known as the Casimir operators, and finally for algebras without proper subalgebras, they serve to build a one-to-one matrix representation. A map between an algebra to another that is not one-to-one is called a *homomorphism.*

4.2 The exponentiation of Lie algebras

We are now going to use a few well-known calculus results to connect the material in this chapter with the previous ones. The calculus theorems we use are convergent Taylor series expansion of the following three analytical functions:

$$e^x = \sum_{k=0}^{\infty} \frac{x^k}{k!}, \qquad \sin(x) = \sum_{k=0}^{\infty} \frac{(-1)^k x^{2k+1}}{(2k+1)!},$$
$$\cos(x) = \sum_{k=0}^{\infty} \frac{(-1)^k x^{2k}}{(2k)!}.$$

Other formulas can be easily derived from these by making appropriate substitutions, namely

$$e^{-x} = \sum_{k=0}^{\infty} \frac{(-x)^k}{k!} = \sum_{k=0}^{\infty} \frac{(-1)^k x^k}{k!}.$$

Combining this result with the expansion of e^x, we can get the hyperbolic sine and cosines,

$$\cosh(x) = \frac{e^x + e^{-x}}{2} = \frac{1}{2}\sum_{k=0}^{\infty} \frac{x^k}{k!} + \frac{1}{2}\sum_{k=0}^{\infty} \frac{(-1)^k x^k}{k!}.$$

We can recombine the two sums into a single one if we realize that all the add terms cancel identically since for k odd $1 + (-1)^k$ is zero. Therefore, after some trivial steps, we end up with

$$\cosh(x) = \frac{e^x + e^{-x}}{2} = \sum_{k=0}^{\infty} \frac{x^{2k}}{(2k)!}$$

and a similar comment applies to the sinh expansion,

$$\sinh(x) = \frac{e^x - e^{-x}}{2} = \sum_{k=0}^{\infty} \frac{x^{2k+1}}{(2k+1)!}$$

where now the sum is over odd terms since when k is even $1-(-1)^k$ vanishes. The power series expansion of e^{ix} for $x \in \mathbb{R}$ is

$$e^{ix} = \sum_{k=0}^{\infty} \frac{i^k x^k}{k!}$$

where i^k for $k = 0, 1, 2, 3, 4, 5, 6, \ldots$ evaluates as $1, i, -1, -i, 1, i, -1, \ldots$, and a pattern emerges. All odd terms are proportional to i with alternating signs, and all even ones are proportional to 1 with alternating signs. Inspection of the resulting even and odd power series proves Euler's theorem, $e^{ix} = \cos x + i \sin x$, etc. The group property of the exponential function, namely $e^a e^b = e^{a+b}$ can be exploited to derive all the multiple and fractional angle trigonometric identities. More importantly, it gives us a map from a Lie algebra to a corresponding group, and vice versa. These correspondences are of importance, so we will spend some time developing the strategies involved.

First, we note that all the above mentioned expansions can be extended to matrices with the matrix product with impunity since the radius of convergence for all of them is the entire number line (complex plane). Thus, an equation such as e^A with A a $n \times n$ matrix is defined as a power series identical to the exponential function of a real number,

$$e^A = I + A + \frac{1}{2}A^2 + \cdots + \frac{1}{n!}A^n.$$

Generally, these expressions create a map from a vector space $\mathcal{V}$ of $n \times n$ matrices back to $\mathcal{V}$. The following theorems provide numerous options for the computation of e^A.

Theorem 11. *The exponential $B = \exp \Lambda$ of a diagonal $n \times n$ matrix Λ is itself diagonal. Moreover, the elements of B are the corresponding elements of Λ, exponentiated:*

$$B = \exp \begin{pmatrix} \lambda_1 & 0 & 0 & \cdots & 0 \\ 0 & \lambda_2 & 0 & \cdots & 0 \\ \vdots & & & & \vdots \\ 0 & 0 & 0 & 0 & \lambda_n \end{pmatrix} = \begin{pmatrix} e^{\lambda_1} & 0 & 0 & \cdots & 0 \\ 0 & e^{\lambda_2} & 0 & \cdots & 0 \\ \vdots & & & & \vdots \\ 0 & 0 & 0 & 0 & e^{\lambda_n} \end{pmatrix}.$$

Proof. Note that the first diagonal element in the power expansion of B_{ii} is 1, whereas the second is λ_i. Consider next the square of Λ. The result

$$\Lambda^2 = \begin{pmatrix} \lambda_1^2 & 0 & 0 & \cdots & 0 \\ 0 & \lambda_2^2 & 0 & \cdots & 0 \\ \vdots & & & & \vdots \\ 0 & 0 & 0 & 0 & \lambda_n^2 \end{pmatrix}$$

can be shown using the definition of the matrix product. The $n = 1$ case is trivial, for any n,

$$(\Lambda^2)_{ij} = \sum_{k=0}^{n} \lambda_{ik}\lambda_{kj}$$

but $\lambda_{ik} = \lambda_i\delta_{ki}$ and $\lambda_{kj} = \lambda_j\delta_{kj}$. Therefore, if $i \neq j$ $\delta_{ki}\delta_{kj}$ cannot both be satisfied and the result is zero, therefore, the matrix is diagonal. Moreover, when $i = j$,

$$(\Lambda^2)_{ii} = \sum_{k=0}^{n} \lambda_{ik}\lambda_{ki} = \sum_{k=0}^{n} \lambda_{ik}\lambda_{ki}\delta_{ki} = \lambda_i^2.$$

It follows by induction on k that

$$\Lambda^k = \begin{pmatrix} \lambda_1^k & 0 & 0 & \cdots & 0 \\ 0 & \lambda_2^k & 0 & \cdots & 0 \\ \vdots & & & & \vdots \\ 0 & 0 & 0 & 0 & \lambda_n^k \end{pmatrix}$$

and the theorem follows by inserting the result into $e^{\Lambda} = I + \Lambda + \frac{1}{2}\Lambda^2 + \cdots + \frac{1}{k!}\Lambda^k$. □

This result gives us a quick way to evaluate the exponential of any matrix that can be diagonalized.

Theorem 12. *Consider the generic similarity transform P on a matrix A of shape $n \times n$,*

$$P^{-1}AP = \Lambda.$$

Then

$$e^A = Pe^{\Lambda}P^{-1}.$$

Proof. Note that $A = P\Lambda P^{-1}$ and $A^2 = P\Lambda P^{-1}P\Lambda P^{-1} = P\Lambda^2P^{-1}$ and, therefore, $A^k = P\Lambda^kP^{-1}$. Inserting this into the expansion, we get

$$e^x = \sum_{k=0}^{\infty} \frac{1}{k!}A^k = \sum_{k=0}^{\infty} \frac{1}{k!}P\Lambda^kP^{-1} = P\left(\sum_{k=0}^{\infty} \frac{1}{k!}\Lambda^k\right)P^{-1},$$

and the theorem follows. □

Example 20. The eigenvalues of

$$A = \begin{pmatrix} 1 & 1 & 0 \\ 0 & 2 & 1 \\ 0 & 0 & 3 \end{pmatrix}$$

are $\lambda_1 = 1$, $\lambda_2 = 2$, and $\lambda_3 = 3$. The similarity transform matrices are

$$P = \begin{pmatrix} 1 & 1 & 0 \\ 0 & 1 & 2 \\ 0 & 0 & 2 \end{pmatrix}, \quad P^{-1} = \frac{1}{2}\begin{pmatrix} 2 & -2 & 1 \\ 0 & 2 & -2 \\ 0 & 0 & 3 \end{pmatrix}.$$

Therefore,

$$e^A = P\begin{pmatrix} e & 0 & 0 \\ 0 & e^2 & 0 \\ 0 & 0 & e^3 \end{pmatrix}P^{-1} = \frac{1}{2}\begin{pmatrix} 2e & -2e + 2e^2 & e - 2e^2 + e^3 \\ 0 & 2e^2 & -2e^2 + 2e^3 \\ 0 & 0 & 2e^3 \end{pmatrix}.$$

Note how in this example the upper triangular shape of A is preserved by exponentiation. This is a general result.

Theorem 13. *If A is upper (or lower) triangular $n \times n$ matrix, e^A is upper (or, respectively, lower) triangular also.*

Proof. Consider an upper triangular matrix A, then

$$a_{ij} = 0$$

when $j < i$ in the notation $a_{\text{row,column}}$ we have adopted. Now consider A^2 with the matrix product. A number of terms in the sum vanish

$$(A^2)_{ij} = \sum_{k=0}^{n} a_{ik}a_{kj} = \sum_{k=i}^{j} a_{ik}a_{kj}$$

only if $j > i$, where the sum must begin with $k = i$, otherwise for $k < i$ $a_{ik} = 0$ since A is upper triangular, and it must terminate when $k = j$, else if $k > j$ $a_{kj} = 0$ for the same reason. Conversely, if $j < i$ $(A^2)_{ij} = 0$, k must begin at j otherwise a_{kj} vanish, and must also begin with $k = i$ else a_{ik} vanishes. But this is a contradiction. Therefore,

$$(A^2)_{ij} = \sum_{k=0}^{n} a_{kj}a_{ik} = 0.$$

Namely, A^2 is upper triangular and by induction so is A^k. Consequently, all terms in the power expansion of a^A are upper triangular and the theorem follows. □

An important class of algebras are the so-called *nilpotent algebras*.

Definition 26 (Nilpotent). A matrix is nilpotent if there exist a power p such that $A^k = 0$, the null matrix for $k \geq p$.

Example 21. The matrix

$$A = \begin{pmatrix} 0 & 1 & 3 \\ 0 & 0 & 2 \\ 0 & 0 & 0 \end{pmatrix}$$

is nilpotent with $p = 3$. In fact,

$$A^2 = \begin{pmatrix} 0 & 0 & 2 \\ 0 & 0 & 0 \\ 0 & 0 & 0 \end{pmatrix}$$

and $A^3 = A^4 = \cdots = 0$ are null matrices. This statement can be generalized as well.

Theorem 14. *If A is $n \times n$, upper triangular and has zeros along the main diagonal, then it is nilpotent and all powers of A^k for $k \geq n$ yield the null matrix.*

Proof. Following the same argument for upper triangular matrices, the i, j element of the square of A is

$$(A^2)_{ij} = \sum_{k=i}^{j} a_{ik}a_{kj} = \sum_{k=i+1}^{j-1} a_{ik}a_{kj}$$

where the last sum on the right must be true because if $k = i$ then $a_{ik}a_{kj}$ must be zero since by definition A has zeros along the main diagonal. Let us now consider the $(A^2)_{i,i+1}$ element. If all the elements in the main diagonal of A are zero, then

$$(A^2)_{i,i+1} = \sum_{k=i+1}^{i} a_{ik}a_{kj}$$

a clear contradiction. Therefore, all entries b_{ii} of $B = A^2$ on the main diagonal are zero and so are the elements $b_{i,i+1}$ immediately above the main diagonal. Now let us inspect A^3 using the same approach,

$$(A^3)_{ij} = (BA)_{ij} = \sum_{k=i+1}^{j-1} b_{ik}a_{kj} = \sum_{k=i+2}^{j-1} b_{ik}a_{kj}$$

where the sum on the right-hand side follows from the property of B we uncovered earlier. Therefore, the elements c_{ij} of $C = A^3$ must satisfy $c_{ii} = c_{i,i+1} = c_{i,i+2} = 0$. The theorem follows by induction on k, the power of A. □

This means that the exponential of a nilpotent matrix has a finite power series expansion.

Example 22. Consider the nilpotent Lie algebra of dimension 3 over the real numbers. That corresponds to the set of 3×3 upper triangular matrices with zeros on the main diagonal. A valid basis set is

$$a_1 = \begin{pmatrix} 0 & 1 & 0 \\ 0 & 0 & 0 \\ 0 & 0 & 0 \end{pmatrix}, \quad a_1 = \begin{pmatrix} 0 & 0 & 1 \\ 0 & 0 & 0 \\ 0 & 0 & 0 \end{pmatrix}, \quad a_1 = \begin{pmatrix} 0 & 0 & 0 \\ 0 & 0 & 1 \\ 0 & 0 & 0 \end{pmatrix}.$$

The exponential of an arbitrary element of this algebra only requires three terms:

$$\exp\begin{pmatrix} 0 & a & b \\ 0 & 0 & c \\ 0 & 0 & 0 \end{pmatrix} = \begin{pmatrix} 1 & 0 & 0 \\ 0 & 1 & 0 \\ 0 & 0 & 1 \end{pmatrix} + \begin{pmatrix} 0 & a & b \\ 0 & 0 & c \\ 0 & 0 & 0 \end{pmatrix} + \begin{pmatrix} 0 & 0 & \frac{ac}{2} \\ 0 & 0 & 0 \\ 0 & 0 & 0 \end{pmatrix}$$
$$= \begin{pmatrix} 1 & a & b+\frac{ac}{2} \\ 0 & 1 & c \\ 0 & 0 & 1 \end{pmatrix}.$$

We are now ready to prove the following useful results.

Theorem 15. *The exponential of an antisymmetric Lie algebra* $\mathrm{so}(n,\mathbb{R})$ *is a member of the* $\mathrm{SO}(n,\mathbb{R})$ *group.*

Proof. If $A \in \mathrm{so}(n,\mathbb{R})$, $A^T = -A$. Whereas if $B = \exp(A)$ is in $\mathrm{SO}(nR)$, then $B^T B = BB^T = I$, must hold. I is the $n \times n$ unit matrix. This is the case since

$$B^T = \exp(A)^T = \exp(A^T) = \exp(-A)$$

and

$$B^T B = \exp(-A)\exp(A) = \exp(0) = I$$ □

In the last line, we have made use of the fact that $\exp(A)\exp(B) = \exp(A+B)$ However, this relationship only holds if $[A,B] = 0$. In this case, A commutes with itself; therefore, we are justified to merge the arguments. Recall that the dimensionality of this algebra is $n(n-1)/2$, which means that there are that many free parameters in the entries of both A and B. Let $A \in \mathrm{so}(n,\mathbb{R})$ and $B \in \mathrm{SO}(n,\mathbb{R})$. Then, in terms of the basis set $\{x_1, x_2, \ldots, x_{n(n-1)/2}\}$ of $\mathrm{so}(n,\mathbb{R})$, we write

$$A = \sum_{i=1}^{n(n-1)/2} a_i x_i$$
$$B = \exp\left\{\sum_{i=1}^{n(n-1)/2} a_i x_i\right\}.$$

The set $\{x_1, x_2, \ldots, x_{n(n-1)/2}\}$ is called the set of generators of the group $\mathrm{SO}(n,\mathbb{R})$. An example should clarify this point further.

Example 23. Consider the matrix,

$$A = t\begin{pmatrix} 0 & 1 \\ -1 & 0 \end{pmatrix},$$

where $t \in \mathbb{R}$ is a free parameter. Then

$$A^2 = -It, \quad A^3 = -At,$$

where I is the 2×2 unit matrix. By induction, all odd powers of A are proportional to A and all even powers of A are proportional to the unit matrix. More precisely, for $k = 0, 1, 2, \ldots,$

$$A^{2k+1} = (-1)^k t^{2k+1} \begin{pmatrix} 0 & 1 \\ -1 & 0 \end{pmatrix}$$
$$A^{2k} = (-1)^k t^{2k} I.$$

Then

$$\exp t \begin{pmatrix} 0 & 1 \\ -1 & 0 \end{pmatrix} = \left\{ \sum_{k=0}^{\infty} (-1)^k \frac{t^{2k+1}}{(2k+1)!} \right\} \begin{pmatrix} 0 & 1 \\ -1 & 0 \end{pmatrix} + \left\{ \sum_{k=0}^{\infty} (-1)^k \frac{t^{2k}}{(2k)!} \right\} I = \begin{pmatrix} \cos t & \sin t \\ -\sin t & \cos t \end{pmatrix}.$$

This is a clockwise rotation matrix. Theorem 15 can be generalized to *all Lie algebras*. Here is its complex counterpart.

Theorem 16. *The exponential* $\exp(A)$ *where* $A \in \mathrm{su}(n, \mathbb{C})$ *a is a member of the special unitary group* $\mathrm{SU}(n, \mathbb{R})$.

Proof. If $A \in \mathrm{su}(n, \mathbb{R})$, $A^\dagger = -A$. Whereas if $B = \exp(A)$ is in $\mathrm{SU}(n, \mathbb{C})$, then $B^\dagger B = BB^\dagger = I$ and must hold. I is the $n \times n$ unit matrix. This is the case since

$$B^\dagger = \exp(A)^\dagger = \exp(A^\dagger) = \exp(-A)$$

and

$$B^\dagger B = \exp(-A)\exp(A) = \exp(0) = I.$$ □

4.3 The cross-product

Consider two regular vectors, $\mathbf{a}, \mathbf{b} \in \mathbb{R}^3$. The classical notation for such objects is expressed in terms of components and basis,

$$\mathbf{a} = a_x \mathbf{i} + a_y \mathbf{j} + a_z \mathbf{k}, \quad \mathbf{b} = b_x \mathbf{i} + b_y \mathbf{j} + b_z \mathbf{k}.$$

The geometric meaning of the cross-product is the oriented area between the two vectors, and it can be defined with the following expression:

$$\mathbf{c} = \mathbf{a} \times \mathbf{b} = (a_y b_z - a_z b_y)\mathbf{i} - (a_x b_z - a_z b_x)\mathbf{i} + (a_x b_y - a_y b_x)\mathbf{k}. \tag{4.4}$$

This formula is likened to, and can be derived from a type of determinant computation

$$\mathbf{c} = \mathbf{a} \times \mathbf{b} = \det \begin{vmatrix} \mathbf{i} & \mathbf{j} & \mathbf{k} \\ a_x & a_y & a_z \\ b_x & b_y & b_z \end{vmatrix} .$$

Alternatively, one can use the following multiplication table for the $\mathbb{R}^3$ basis set:

×	i	j	k
i	0	k	–j
j	–k	0	i
k	j	–i	0

(4.5)

For the subject of gauge theory, the fundamental importance of the conservation of angular momentum in the n body system cannot be understated. The reader may recall that angular momentum is defined by the cross-product of the spacial coordinates with their respective conjugate momenta, namely $\mathbf{l} = \mathbf{r} \times \mathbf{p}$. It is a less known fact, however, that the cross-product over $\mathbb{R}^3$ is Lie algebra. The closure property of the cross-product is manifested clearly in its multiplication table. The distributive properties can be gleaned by carrying out the following operation longhand:

$$\mathbf{c} = \mathbf{a} \times \mathbf{b} = (a_x\mathbf{i} + a_y\mathbf{j} + a_z\mathbf{k}) \times (b_x\mathbf{i} + b_y\mathbf{j} + b_z\mathbf{k})$$

distributing the two expressions in the parenthesis generates nine distinct terms

$$\begin{aligned} \mathbf{a} \times \mathbf{b} = {} & (a_x b_x)\mathbf{i} \times \mathbf{i} + (a_x b_y)\mathbf{i} \times \mathbf{j} + (a_x b_z)\mathbf{i} \times \mathbf{k} \\ & + (a_y b_x)\mathbf{j} \times \mathbf{i} + (a_y b_y)\mathbf{j} \times \mathbf{j} + (a_y b_z)\mathbf{j} \times \mathbf{k} \\ & + (a_z b_x)\mathbf{k} \times \mathbf{i} + (a_z b_y)\mathbf{k} \times \mathbf{j} + (a_z b_z)\mathbf{k} \times \mathbf{k}. \end{aligned}$$

Application of the rules in equation (4.5), while taking care to preserve the correct order of multiplication, yields the expression in equation (4.4) after the six surviving terms are regrouped into components.

Finally, using the rules in equation (4.5), it should be straightforward to prove the Jacobi identity:

$$\begin{aligned} & \mathbf{i} \times \mathbf{j} \times \mathbf{k} + \mathbf{k} \times \mathbf{i} \times \mathbf{j} + \mathbf{j} \times \mathbf{k} \times \mathbf{i} \\ & = \mathbf{i} \times \mathbf{i} + \mathbf{k} \times \mathbf{k} + \mathbf{j} \times \mathbf{j} = 0 \end{aligned}$$

The second line results regardless of the chosen order of operations in each of the three terms.

Therefore, we have shown that the cross-product over a vector space creates a Lie algebra. To generate the corresponding Lie group, we can begin by exploiting the isomorphism between the **i**, **j**, **k**, and the three 3 × 3 matrices of example 19 by inspecting the structure constants:

$$\mathbf{i} \to J_x, \quad \mathbf{j} \to J_y, \quad \mathbf{k} \to J_z.$$

We now find two related but different matrix representation of the corresponding group. As proved earlier, the exponential of a Lie algebra of skew symmetric (or anti-symmetric) matrices is a map from the Lie algebra so$(n, \mathbb{R})$ to the SO$(n, \mathbb{R})$ group,

$$\mathcal{G} = \exp(\mathcal{A}).$$

Clearly, our three matrices are skew symmetric, so we are in fact presented with the so$(3, \mathbb{R})$ Lie algebra and we want to exponentially map it into members of the SO$(3, \mathbb{R})$ group. We can do this in a number of ways, depending on the actual geodesic we choose in the associated three-dimensional space mapped with parameters α_x, α_y, and α_z:

$$\mathcal{G}_{x,y,z} = e^{\alpha_x J_x + \alpha_y J_y + \alpha_z J_z}.$$

For instance, suppose we choose the line $\alpha_y = \alpha_z = 0$, then the matrix representation for the SO$(3, \mathbb{R})$ group member is given by the following result.

Theorem 17. *The members of the Lie group $\mathcal{G}_{x,y,z}$ along the parameter space $\alpha_y = \alpha_z = 0$ geodesic takes the following form:*

$$e^{\alpha_x J_x} = \begin{pmatrix} 1 & 0 & 0 \\ 0 & \cos(\alpha_x) & \sin(\alpha_x) \\ 0 & -\sin(\alpha_x) & \cos(\alpha_x) \end{pmatrix}.$$

In other words, $e^{\alpha_x J_x}$ is a matrix representation of a clockwise rotation about the x axis, by an angle α_x.

Proof. By the direct matrix product, it follows that

$$J_x^2 = \begin{pmatrix} 0 & 0 & 0 \\ 0 & -1 & 0 \\ 0 & 0 & -1 \end{pmatrix}$$

and

$$J_x^3 = -J_x.$$

By induction, we prove that

$$J_x^5 = J_x, \quad J_x^7 = -J_x, \quad \ldots, \quad J_x^{2n+1} = (-1)^n J_x$$
$$J_x^4 = -J_x^2, \quad J_x^6 = J_x^2, \quad \ldots, \quad J_x^{2n} = -(-1)^n J_x^2.$$

Using Taylor's theorem about the origin, we expand the exponential function,

$$e^{\alpha_x J_x} = \sum_{n=0}^{\infty} \frac{(\alpha_x J_x)^n}{n!}.$$

The division into an even and an odd series gives

$$e^{\alpha_x J_x} = I + \sum_{n=1}^{\infty} \frac{(\alpha_x J_x)^{2n}}{(2n)!} + \sum_{n=0}^{\infty} \frac{(\alpha_x J_x)^{2n+1}}{(2n+1)!},$$

where I is the 3×3 unit matrix. Using the expressions for the odd and even powers of J_x,

$$e^{\alpha_x J_x} = \mathbb{I} - J_x^2 \sum_{n=1}^{\infty} (-1)^n \frac{(\alpha_x)^{2n}}{(2n)!} + J_x \sum_{n=0}^{\infty} (-1)^n \frac{(\alpha_x)^{2n+1}}{(2n+1)!}.$$

The two Taylor's series of the last line are recognized as those of the cosine and the sine of α_x, respectively,

$$e^{\alpha_x J_x} = \mathbb{I} - J_x^2 \cos(\alpha_x) + J_x \sin(\alpha_x).$$

The theorem follows trivially. □

The following is the group representation along any trajectory of the space.

Theorem 18. *The exponential map of*

$$A = \begin{pmatrix} 0 & \alpha_z & -\alpha_y \\ -\alpha_z & 0 & \alpha_x \\ \alpha_y & -\alpha_x & 0 \end{pmatrix}$$

is

$$\mathcal{G} = e^A = I + A\,\alpha^{-1} \sin(\alpha) - A^2\,\alpha^{-2} \cos(\alpha)$$

where I is the unit matrix and $\alpha^2 = \alpha_x^2 + \alpha_y^2 + \alpha_z^2$.

Proof. By direct matrix multiplication, one obtains

$$A^2 = \begin{pmatrix} -\alpha_z^2 - \alpha_y^2 & \alpha_x \alpha_y & \alpha_z \alpha_x \\ \alpha_y \alpha_x & -\alpha_x^2 - \alpha_z^2 & \alpha_y \alpha_z \\ \alpha_z \alpha_x & \alpha_z \alpha_y & -\alpha_x^2 - \alpha_y^2 \end{pmatrix} \tag{4.6}$$

$$A^3 = -a^2 A. \tag{4.7}$$

By induction, we have

$$A^3 = -a^2 A, \quad A^5 = a^4 A, \quad A^7 = -a^6 A, \quad \ldots, \quad A^{2n+1} = (-1)^n a^{2n} A$$
$$A^4 = -a^2 A^2, \quad A^6 = a^4 A^2, \quad A^8 = -a^6 A^2, \quad \ldots, \quad A^{2n} = -(-1)^n a^{2(n-1)} A.$$

Inserting these expressions into the odd and even terms of the power series expansion of e^A,

$$e^A = I + \sum_{n=0}^{\infty} \frac{A^{2n+1}}{(2n+1)!} + \sum_{n=1}^{\infty} \frac{A^{2n}}{(2n)!},$$

gives

$$e^A = I + A \sum_{n=0}^{\infty} \frac{(-1)^n a^{2n}}{(2n+1)!} - A^2 \sum_{n=1}^{\infty} \frac{(-1)^n a^{2(n-1)}}{(2n)!}.$$

The theorem follows by multiplying the first sum by a/a and the second by a^2/a^2. □

4.4 Groups as solutions

We are interested in Lie groups and their corresponding Lie algebras for the following fundamental reason. Lie groups solve coupled differential equations, whereas their corresponding Lie algebras are the so-called generators of the group. Lie's theorems and their inverses are the tools that allow us to find the infinitesimal generators if a group is given. Moreover, given the infinitesimal generators, one can find the corresponding group by exponentiation. The best way to see how this powerful machinery works is to look at some examples.

Example 24. Consider the following coupled set of linear first-order differential equations:

$$\frac{\partial x}{\partial t} = \alpha y + \beta z \quad \frac{\partial y}{\partial t} = \gamma z$$
$$\frac{\partial z}{\partial t} = 0 \quad x, y, z(t=0) = x_0, y_0, z_0$$

where x, y, z are the dependent variables and $\alpha, \beta, \gamma \in \mathbb{R}$ are constant parameters. We can write this set into matrix-vector form,

$$\frac{\partial}{\partial t}\begin{pmatrix} x \\ y \\ z \end{pmatrix} = \begin{pmatrix} 0 & \alpha & \beta \\ 0 & 0 & \gamma \\ 0 & 0 & 0 \end{pmatrix}\begin{pmatrix} x \\ y \\ z \end{pmatrix} \quad \lim_{t \to 0}\begin{pmatrix} x \\ y \\ z \end{pmatrix} = \begin{pmatrix} x_0 \\ y_0 \\ z_0 \end{pmatrix}.$$

Note that this is analogous to the first-order linear differential equation $df/dt = af$, $f(0) = f_0$, which is solved with $f(t) = f_0 e^{at}$. Therefore, the solution of the coupled set of linear differential equations is

$$\begin{pmatrix} x \\ y \\ z \end{pmatrix} = \left\{ \exp \begin{pmatrix} 0 & \alpha t & \beta t \\ 0 & 0 & \gamma t \\ 0 & 0 & 0 \end{pmatrix} \right\} \begin{pmatrix} x_0 \\ y_0 \\ z_0 \end{pmatrix},$$

where the exponentiation of this nilpotent matrix has been shown earlier,

$$\begin{pmatrix} x \\ y \\ z \end{pmatrix} = \left\{ \begin{matrix} 1 & \alpha t & (\beta + \alpha\gamma/2)t \\ 0 & 1 & \gamma t \\ 0 & 0 & 1 \end{matrix} \right\} \begin{pmatrix} x_0 \\ y_0 \\ z_0 \end{pmatrix}.$$

Example 25. To see how, given an expression of a group, one can find the generators, consider the following three parameter Lie group:

$$B = \begin{pmatrix} e^{\alpha}(1 + \beta\alpha) & \beta e^{\alpha} \\ \gamma e^{-\alpha} & e^{-\alpha} \end{pmatrix}.$$

This means that there are three distinct generators X_α, X_β, and X_γ and these are found as follows:

$$\begin{aligned} X_\alpha &= \frac{\partial}{\partial \alpha} B \bigg|_{\alpha,\beta,\gamma=0} &= \begin{pmatrix} 1 & 0 \\ 0 & -1 \end{pmatrix} \\ X_\beta &= \frac{\partial}{\partial \beta} B \bigg|_{\alpha,\beta,\gamma=0} &= \begin{pmatrix} 0 & 1 \\ 0 & 0 \end{pmatrix} \\ X_\gamma &= \frac{\partial}{\partial \gamma} B \bigg|_{\alpha,\beta,\gamma=0} &= \begin{pmatrix} 0 & 0 \\ 1 & 0 \end{pmatrix}. \end{aligned}$$

There is a great deal of nuance that we are leaving out of this discussion regarding these last two results at this point. For now, regarding applications to quantum mechanics, it suffices to say that the solution of a coupled set of differential equations can rarely be solved with a single exponentiation. Rather, the integration is carried out by an iterative process taking small steps along a carefully chosen trajectory in parameter space. This is because in general the generators (such as, e. g., the Hamiltonian operator) evaluated at a point in parameter space does not commute with the same generator evaluated at a different point. Moreover, parameter spaces are seldom Euclidean in structure. They have a tendency instead to be smooth *manifolds*. For this reason, we differ the discussion to a later chapter where these issues can be addressed a little more formally after differential geometry is introduced. Nonetheless, those algorithms that make use of group theory to preserve the fundamental symmetries are usually much more stable than those that do not. Therefore, symmetry, group operations, and Lie algebras are very important objects in chemical physics.

Important applications of the commutator

The commutator of two quantum operators is closely related to the Heisenberg uncertainty principle. Consider the simplest example of two *incompatible physical properties*, the position and the momentum. The measurement of one disturbs the sharpness of the other. This is typically explained in the following way. When the particle is in a "infinitely sharp" momentum state $\psi_p = \delta(p - p_0)$, and we Fourier transform this into the position representation, we end up with an infinitely broad position wave function $\psi_x \backsim e^{ixp/\hbar}$ and the converse is true if we begin with a "infinitely sharp" position state. In these theoretical states, the uncertainty of one variable is zero, while the uncertainty of the other approaches infinity. The general statement for the product of the uncertainty of two physical properties a, b with A, B their respective quantum operators is

$$\Delta A \Delta B \geq \frac{1}{2}\big|[A, B]\big|.$$

Therefore, position and momentum are incompatible because $[x, p] = i\hbar$ is not zero. Let us review how this result comes about using x as the position operator and $-i\hbar d/dx$ for the momentum operator. Then note that

$$\begin{aligned}
\langle [x,p] \rangle &= \int_{\mathcal{D}} \psi^* [x,p] \psi \, dx \\
&= -i\hbar \left\{ \int_{\mathcal{D}} \psi^* \left(x \frac{d}{dx} \psi \right) dx - \int_{\mathcal{D}} \psi^* \left(\frac{d}{dx} x\psi \right) dx \right\} \\
&= -i\hbar \left\{ \int_{\mathcal{D}} \psi^* \left(x \frac{d}{dx} \psi \right) dx - \int_{\mathcal{D}} \psi^* \left(\psi + \frac{d}{dx} x\psi \right) dx \right\} = i\hbar
\end{aligned}$$

where we have used the chain rule to get to the last line.

The commutator of two variables also plays a role in determining the time evolution of the expectation of variables. For instance, consider

$$\frac{d}{dt}\langle x \rangle = \frac{d}{dt} \int_{\mathcal{D}} \psi^* x \psi \, dx.$$

We are going to work here in the Schrödinger picture, meaning operators are time independent, and the time dependence is carried by the wave function. Then

$$\frac{d}{dt}\langle x \rangle = \int_{\mathcal{D}} \psi^* x \left(\frac{d\psi}{dt} \right) dx + \int_{\mathcal{D}} \left(\frac{d\psi^*}{dt} \right) x\psi \, dx.$$

We then use the time dependent Schrödinger equation:

$$i\hbar \frac{d\psi}{dt} = H\psi \tag{4.8}$$

where

$$H = -\frac{\hbar^2}{2m}\frac{d^2}{dx^2} + U \tag{4.9}$$

is the Hamiltonian operator, and to write this as

$$\begin{aligned}\frac{d}{dt}\langle x\rangle &= -\frac{i}{\hbar}\left\{\int_{\mathcal{D}} \psi^*(xH)\psi\,dx - \int_{\mathcal{D}} \psi^*(Hx)\psi\,dx\right\}\\ \frac{d}{dt}\langle x\rangle &= \frac{i}{\hbar}\int_{\mathcal{D}} \psi^*[H,x]\psi\,dx.\end{aligned} \tag{4.10}$$

Therefore, if we know the commutator between the Hamiltonian and the position x, we can compute the trajectory as a function of time by solving differential equation,

$$\langle x\rangle_t = \langle x\rangle_{t=0} e^{i\langle [x,H]\rangle t/\hbar}. \tag{4.11}$$

Note that this equation actually works for any arbitrary operator including, e. g., the momentum p. Equation (4.11) is one example that explains the claim “solutions of the Schrödinger equations are members of continuous Lie groups.” Note that the argument of the exponential is a particular member of the Lie algebra.

The commutator between p and H is more interesting for the following reason:

$$\begin{aligned}\langle [H,p]\rangle = &+i\hbar \int_{\mathcal{D}} \psi^*\left[\left(-\frac{\hbar^2}{2m}\frac{d^2}{dx^2} + U\right)\frac{d}{dx}\right]\psi\,dx\\ &- i\hbar \int_{\mathcal{D}} \psi^* \frac{d}{dx}\left(-\frac{\hbar^2}{2m}\frac{d^2}{dx^2} + U\right)\psi\,dx,\end{aligned}$$

using the chain rule inside the first of the two integrals,

$$\begin{aligned}\langle [H,p]\rangle = &+i\hbar \int_{\mathcal{D}} \psi^*\left[-\frac{\hbar^2}{2m}\frac{d^3\psi}{dx^3} + U\frac{d\psi}{dx}\right]dx\\ &- i\hbar \int_{\mathcal{D}} \psi^*\left[-\frac{\hbar^2}{2m}\frac{d^3\psi}{dx^3} + \left(\frac{dU}{dx}\right)\psi + U\frac{d\psi}{dx}\right]dx\end{aligned}$$

and canceling yields

$$\langle [H,p]\rangle = i\hbar\left\langle \frac{dU}{dx}\right\rangle.$$

Let us now use (4.10) and the fact that dU/dx is $-F$ the force, and we get

$$\frac{d}{dt}\langle p\rangle = \langle F\rangle. \tag{4.12}$$

Newton's second law, but for the expectation of variables rather than the variables themselves. Here, we can begin to see how the classical limit of the quantum theory emerges. When particles are sufficiently massive, the deBroglie wavelength becomes smaller and the expectation of variables along a trajectory that solve Newton's equations collapse to their classical variables. This is an important observation that is needed later on when the Feynman path integral is introduced. Equation (4.12) is the first of the two parts of Ehrenfest's theorem, named after Austrian theoretical physicist Paul Ehrenfest. The proof of the second part,

$$m\frac{d}{dt}\langle x\rangle = \langle p\rangle \tag{4.13}$$

is left to the reader as an exercise.

4.5 Exercises

1. Consider the algebra $\mathcal{A}$ built over the set of real $n \times n$ matrices with the regular matrix product. Show that the distributive property is satisfied.
2. Show that the symmetrized matrix product of two real $n \times n$ symmetric matrices,

$$\{A, B\}_{ij} = \sum_{k=1}^{n} a_{ik}b_{kj} + b_{ij}a_{kj}$$

   satisfies closure, and is distributive.
3. Verify the result in equation (4.1).
4. Verify the Jacobi identity for the commuted product.
5. Demonstrate that the following basis set:

$$a_0 = \begin{pmatrix} 1 & 0 \\ 0 & 1 \end{pmatrix}, \quad a_1 = \begin{pmatrix} 0 & 1 \\ 1 & 0 \end{pmatrix},$$
$$a_2 = \begin{pmatrix} 0 & -1 \\ 1 & 0 \end{pmatrix}, \quad a_3 = \begin{pmatrix} 1 & 0 \\ 0 & -1 \end{pmatrix}.$$

   is a Lie algebra under the commutation product by finding the structure constants. Then show that it is isomorphic to the basis set of 2×2 matrices over the real numbers.
6. Find the structure constants for su$(2, \mathbb{C})$ with the Pauli spin matrices in example 16.

7. Prove that the set of $n \times n$ Hermitian matrices satisfies the following requirements:

$$\mathrm{Im}(a_{ii}) = 0, \quad \mathrm{Re}(a_{ij}) = \mathrm{Re}(a_{ji}), \quad \mathrm{Im}(a_{ij}) = -\,\mathrm{Im}(a_{ji})$$

where $\mathrm{Im}(a + ib) = b$ is the imaginary part of a complex number and $\mathrm{Re}(a + ib) = a$ is the real part.
8. Propose a basis set for the Lie algebra of 3×3 Hermitian matrices, then use it to find the structure constants.
9. Let

$$A = \begin{pmatrix} 2 & -3 \\ 1 & -1 \end{pmatrix}, \quad P = \begin{pmatrix} 2 & 1 \\ 1 & 1 \end{pmatrix}.$$

Show that the characteristic polynomial of A and $P^{-1}AP$ are both equal to $\lambda^2 - \lambda + 1$. This means that similar matrices have the same eigenvalues.
10. Prove that $\det A = \det B$ if A and B are similar matrices.
11. Verify that

$$T = \frac{1}{2}\begin{pmatrix} 1+i & -1+i \\ 1+i & 1-i \end{pmatrix}$$

is unitary.
12. Find

$$\exp\begin{pmatrix} 2 & -1 & 0 \\ -1 & 5 & -1 \\ 0 & -1 & 2 \end{pmatrix}$$

by diagonalization.
13. Find

$$\exp\begin{pmatrix} 0 & -1 & 5 \\ 0 & 0 & -2 \\ 0 & 0 & 0 \end{pmatrix}.$$

14. Using P, P^{-1} and A in example 20 confirm that $P^{-1}P = I$, the 3×3 unit matrix, and that

$$P^{-1}AP = \begin{pmatrix} 1 & 0 & 0 \\ 0 & 2 & 0 \\ 0 & 0 & 3 \end{pmatrix}.$$

15. Confirm that for the matrix A in example 21, A^3 is the null matrix.
16. Verify equations (4.6) and (4.7).

17. Prove that

$$\exp\begin{pmatrix} 0 & x \\ x & 0 \end{pmatrix} = \begin{pmatrix} \cosh x & \sinh x \\ \sinh x & \cosh x \end{pmatrix}.$$

18. Exponentiate $\left(\begin{smallmatrix} a & 0 \\ 0 & -a \end{smallmatrix}\right)$ for some constant a.
19. Show that

$$B^T \begin{pmatrix} 0 & 1 \\ -1 & 0 \end{pmatrix} B = \begin{pmatrix} 0 & 1 \\ -1 & 0 \end{pmatrix}$$

where B is the matrix in example 25.
20. In this exercise, you will compute the exponential of a $n \times n$ matrix A using two approaches. The first is a recursion scheme that initiates a matrix B as a unit matrix,

$$B_0 = I,$$

and updates using

$$B_{n+1} = \frac{1}{n+1} A B_n.$$

(a) Prove by induction that

$$\sum_{k=0}^{m} B_k = \sum_{k=0}^{m} \frac{1}{k!} A^k.$$

(b) The following is a minimal Python code for the exponentiation of a 3×3 matrix. The example loads a symmetric matrix.

```
import numpy as np
a = np.array([[3, -1, 0], [-1, 2, -1], [0, -1, 3]])
bn = np.identity((3),np.float64)
sume = np.identity((3),np.float64)
for k in range(30):
    bnp1 = np.matmul(a,bn)/np.float64(k+1)
    sume += bnp1
    bn = bnp1
print(sume)
```

The command `np.float64(k+1)` converts the integer $k + 1$ into a double precision floating point number. Run the code, then write code to diagonalize the matrix A using the `np.linalg.eig(a)` to get the eigenvalues λ_i and eigenvectors P,

$$P^T AP = P^T \begin{pmatrix} 3 & -1 & 0 \\ -1 & 2 & -1 \\ 0 & -1 & 3 \end{pmatrix} P = \begin{pmatrix} 1 & 0 & 0 \\ 0 & 3 & 0 \\ 0 & 0 & 4 \end{pmatrix}$$

exponentiate the eigenvalues, and compute

$$e^A = P \begin{pmatrix} e^1 & 0 & 0 \\ 0 & e^3 & 0 \\ 0 & 0 & e^4 \end{pmatrix} P^T.$$

The answer should match the computation by the series expansion.

21. In the previous exercise, you observed that the exponential of a symmetric matrix is also symmetric. Use the definition of the matrix product to prove the general case.
22. Prove that for any $n \times n$ matrix A over the real numbers, $e^A \in \mathrm{Gl}(n, \mathbb{R})$.
23. Show that

$$\begin{pmatrix} \cos\theta & -\sin\theta \\ \sin\theta & \cos\theta \end{pmatrix}^T \begin{pmatrix} 0 & 1 \\ -1 & 0 \end{pmatrix} \begin{pmatrix} \cos\theta & -\sin\theta \\ \sin\theta & \cos\theta \end{pmatrix} = \begin{pmatrix} 0 & 1 \\ -1 & 0 \end{pmatrix}.$$

Therefore, there are transformations in the intersection of the symplectic group and the special orthogonal group.

24. First, show that

$$[H, x] = -i\frac{\hbar}{m}p,$$

then use it to derive equation (4.13):

$$m\frac{d}{dt}\langle x \rangle = \langle p \rangle.$$

5 The diffusion Monte Carlo method

> This is a sad but necessary chapter. It is sad because we have reached the point where the hope of finding exact solutions to the problem is set aside, and we begin to look for methods of approximation.
>
> Peter W. Atkins

We indeed have reached a point where, finding *analytical* solutions to the problem is set aside, but with the tools in this chapter we can compute the ground state of the Schrödinger equation in one dimension (for now) for any reasonable potential that permits bound states. The simulation tool we introduce is one of two powerful research methods that we discuss in this book. Both are heavily used in computational quantum mechanics. The diffusion Monte Carlo (DMC) method has been used for research in nuclear, electronic, and molecular quantum physics. There are two flavors of the method created independently and nearly simultaneously. The first was published in 1974 by M. H. Kalos, D. Levesque, and L. Verlet [3, 15] and the second by J. B. Anderson in 1975 [26]. Both methods represent the propagated wave function with populations of pseudo-particles. Both methods take advantage of the isomorphism between the imaginary time-dependent Schrödinger equation and the diffusion equation with sources and sinks. Both methods model the kinetic energy portion of the propagator by changing the position of pseudo-particles with a Gaussian random number with an appropriate standard deviation, and use the potential U to model the sources and sinks. The former approach makes use of a continuous rate term to modify the weights of each individual pseudo-particle, whereas the latter uses a random integer to decide if a particular replica will have zero, one or two copies created at the location of the pseudo-particle. The present chapter is designed to explain in detail the algorithms and the theory behind it. We start with the notion of the time evolution propagator as a member of the unitary group, one of the Lie groups we introduced in Chapter 4. The first section contains the theoretical underpinning. The second one derives a simple estimator of the energy. The third presents the algorithm and some test data.

5.1 The ground state projection operator

As we have seen, to study quantum dynamics one needs to introduce the time variable into the Schrödinger equation,

$$i\hbar\frac{\partial\psi}{\partial t} = -\frac{\hbar^2}{2m}\frac{d^2}{dx^2}\psi + U\psi.$$

If the potential is time independent, namely $\partial U/\partial_t = 0$, this partial differential equation is separable in the time and space,

https://doi.org/10.1515/9783111610207-005

$$-\frac{\hbar^2}{2m}\frac{d^2}{dx^2}\psi_n + U\psi = E_n\psi_n \quad i\hbar\frac{\partial\psi_n}{\partial t} = E_n\psi_n.$$

The time dependent part can be solved readily:

$$\psi_n(x,t) = \psi_n(x,0)e^{-iE_nt/\hbar}.$$

Let $\mathcal{V} = \text{span}\{\psi_n(x,0)\}$ be the vector space of eigenfunctions of the Hamiltonian operator $\hat{H}$ at $t = 0$, i. e., $\hat{H}\psi_n(x,0) = E_n\psi_n(x,0)$, where

$$\hat{H} = -\frac{\hbar^2}{2m}\frac{d^2}{dx^2} + U\psi.$$

In $\mathcal{V}$, the Hamiltonian matrix we introduce in Chapter 2 is by definition diagonal with E_i the diagonal elements. Moreover, H is Hermitian (real valued in this case). If we arrange the coefficients $c_n(0)$ as a column vector, and use the exponential of $iHt/\hbar$, equation (5.1) can be understood as the operation of a member of the SU group transforming the wave vector ψ_T. We first show that the eigenstates of $\hat{H}$ are stationary states since the matter density $\rho(x,t)$ is time independent:

$$\begin{aligned}\rho(x,t) &= \psi_n(x,t)^*\psi_n(x,t) = \psi_n(x,0)^* e^{iE_nt/\hbar}\psi_n(x,0)e^{-iE_nt/\hbar} \\ &= \rho(x,0).\end{aligned}$$

5.1.1 Wave packets

Generally, we do not know the set ψ_n before hand. To simulate a quantum system, one uses some educated guess for the wave function ψ_T. Of course, no matter how good our guess may be, we cannot expect ψ_T to equal any of the ψ_n. However, if ψ_T satisfies the same boundary conditions as the exact solutions $\{\psi_n(x,0)\}$, it can always be expanded into $\mathcal{V}$,

$$\psi_T = \sum_{n=0}^{\infty} c_n\psi_n.$$

However, ψ_T is no longer a stationary state $\partial\rho(x,t)/\partial t \neq 0$, and it is straightforward to prove such claim:

$$\psi_T(x,t) = \sum_{n=0}^{\infty} c_n(t)\psi_n(x).$$

Each coefficient of the expansion gains a complex phase at $t > 0$,

$$c_n(t) = c_n(0)e^{-iE_nt/\hbar}.$$

Inserting this form into the expansion, we get the real time evolution,

$$\psi_T(x,t) = \sum_{n=0}^{\infty} c_n(0)e^{-iE_n t/\hbar}\psi_n(x). \tag{5.1}$$

Evaluating the complex conjugate of Ψ_T,

$$\psi_T^*(x,t) = \sum_{m=0}^{\infty} c_m(0)e^{iE_m t/\hbar}\psi_m^*(x),$$

and using the result to find an expression for the matter density, we end up with function of both space and time,

$$\rho(x,t) = \sum_{m=0}^{\infty}\sum_{n=0}^{\infty} c_m(0)c_n(0)e^{i\omega_{n,m}t}\psi_m^*(x)\psi_n(x),$$

where $\omega_{n,m} = (E_m - E_n)/\hbar$ are the Bohr frequencies of the operator $\hat{H}$. This means that wave packets evolve in time.

5.1.2 The imaginary time evolution of the wave packet

Without knowing the solutions ψ_n, E_n a priori, it is challenging to formulate and numerically compute the time evolution operators and get the desired properties of the physical system. Switching to imaginary time $t = -i\tau$, one can show that the time evolution leads to the ground state. To see this, let us first consider the evolution of the wave packet under these new conditions,

$$\psi_T(x,\tau) = \sum_{n=0}^{\infty} c_n(\tau)\psi_n(x),$$

where

$$c_n(\tau) = c_n(0)e^{-(E_n-E_0)\tau/\hbar},$$

and E_0 is the ground state energy. Inserting this into the expression above, and taking the limit as $\tau \to \infty$, give the following result:

$$\lim_{\tau\to\infty}\psi_T(x,\tau) = \lim_{\tau\to\infty}\sum_{n=0}^{\infty} c_n(0)e^{-(E_n-E_0)\tau/\hbar}\psi_n(x) \to \psi_0(x),$$

since all states other than the ground state ($n > 0$) are multiplied by an exponentially decaying function

$$e^{-(E_n-E_0)\tau/\hbar}. \tag{5.2}$$

Therefore, any arbitrary wave function that can be expanded in $\mathcal{V}$ and subjected to the imaginary time projection operator will yield the ground state. E_0 may not be known a priori, however, knowing its value beforehand is not strictly necessary as we soon show.

The time dependent Schrödinger equation under the transformation to imaginary time $t = -i\tau$ becomes a diffusion equation with sources and sinks ($\tau = it$),

$$i\hbar\frac{\partial\psi}{\partial\tau}\frac{\partial\tau}{\partial t} = -\frac{\hbar^2}{2m}\frac{d^2}{dx^2}\psi + (U - E_0)\psi.$$

The chain rule is used on the right and U can be shifted by a constant. E_0 in this case, without any consequences,

$$-\hbar\frac{\partial\psi}{\partial\tau} = -\frac{\hbar^2}{2m}\frac{d^2}{dx^2}\psi + (U - E_0)\psi$$

or

$$\hbar\frac{\partial\psi}{\partial\tau} = \frac{\hbar^2}{2m}\frac{d^2}{dx^2}\psi - (U - E_0)\psi.$$

This is still a formidable equation to solve analytically, but as we show later on, can be computed in most cases. Before discussing the general case, it pays to study the potential free case. For a free particle in imaginary time,

$$\frac{\hbar^2}{2m}\left(\frac{\partial^2\psi}{\partial x^2}\right) = \hbar\frac{\partial\psi}{\partial\tau}, \tag{5.3}$$

the differential equation can be derived from the equation of continuity, an important conservation law. The physically acceptable solution satisfying the initial conditions

$$\psi(x, \tau = 0) = 1,$$

is

$$\psi(x,\tau) = N\exp\left(-\frac{mx^2}{2\hbar\tau}\right), \tag{5.4}$$

where N is some normalization constant. It is useful to check that this expression for ψ does in fact satisfy equation (5.3), as it validates the choice made for t. Here is the proof in detail:

$$\frac{\partial\psi}{\partial x} = -\frac{mx}{\hbar\tau}\psi$$

$$\frac{\partial^2\psi}{\partial x^2} = \left[-\frac{m}{\hbar\tau} + \left(\frac{mx}{\hbar\tau}\right)^2\right]\psi$$

$$\frac{\partial \psi}{\partial \tau} = \left[-\frac{1}{2\tau} + \left(\frac{mx^2}{2\hbar\tau^2}\right)\right]\psi.$$

Inserting these into equation (5.3), and canceling ψ on both sides, leaves

$$\frac{\hbar^2}{2m}\left[-\frac{m}{\hbar\tau} + \left(\frac{mx}{\hbar\tau}\right)^2\right] = \hbar\left[-\frac{1}{2\tau} + \left(\frac{mx^2}{2\hbar\tau^2}\right)\right]$$
$$-\frac{\hbar}{2\tau} + \frac{mx^2}{2\tau^2} = -\frac{\hbar}{2\tau} + \frac{mx^2}{2\tau^2},$$

therefore, the solution in equation (5.4) is validated.

It is highly instructive to see how equation (5.4) is actually derived. The approach we choose is the Fourier transform method. The Fourier transform of a function is defined as

$$\mathscr{F}[f(x)] = \tilde{f}(a) = \int_{-\infty}^{\infty} f(x)e^{-iax}dx.$$

The solution of differential equations by the Fourier transform requires three theorems, the proofs of which are left to the reader,

$$\mathscr{F}\left[\frac{\partial}{\partial t}f(x,t)\right] = \frac{\partial}{\partial t}\tilde{f}(x,t), \quad \mathscr{F}\left[\frac{\partial}{\partial x}f(x,t)\right] = ia\tilde{f}(x,t),$$
$$\mathscr{F}\left[\frac{\partial^2}{\partial x^2}f(x,t)\right] = -a^2\tilde{f}(x,t).$$

The Fourier transform of a partial differential equation like the diffusion equation becomes an ordinary separable first-order differential equation satisfied by $\tilde{f}(a,t)$ with time as the independent variable. In particular, the diffusion problem,

$$\frac{\partial}{\partial t}\psi(x,t) = D\frac{\partial}{\partial x^2}\psi(x,t),$$

subject to the initial condition,

$$\psi(x,t=0) = \delta(x),$$

is transformed into

$$\frac{\partial}{\partial t}\tilde{\psi}(a,t) = -Da^2\tilde{\psi}(a,t).$$

The solution in the frequency domain with the initial condition,

$$\tilde{\psi}(a,t=0) = 1$$

becomes

$$\widetilde{\psi}(a,t) = e^{-Dta^2},$$

and this can be transformed back into the position domain by the inverse integral,

$$\psi(x,t) = \frac{1}{2\pi}\int_{-\infty}^{\infty} da\widetilde{\psi}(a,t)e^{iax},$$

$$\psi(x,t) = \frac{1}{2\pi}\int_{-\infty}^{\infty} da\, e^{-Dta^2}\, e^{iax}.$$

This can be turned into a standard Gaussian integral [104] by combining exponents and completing the square,

$$-Dta^2 + iax = -Dt(a+b)^2 + Dtb^2$$

where b is found by expanding the square on the right $-Dta^2 + iax = -Dta^2 - 2Dtab$ or $b = -iax/2Dt$. Then

$$\psi(x,t) = \frac{1}{2\pi}e^{-x^2/4Dt}\int_{-\infty}^{\infty} da\, e^{-Dt(a-ix/2Dt)^2}.$$

We finally arrive at the answer,

$$\psi(x,t) = \frac{1}{2\sqrt{\pi Dt}}e^{-x^2/4Dt}$$

where the diffusion constant is $D = \hbar/2m$.

It is evident that the preexponential factor normalizes ψ instead of ψ^2. Imaginary time propagation does not conserve the norm of quantum wave functions. This fact, however, does not impact our ability to simulate ground states. It is important at this point to note that the differential equation in this section is solved exactly. Generally, this is not possible primarily because the potential energy surface U depends on x, and this prevents us from finding an analytical solution. Then the most rigorous way to solve the problem computationally is to derive a time evolution propagator and apply it to the system to extract the properties of interest. Rather than trying to solve the differential equation in the frequency domain for all times, we choose to do so only for a small interval Δt. In the limit $\Delta t \to 0$, it is sufficient to expand U into a power series in x about the initial position x_n and just keep the constant term. In this case, $U(x) \approx U(x_n)$. The resulting diffusion equation,

$$\frac{\partial}{\partial t}\psi(x,t) = D\frac{\partial^2}{\partial x^2}\psi(x,t) - U\psi(x,t)$$

subject to the initial condition,

$$\psi(x, t = 0) = \delta(x)$$

yields the following Fourier transform:

$$\frac{\partial}{\partial t}\tilde{\psi}(a, t) = -Da^2\tilde{\psi}(a, t) - U\tilde{\psi}(a, t)$$

with

$$\tilde{\psi}(a, t = 0) = 1$$

as the initial condition. The solution in the frequency domain is

$$\tilde{\psi}(a, t) = e^{-Dta^2 - Ut}.$$

The inverse integral,

$$\mathcal{K}(x, 0, t) = \frac{1}{2\pi}e^{-Ut}\int_{-\infty}^{\infty} da\, e^{-Dta^2}\, e^{iax},$$

can also be turned into a standard Gaussian integral once more by combining exponents and completing the square in a manner analogous to the free particle case. Formally, the wave function is updated with

$$\psi(x, t + \Delta t) = \int_{-\infty}^{\infty} \mathcal{K}(x, x', t)\psi(x', t)dx' \tag{5.5}$$

where the propagator from x', t to $x, t + \Delta t$ is derived from the inverse Fourier transform using $x(t) = x'$ as the initial condition,

$$\mathcal{K}(x, x', t) = \frac{1}{2\sqrt{\pi D t}}e^{-U(x)t}\, e^{-(x-x')^2/4Dt}. \tag{5.6}$$

We can clearly see that the $U\psi$ term in the diffusion equation changes the amplitude of the wave. If U is uniform over all space, then the amplitude change is also uniform, and its contribution becomes irrelevant, since one always works with normalized wave functions in the end. When U is not uniform, the e^{-Ut} term is handled by a combination of branching and feedback to maintain a target number of replicas in the population. Note that in order to derive equation (5.6), we approximate the function $U(x)$ with a constant, $U(x) \approx U(x' + \eta)$ where η is some random number sampled from the $e^{-(x-x')^2/4Dt}$ distribution. More about the algorithm is discussed later in the chapter. The next two sections are dedicated to the explanation of the implementation in detail.

5.2 The basic diffusion Monte Carlo algorithm

We begin with a special type of trial wave function constructed using an ensemble of M Dirac delta functions centered at a set of positions $\{x_1, x_2, \ldots, x_M\}$,

$$\psi_T = A \sum_{i=1}^{M} w_i \delta(x - x_i), \tag{5.7}$$

where the coefficients w_i are the weights of each of these functions. The interpretation of equation (5.7) is to create a representation of ψ_T by placing each pseudo-particle at position x_i. We will occasionally refer to a pseudo-particle as a "replica," both carry the same meaning. The idea behind the propagation algorithm is to move each replica position x_i by simply using

$$x_i' \leftarrow x_i + \eta \tag{5.8}$$

where η is a Gaussian random numbers with zero mean and variance $S^2 = 2D\Delta\tau$. The weight for the new position is computed using

$$w_i^2 = e^{-(U(x_i') - E_{\text{ref}})\Delta\tau/\hbar}. \tag{5.9}$$

E_{ref} is the best estimate for the ground state energy, and the choice of squaring w_i becomes clear later in this section. If the position of the pseudo-particle is such that $U > \hbar^{-1} E_{\text{ref}}$, the corresponding weight is small and the particle is removed by setting w_i to zero. Conversely, if $U < \hbar^{-1} E_{\text{ref}}$, its weight is large and the particle reproduces $w_i^2 = 2$. Critically, the feedback part of the algorithm decreases the value of E_{ref} if too many replicas reproduce, and increases the value of E_{ref} if too many replicas are annihilated. This step accomplishes two important goals. It maintains a population size M around a predetermined target, and systematically improves the estimate of E_{ref}.

The resemblance between equations (5.9) and (5.2) should be clear. They are two different representations of the same object, the imaginary time evolution operator. In equation (5.2), it is represented by projecting it into the vector space of solutions of the time independent Schrödinger equation, whereas here it is represented approximately (for small time intervals) using a type of position representation. Clearly, the second representation is much more amenable to computation when the eigenstates of the Hamiltonian are not known a priori.

To develop the basic diffusion Monte Carlo algorithm, we need to derive an estimator of the ground state energy. This exercise requires a careful analysis using delta calculus, but the process is valuable for the interpretation of the results we obtain with the introduction of gauged fields. The Dirac delta function is technically not a function, but can be defined carefully in terms of a limit over a class of analytical functions. There are a number of options but for our purpose the series of increasingly narrow Gaussian functions is sufficient. In terms of these, we can carefully define this limiting process,

$$\delta(x - x_i) = \lim_{\sigma \to 0} \Phi_\sigma(x, x_i),$$

where

$$\Phi_\sigma(x, x_i) = N_\sigma e^{-(x-x_i)^2/2\sigma^2}.$$

The normalization constant N_σ follows from the regular interpretation of the wave function,

$$\langle i|i\rangle = \int_{-\infty}^{\infty} \Phi_\sigma(x, x_i)^2 dx = 1$$

$$\langle i|i\rangle = N_\sigma^2 \int_{-\infty}^{\infty} e^{-(x-x_i)^2/\sigma^2} dx = 1$$

or

$$N_\sigma = \frac{1}{\sqrt{\sigma\sqrt{\pi}}}.$$

Therefore, a monotonically decreasing sequence of σ values that tends toward zero as, e. g., $\{\sigma_n\}$ $n \in \mathbb{N} = \{\sigma_0, \sigma_0/2, \sigma_0/3, \ldots, \sigma_0/n, \ldots\}$, produces a set of increasingly sharper Gaussian distributions centered at x_i. The distribution evaluated at $x = x_i$ approaches infinity as n tends to infinity, and approaches zero from above at all other values of x.

Given that deriving the matrix elements of $\hat{H}$ with the wave packet in equation (5.7) requires some care, we prove in what follows some important properties for the Dirac delta distribution. First, we show that the Hilbert space basis set used in equation (5.7) has zero overlap in the $\{\sigma_n\}$ sequence limit.

Theorem 19. *The orthogonality relation for the basis set is,*

$$\langle i|i\rangle = \delta_{ij}.$$

Proof. For any σ in the monotonic sequence,

$$\langle i|j\rangle_n = N_\sigma^2 \int_{-\infty}^{\infty} e^{-n^2(x-x_i)^2/2\sigma_0^2} e^{-n^2(x-x_j)^2/2\sigma_0^2} dx.$$

Combining both exponents and completing the square,

$$(x - x_i)^2 + (x - x_j)^2 = 2(x - \alpha_{ij})^2 + \beta_{ij}$$

where

$$\alpha_{ij} = \frac{x_i + x_j}{2}, \quad \beta_{ij} = \frac{(x_i - x_j)^2}{2},$$

gives

$$\langle i|j\rangle = \frac{n}{\sigma_0\sqrt{\pi}} e^{-n^2\beta_{ij}/2\sigma_0^2} \int_{-\infty}^{\infty} e^{-n^2(x-\alpha_{ij})^2/\sigma_0^2} dx.$$

The integral on the left is a standard Gaussian integral

$$\langle i|j\rangle_n = e^{-n^2\beta_{ij}/2\sigma_0^2}.$$

The theorem follows trivially as n becomes large. The case when $i = j$ simply produces N_σ^{-2} for all values of σ_n. □

With this result, we can normalize the wave function in equation (5.7),

$$A^2 \int_{-\infty}^{\infty} \psi_T^* \psi_T \, dx = 1,$$

$$A^2 \int_{-\infty}^{\infty} \sum_{i,j=1}^{M} w_i w_j \delta(x - x_i)\delta(x - x_j) dx = 1,$$

exchanging the double summation with the integration and using Theorem 19 gives the normalization constant,

$$A = 1 \Big/ \sqrt{\sum_{i=1}^{M} w_i^2}.$$

Next, let us consider the δ matrix elements for the momentum operator.

Theorem 20. *The expectation value for the momentum operator vanishes:*

$$\langle i|\hat{p}|j\rangle = 0.$$

Proof. Since $\hat{p} = -i\hbar d/dx$, it is sufficient to prove that

$$\left\langle i \left| \frac{d}{dx} \right| j \right\rangle_n$$

approaches 0 as $n \to \infty$, using

$$\frac{d}{dx}\Phi_\sigma(x, x_i) = N_\sigma \frac{d}{dx} e^{-n^2(x-x_i)^2/2\sigma_0^2} = -n^2 N_\sigma \frac{x - x_i}{\sigma_0^2} e^{-n^2(x-x_i)^2/2\sigma_0^0}.$$

With this expression, we are ready to consider two cases. First, the $i = j$ case produces the following expression:

$$\left\langle i\left|\frac{d}{dx}\right|j\right\rangle_n = -\frac{n^2}{\sigma_0^2}N_\sigma^2\int_{-\infty}^{\infty}(x-x_i)e^{-n^2(x-x_i)^2/\sigma_0^2}dx = 0,$$

where the $u = x - x_i$ yields a vanishing integral by symmetry. The case $i \neq j$ requires the same completion of the square in the exponent as in the previous proof,

$$\left\langle i\left|\frac{d}{dx}\right|j\right\rangle_n = -\frac{n^2}{\sigma_0^2}N_\sigma^2 e^{-n^2\beta_{ij}/2\sigma_0^2}\int_{-\infty}^{\infty}(x-x_i)e^{-n^2(x-a_{ij})^2/\sigma_0^2}dx.$$

In this case, the natural substitution $u = x - a_{ij}$ produces two integrals,

$$\left\langle i\left|\frac{d}{dx}\right|j\right\rangle_n = -\frac{n^2}{\sigma_0^2}N_\sigma^2 e^{-n^2\beta_{ij}/2\sigma_0^2}\left[\int_{-\infty}^{\infty}ue^{-n^2u^2/\sigma_0^2}dx + \frac{(x_j-x_i)}{2}\int_{-\infty}^{\infty}e^{-n^2u^2/\sigma_0^2}dx\right].$$

This first integral vanishes by symmetry, and the second is a standard integral,

$$\left\langle i\left|\frac{d}{dx}\right|j\right\rangle_n = \frac{n^2}{\sigma_0^2}\frac{(x_i-x_j)}{2}e^{-n^2\beta_{ij}/2\sigma_0^2}.$$

The theorem follows in the limit as $n \to \infty$ by a straightforward application of l'Hôpital's rule. □

The kinetic energy operator is considered next. The relevant result is the following.

Theorem 21. *The expectation value for the second derivative operator satifies the following:*

$$\left\langle i\left|\frac{d^2}{dx^2}\right|j\right\rangle_n = \begin{cases} 0 & i \neq j \\ -\frac{n^2}{2\sigma_0^2} & i = j. \end{cases}$$

Proof. The second derivative of Φ_σ is

$$\frac{d^2}{dx^2}\Phi_\sigma(x, x_i) = N_\sigma\left[\frac{n^4}{\sigma_0^4}(x-x_i)^2 - \frac{n^2}{\sigma_0^2}\right]e^{-n^2(x-x_i)^2/2\sigma_0^2}.$$

For the $i = j$ case, we get the following two integrals:

$$\left\langle i\left|\frac{d^2}{dx^2}\right|i\right\rangle_n = N_\sigma^2\left[\frac{n^4}{\sigma_0^4}\int_{-\infty}^{\infty}(x-x_i)^2 e^{-n^2(x-x_i)^2/\sigma_0^2}dx - \frac{n^2}{\sigma_0^2}\int_{-\infty}^{\infty}e^{-n^2(x-x_i)^2/\sigma_0^2}dx\right].$$

Both of these can be brought into standard form with the $u = x - x_i$ substitution,

$$\left\langle i\left|\frac{d^2}{dx^2}\right|j\right\rangle_n = \frac{n}{\sigma_0}\left[\frac{n}{2\sigma_0} - \frac{n}{\sigma_0}\right] = -\frac{n^2}{2\sigma_0^2}.$$

Whereas, for the $i \neq j$ case, we get

$$\left\langle i\left|\frac{d^2}{dx^2}\right|j\right\rangle_n = N_\sigma^2 e^{-n^2\beta_{ij}/2\sigma_0^2}\left[\frac{n^4}{\sigma_0^4}\int_{-\infty}^{\infty}(x-x_i)^2 e^{-n^2(x-a_{ij})^2/2\sigma_0^2}dx - \frac{n^2}{\sigma_0^2}\int_{-\infty}^{\infty}e^{-n^2(x-a_{ij})^2/2\sigma_0^2}dx\right].$$

Letting $u = x - a_{ij}$ into the first integral gives

$$\int_{-\infty}^{\infty}(x-x_i)^2 e^{-n^2(x-a_{ij})^2/2\sigma_0^2}dx = \int_{-\infty}^{\infty}\left(u + \frac{x_j - x_i}{2}\right)^2 e^{-n^2u^2/2\sigma_0^2}dx.$$

We expand the square

$$\left(u + \frac{x_j - x_i}{2}\right)^2 = u^2 + (x_j - x_i)u + \frac{(x_j - x_i)^2}{4},$$

and insert this expression into the matrix element $\langle i|d^2/dx^2|j\rangle_n$. Evaluating the standard Gaussian integrals, after some cancellations gives,

$$\left\langle i\left|\frac{d^2}{dx^2}\right|j\right\rangle_n = \left[\frac{n^4(x_j - x_i)^2}{4\sigma_0^4} - \frac{n^2}{2\sigma_0^2}\right]e^{-n^2\beta_{ij}/2\sigma_0^2}.$$

As with the previous results, taking the limit as $n \to \infty$ proves the theorem. □

To complete the analysis, we need one additional result.

Theorem 22. *The expectation value for any continuous function $f(x)$ is*

$$\langle i|f(x)|j\rangle = f.(x_i)\delta_{ij}.$$

This result is closely related to the sifting property of the delta distribution and it is shown with the following argument.

Proof. Let us begin with the $i = j$ case, then

$$\langle i|f(x)|i\rangle_n = \frac{n}{\sigma_0\sqrt{\pi}} \int_{-\infty}^{\infty} f(x)\, e^{-n^2(x-x_i)^2/\sigma_0^2} dx$$

where, for any given value of n, one can define ϵ such that

$$\langle i|f(x)|i\rangle_n \approx \frac{n}{\sigma_0\sqrt{\pi}} \int_{x_i-\epsilon}^{x_i+\epsilon} f(x)\, e^{-n^2(x-x_i)^2/\sigma_0^2} dx$$

because outside of the $x_i - \epsilon < x < x_i + \epsilon$ interval the value of Φ_σ is sufficiently small. Moreover, if f is continuous,

$$\langle i|f(x)|i\rangle_n \approx f(x_i) \frac{n}{\sigma_0\sqrt{\pi}} \int_{x_i-\epsilon}^{x_i+\epsilon} e^{-n^2(x-x_i)^2/\sigma_0^2} dx = f(x_i),$$

since Φ_σ^2 is normalizable. The approximation becomes exact in the limit as n approaches infinity.

When $i \neq j$,

$$\begin{aligned}\langle i|f(x)|j\rangle_n &= \frac{n}{\sigma_0\sqrt{\pi}} e^{-n^2\beta_{ij}/2\sigma_0^2} \int_{-\infty}^{\infty} f(x) e^{-n^2(x-a_{ij})^2/\sigma_0^2} dx \\ &\approx \frac{n}{\sigma_0\sqrt{\pi}} e^{-n^2\beta_{ij}/2\sigma_0^2} f(a_{ij})\end{aligned}$$

by the previous argument. Since $f(a_{ij})$ is finite at a_{ij}, the element $\langle i|f(x)|j\rangle$ tends to zero in the limit. □

With the last four theorems, we can find an expression for the elements of the Hamiltonian matrix,

$$\langle i|\hat{H}|j\rangle = -\frac{\hbar^2}{2m}\left\langle i\left|\frac{d^2}{dx^2}\right|j\right\rangle + \langle i|U(x)|j\rangle = U(x_i)\delta_{ij}.$$

For a trial wave function ψ_T, the energy is

$$E_0 = \int_{-\infty}^{\infty} \psi_T^* \hat{H}\psi_T\, dx = A^2 \int_{-\infty}^{\infty} \sum_{i,j=1}^{M} w_i w_j \langle i|\hat{H}|j\rangle$$

and it follows from Theorems 21 and 22 that

$$E_0 = \sum_{i=1}^{M} w_i^2\, U(x_i) \Big/ \sum_{i=1}^{M} w_i^2. \tag{5.10}$$

Equation (5.10) estimates the energy of any trial wave function ψ_T. When the diffusion-branching processes of the random walk reaches equilibrium, it estimates the ground state energy.

Theorem 21 is a worrisome result, the kinetic energy matrix has divergent diagonal elements. At first, it may seem something is seriously wrong. However, these divergent energies can be safely ignored, as we have here, because for any value of n they are uniform, meaning they do not depend on x. Consequently, they are immeasurable. The situation is analogous to the quantum field theory infinite zero point energy issue. Eliminating constant energy terms is perhaps the most trivial example of a fundamental principle upon which gauge theory is founded. We may say at this point that *The laws of physics look exactly the same in a constant (uniform) energy field as they do in its absence.* This first encounter with a seemingly divergent result is perhaps a good example of just how important gauge theory is to us. We will soon encounter a generalized version of the previous observation that states: *The laws of physics look exactly the same after a change of variables that preserves the fundamental symmetries of the system.*

5.3 A Python code for the primitive Diffusion Monte Carlo

```
def pot(x):
    force_constant = 1.0
    return  force_constant*x*x/2.0
import numpy as np
import random as rnd
rnd.seed()    # seed with system time
mass,mu,deltat,eref = 1.0,0.0,0.06,0.0
sigma = np.sqrt(deltat/mass)
target = 10000.0
size = int(target)
x = np.zeros(20000)
y = np.zeros(20000)
m = np.zeros(20000)
for k in range(2000): # perform 2000 steps
    sume = 0.0
    isum = 0.0
    for i in range(size):
        x[i] += rnd.gauss(mu,sigma)
        u = pot(x[i])
        weight = np.exp(-(u - eref)*deltat)
        r = rnd.random()
        iw = int(weight+r)
        m[i] = min(iw,2)
```

```
        sume += m[i]*u
        isum += float(m[i])
    ae = sume/isum
    eref = ae - np.log(isum/target)
    print("%5.4f %10.5f %6d %10.5f" %((k+1)*deltat,ae,size,eref))
    index = 0  # reset the arrays
    for i in range(size):
        j = 1
        while j <= m[i]:
            j += 1
            y[index] = x[i]
            index += 1
    size = int(isum)
    for i in range(size):
        x[i] = y[i]
        m[i] = 1.0
```

In the top three lines, we have used a function definition for the computation of the potential energy U, which is that of a harmonic oscillator with a force constant equal to one. After assigning values to the mass m, the time increment Δt, and a first estimate of E_0, the code computes, $\sigma = \sqrt{\Delta t/m}$, the standard deviation of the Gaussian random number η in equation (5.8). Figure 5.1 contains the graph of the value of ae, the estimate

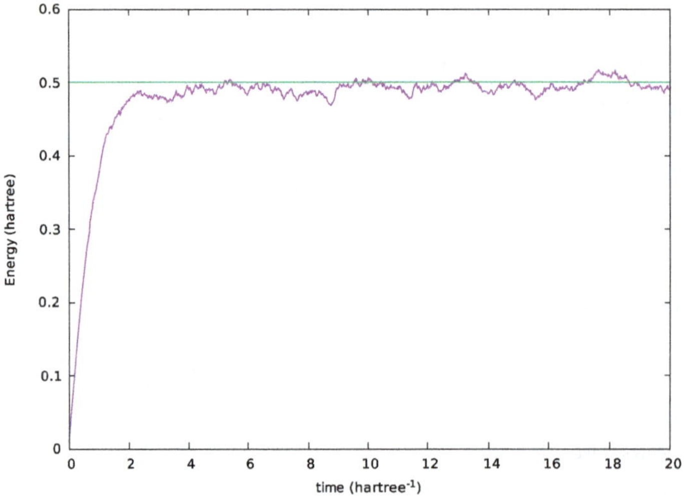

Figure 5.1: The ground state energy over simulation time for a particle on unit mass in a harmonic oscillator with unit force constant as estimated by equation (5.10). The horizontal line (green) at 0.5 hartree represents the exact ground state energy E_0. The purple line begins near zero, and rapidly increases toward the expected value. At around $\tau = 5$ hartree^{-1}, the line begins to fluctuate around the correct result.

of E_0 as a function of time computed with `((k+1)*deltat`. The array `x[]` contains the position of the pseudo-particles with a population size that fluctuates around the value assigned to the variable `target`. The outer loop over `k` steps through the simulation time, while the loop over `i` handles each of the members of the populations. The branching weight is computed according to equation (5.9) and the feedback step that updates the estimate of E_{ref} takes place after the loop over `i` is completed in the line `eref = ae - np.log(isum/target)`, where `ae` is the ground state energy of equation estimated with equation (5.10). If the ratio inside the logarithm is less than one, too many replicas were annihilated, and as a result E_{ref} is moved slightly above the running ground state energy estimate. The converse occurs if the ratio inside the logarithm is greater than one. When the value of E_{ref} and E_0 begin to fluctuate about some constant value the population of pseudo-particles has reached the equilibrium shape and samples the unnormalized ground state wave function φ_0 with some bias that can, in principle, be controlled systematically. The lines of code that follow the `index = 0` statement are performing some housekeeping tasks to account for those positions that have doubled and those that were removed to fit into a contiguous array `x[]`.

For any step k at equilibrium, we can represent the wave function using

$$\psi_T = A \sum_{i=1}^{M} \delta(x - x_i), \tag{5.11}$$

and the estimate of the ground state energy is

$$E_0 \approx \frac{1}{M} \sum_{i=1}^{M} U(x_i). \tag{5.12}$$

Many of the exercises at the end of this chapter are meant to introduce readers into the craft of stochastic simulations including the estimate and management of potential sampling bias.

With the code in this section, we create the data plotted in Figure 5.2 where we compare first- and second-order convergence with respect to the parameter Δt. It is evident that the parameter Δt needs to be carefully tuned. If it is too large, the estimate of the ground state energy will be inaccurate, and if it is too small the algorithm will be less efficient. The basic algorithm has a bias proportional to Δt. However, a small modification [18] to the branching function has been proposed to increase the order of convergence to Δt^2. This means that if Δt is halved, the deviation from the exact result drops by a factor of four. To achieve this result, the branching in equation (5.9) is modified as follows. Let x_i be the position of pseudo-particle i before the diffusion step takes place and $x_i + \eta$ the position after the diffusion move. Then

$$w_i^2 = \exp[-(\overline{U} - E_{\text{ref}})\Delta t], \tag{5.13}$$

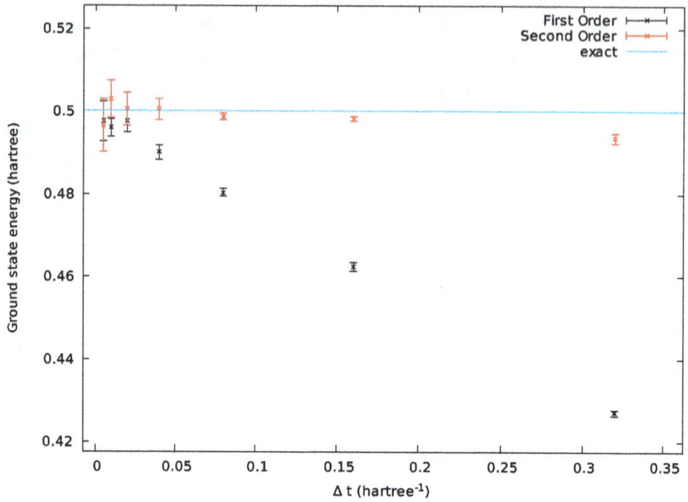

Figure 5.2: The convergence of the ground state energy for several values of the time step Δt. The black data points trend linearly to the exact value from the bottom right where Δt is around 0.3 hartree^{-1} toward the top left where Δt is 0.01. The red data points are systematically closer to the exact value, converge within the statistical error at values of Δt around 0.08 hartree^{-1} and should follow a curved trend, however the curvature is hard to detect at the scale of the graphs. The error bars were estimated using five independent samples consisting of 200 steps.

where $\overline{U} = [U(x_i + \eta) + U(x_i)]/2$ is the arithmetic average between the two values of U. The data in Figure 5.2 demonstrates the clear superiority of the approach. When the computation of U is expensive, one can store the values of $U(x_i)$ in an array, so that simulations with second-order branching do not cost more computer time than the basic algorithm.

5.4 Exercises

1. Prove the following results:

$$\mathcal{F}\left[\frac{\partial}{\partial t} f(x,t)\right] = \frac{\partial}{\partial t}\tilde{f}(x,t),$$

$$\mathcal{F}\left[\frac{\partial}{\partial x} f(x,t)\right] = i\alpha\tilde{f}(x,t),$$

$$\mathcal{F}\left[\frac{\partial^2}{\partial x^2} f(x,t)\right] = -\alpha^2\tilde{f}(x,t),$$

by using the integral definition of the Fourier transform. The order of differentiation and integration can be interchanged if the integrand is bound over the integration domain. For this exercise, you may assume the integrands are properly bounded.

2. Fill in the details of the derivation of equation (5.4) by evaluating the proper inverse Fourier transform.
3. For this exercise, consider a vector space made up of multicentered Gaussian functions

$$\psi = \sum_{i=1}^{n} c_i \chi_i$$

of the following type (in $\mathbb{R}$):

$$\chi_i = N_i \exp\left[-\frac{(x - x_i)^2}{2\sigma_i^2}\right].$$

(a) The normalization constant N_i for χ_i must satisfy

$$1 = N_i^2 \int_{-\infty}^{\infty} \exp\left[-\frac{(x - x_i)^2}{2\sigma_i^2}\right] \exp\left[-\frac{(x - x_i)^2}{2\sigma_i^2}\right] dx.$$

Prove that this becomes

$$N_i = \sqrt{\frac{1}{\sigma_i \sqrt{\pi}}}.$$

(b) The overlap matrix S has the following elements:

$$S_{ij} = N_i N_j \int_{-\infty}^{\infty} \exp\left[-\frac{(x - x_i)^2}{2\sigma_i^2}\right] \exp\left[-\frac{(x - x_j)^2}{2\sigma_j^2}\right] dx,$$

using the group properties of the exponential function we can combine the exponents,

$$\frac{(x - x_i)^2}{2\sigma_i^2} + \frac{(x - x_j)^2}{2\sigma_j^2} = \frac{(x - \alpha_{ij})^2}{2\sigma_{ij}^2} + \gamma_{ij}$$

where

$$\alpha_{ij} = \frac{\sigma_j^2 x_i + \sigma_i^2 x_j}{\sigma_i^2 + \sigma_j^2} \quad \gamma_{ij} = \frac{(x_i - x_j)^2}{2(\sigma_i^2 + \sigma_j^2)} \quad \sigma_{ij}^2 = \frac{\sigma_i^2 \sigma_j^2}{\sigma_i^2 + \sigma_j^2}.$$

Derive these three expressions.

(c) Use the expressions for α_{ij}, γ_{ij}, and σ_{ij} to derive the overlap matrix

$$S_{ij} = \sqrt{\frac{2}{\sigma_i \sigma_j}} \sigma_{ij} \exp(-\gamma_{ij}).$$

Then show that the diagonal elements are $S_{ii} = 1$.
(d) Show that if $\sigma_i = \sigma_j = \sigma_n$ the off diagonal elements of the overlap matrix vanish.
(e) The kinetic energy matrix K has the following elements:

$$K_{ij} = -\frac{\hbar^2}{2m}\sqrt{\frac{1}{\sigma_i\sigma_j\pi}}\int_{-\infty}^{\infty}\exp\left[-\frac{(x-x_i)^2}{2\sigma_i^2}\right]\frac{d^2}{dx^2}\exp\left[-\frac{(x-x_j)^2}{2\sigma_j^2}\right]dx.$$

Derive the following result:

$$\frac{d^2}{dx^2}\exp\left[-\frac{(x-x_j)^2}{2\sigma_j^2}\right] = \left[\frac{(x-x_j)^2}{\sigma_j^4} - \frac{1}{\sigma_j^2}\right]\exp\left[-\frac{(x-x_j)^2}{2\sigma_j^2}\right].$$

(f) To compute the kinetic energy, we have to unravel two integrals,

$$K_{ij} = -\frac{\hbar^2}{2m}\sqrt{\frac{1}{\sigma_i\sigma_j\pi}}\exp(-\gamma_{ij})\int_{-\infty}^{\infty}\left[\frac{(x-x_j)^2}{\sigma_j^4} - \frac{1}{\sigma_j^2}\right]\exp\left[-\frac{(x-a_{ij})^2}{2\sigma_{ij}^2}\right]dx.$$

Use the substitution $u = x - a_{ij}$ to show that

$$K_{ij} = -\frac{\hbar^2}{2m}\sqrt{\frac{2\sigma_i\sigma_j}{\sigma_i^2+\sigma_j^2}}\exp(-\gamma_{ij})\left\{\frac{(x_i-x_j)^2}{(\sigma_i^2+\sigma_j^2)^2} - \frac{1}{\sigma_i^2+\sigma_j^2}\right\}.$$

(g) Show that if $\sigma_i = \sigma_j = \sigma_n$ we get

$$K_{ij} = -\frac{\hbar^2}{2m}\exp(-\gamma_{ij})\left\{\frac{(x_i-x_j)^2}{2\sigma_n^4} - \frac{1}{2\sigma_n^2}\right\}$$

where $i = j$ case grows without bound as the square of n,

$$K_{ii} = \frac{\hbar^2}{4m\,\sigma_n^2},$$

however, the term is uniform, and consequently, it can be ignored.

4. Test the basic diffusion Monte Carlo Python code and reproduce the graph in Figure 5.1. Then modify the basic code so that after the system has equilibrated the estimates of the ground state energy are stored in an array so that the proper statistical analysis can be carried out. For now, simply compute the average over a run consisting of 1000 steps after the system has equilibrated. This may take some explorations at first. The numpy class has a built-in array averaging. For example, the `print("mean = " np.mean(se))` will print the arithmetic mean of the array `se`. Note, however, that zero values are included into the computation of the statistics. You may want to be sure you are computing the average of an array that has no padding zeros if you are using this feature.

5. The basic diffusion Monte Carlo Python code we have included in the text is fine for the necessary initial explorations, however, when optimizing the simulation parameters and for data collection purposes it is more efficient to restart simulations from the ending population of an equilibrated walk. Append the following snippet of code at the end of the basic diffusion Monte Carlo code so that the necessary data to restart the simulation from the data stored in the file `population.txt`.

```
f = open('population.txt','wb')
s1 = str(eref)
f.write(s1+"\n")
for i in range(size):
    s2 = str(x[i])
    space = "    "
    f.write(space+s2+"\n")
f.close()
```

An ACSII file called `population.txt` will be created at the end of the run.

6. Make a copy of the file that contains the basic diffusion Monte Carlo code with a new name, and insert the following lines:

```
f = open('population.txt', 'r')
eref = float(f.readline())
print(eref)
j = 0
for line in f:
    line = line.strip()
    columns = line.split()
    x[j] = float(columns[0])
    print(j,m[j],x[j])
    j += 1
f.close()
print(j, 'lines')
size = j
```

immediately before the `for k in range(200):` line. Note that the line `size = int(target)` should be deleted, and that the range of k is now changed to take 200 steps, since we no longer need to equilibrate and this code should run relatively fast (it takes about 14 seconds on a modern desktop computer), and that is useful for the next exercises.

7. Random numbers are generated by a computer using chaotic equations that are deterministic in nature. These generate sequences of random numbers using an initial value called the *seed*. If we were to restart sampling with the same seed, we would get exactly the same set of random numbers. The technical term for these reproducible sequences of figures is *pseudo-random* numbers. The line in the diffusion Monte Carlo code that reads `rnd.seed()` uses the system clock time as a true random event to prevent the generation of pseudo-random numbers to take place. The

pseudo-random numbers are quite useful when tracking bugs in codes, but are not desirable when many independent simulations are needed to estimate fluctuations. Change the line to read `rnd.seed(715)`. Run two separate simulations and store the energy into separate files but with the same seed. Graph the energy estimates as a function of time from both simulations and observe the result.

8. The diffusion Monte Carlo algorithm is "noisy" and, moreover, the random walks seem to have some amount of "memory," meaning, if at any point in the simulation time the population of pseudo-particles overestimates the energy some successive populations will do so as well. And a similar statement applies if at any point in the simulation time the population of pseudo-particles underestimates the energy. These "cycles" are the result of *correlated sampling*. Because the time increment is typically small, it takes several time steps to reshape the sampling of the ground state energy. It is important when averaging the ground state energy that a sufficient amount of steps are taken to (a) equilibrate the population and (b) include enough correlation cycles. Correlated sampling issues are a permanent headache in stochastic simulations. If we try to estimate the standard deviation of the energy using the energy data from a single run, we would fail to get a meaningful number. A better approach to estimate fluctuations is to run several independent equilibrated runs and find the standard deviation of the means. Run a minimum of five independent simulations with the restarting code (from exercises 5 and 6), then use your favorite method to compute the average and the standard deviation of the means. (Be sure to reset the seed line to `rnd.seed()` in your code.)
9. The standard deviation can be used to estimate the *efficiency* of a stochastic algorithm. Let N be the number of steps in one independent run, σ the standard deviation of the means defined in the previous exercise, and t_C the execution time on a single CPU. Then the relative efficiency ϵ_f of the algorithm can be estimated with

$$\epsilon_f \approx \frac{1}{N\sigma t_C}.$$

   The estimate ϵ_f is not meaningful in isolation, it is a relative number to be used when comparing algorithms. Estimate the efficiency of the diffusion Monte Carlo algorithm consisting of 200 steps assuming it takes around 14.5 seconds to execute.
10. There are several sources of sampling bias. Two of the most important are the size of the time step Δt and the population size M (`target` in the code). As the data in Figure 5.2 demonstrates, the value of Δt should be sufficiently small to converge to the right result, but it cannot be too small or correlated sampling problems will cause the statistical fluctuations to grow substantially. Run five independent samples each 200 steps long, for each of the following values of Δt, 0.005, 0.01, 0.02, 0.04, 0.08, 0.16, and 0.32 hartree^{-1} so as to replicate the data in Figure 5.2. Compute the efficiency of the algorithm at all values of Δt. You should observe that the algorithm is losing efficiency as Δt decreases.

6 Variational methods

> It is my opinion that many physicists, in the last few years, have gone too far in the use of elaborate mathematical methods, often losing sight of the physical reality in the process.
>
> John C. Slater

In previous chapters, we have made occasional use of some educated guesses for a wave function ψ_T, without discussing exactly what makes our guesses "educated." The variational principle offers a way of optimizing the parameters of a trial wave function after some reasonable functional form is assumed. This is the opportune place to demonstrate the process. Generally, one needs optimized wave functions for a number of reasons. In particular, a trial wave function can be used to make diffusion Monte Carlo more efficient by (a) improving the starting distribution of the pseudo-particles and the initial value of E_{ref}, (b) improving the statistical quality of the energy estimators, (c) finding approximate distributions for excited states. The first objective can be accomplished by simply starting the population of pseudo-particles using random numbers distributed according to an optimized ψ_T, when such sampling is possible, and set the initial value of E_{ref} to the variational estimate of the ground state energy. This minor modification alone makes the initial exploratory runs much more efficient. Importance sampling also improves the efficiency of diffusion Monte Carlo sampling and the statistical quality of the energy estimators. It is important to note, however, that the typical importance sampling DMC simulation does not sample φ_0 as the basic algorithm does, but products of the type $\psi_T\varphi_0$ instead [26]. Therefore, the random walk does not solve the Schrödinger equation in imaginary time. Instead, it finds solutions to a modified diffusion expression called the Smoluchowski equation, named after Marian Smoluchowski, a Polish physicist. Two numerical implementations of the importance sampling DMC exist in the literature, one that incorporates the drift term of the Smoluchowski operator into the random move of each pseudo-particle, and another that directly introduces the functional form of ψ_T into the branching term of the Green function propagator.

Importance sampling diffusion Monte Carlo gained popularity shortly after the publication of the basic method for the reasons we articulated earlier, including approximating excited states. However, finding appropriate trials wave functions for excited states is a difficult task. Excited states wave functions have nodes. These are points in one dimension, lines in two, surfaces in three dimensions, and hypersurfaces in multidimensional systems. Generally, one can make educated guesses about such nodal surfaces, however, using random numbers to sample a wave function that changes sign as a pseudo-particle crosses a node is problematic. Random number distributions are probability distributions and these cannot be negative. Many strategies to handle these difficulties have been published, but generally the sampling is not as efficient as it is for the ground state. Moreover, the approach is fundamentally problematic from a theoretical perspective, because the nodal surface cannot be changed in the course of a simulation leading to the so-called *fixed node bias*. Consequently, the approximations made are

https://doi.org/10.1515/9783111610207-006

not controllable, meaning there is no guarantee that reaching numerical convergence implies accuracy of the result.

Before we can address the challenges of simulating excited states with diffusion Monte Carlo, we need some additional background. Therefore, the focus of this chapter is to demonstrate how the variational principle can be used to derive the Schrödinger equation, and how it can be used to optimize parameters for some trial wave functions. Though we do derive the Smoluchowski operator, we do not implement importance sampling simulations in this chapter. Instead, we introduce a strategy published by Bressanini, Cremaschi, Mella, and Morosi [54] in 1997, which allows one to fit a set of orthogonal functions to the ground state wave function sampled by the basic diffusion Monte Carlo. The Fourier expansion converges in the mean to φ_0, provided the correct boundary conditions are satisfied by the orthogonal functions. The resulting analytical expression for φ_0 can be used to estimate with greater efficiency the ground state energy, as well as other properties of the system even when these do not commute with the Hamiltonian operator. Additionally, the approach allows us to graph the ground state wave function in one dimension (or slices of it in higher dimensions) without having to create histograms of the pseudo-particles positions.

The Smoluchowski operator and the expression of its Green function are derived here since the result tells us how to handle terms in the differential equation proportional to the gradient of the wave function. Such terms can appear in the gauged Schrödinger equation, especially when independent variables are transformed. The analysis is also helpful when the interaction terms created by the nonrelativistic gauge theory have to be interpreted. Therefore, such analysis is an important piece of the puzzle.

6.1 The variational principle

The variational principle is the foundation upon which just about all physics theories rest. It plays a central role in gauge theory, and consequently, it of interest to us. In this chapter, we are concerned with a specific application of the variational principle applied to the solution of the Schrödinger equation. So, that is where we begin. Suppose we are trying to find the ground state of a Hamiltonian operator H. We can start, as in previous chapters, with some arbitrary function ψ_T and vary it in some way, typically by manipulating some parameter α to minimize the error between the average

$$\langle H\rangle_\alpha = \frac{\int_{\mathcal{D}} \psi_T^* H \psi_T \, dx}{\int_{\mathcal{D}} \psi_T^* \psi_T \, dx}, \tag{6.1}$$

and the actual ground state energy. Let us begin with a concrete example that we can work out analytically. Suppose we are trying to solve the Schrödinger equation for a particle of mass m in a quartic potential,

$$H = -\frac{\hbar^2}{2m}\frac{\partial^2}{\partial x^2} + D(x - x_0)^4.$$

To rescale the energy, we simply multiply by $2m/(\hbar^2)$ on both sides and get

$$-\frac{\partial^2}{\partial x^2}\varphi_0 + \lambda^2(x - x_0)^4\varphi_0 = \epsilon_0\varphi_0$$

where clearly $\epsilon_0 = 2mE_0/(\hbar^2)$, and $\lambda = \sqrt{2mD}/\hbar$.

We now introduce a trial wave function with one parameter

$$\psi_T = e^{-\beta(x-x_0)^2} \quad x \in (-\infty, \infty).$$

The idea is to solve this equation for E_0 as a function of the parameter $\beta > 0$. Then, to find the best value of the parameter, we vary β until the slope of $E_0(\beta)$ vanishes at $\beta = \beta_{\text{min}}$. The value $E_0(\beta_{\text{min}})$ is the best estimate of the ground state energy attainable with our guess. A sketch of the process using the rescaled version of equation (6.1) is below. The normalization integral is

$$\mathcal{I}_0 = \int_{-\infty}^{\infty} \psi_T^* \psi_T dx = \sqrt{\frac{\pi}{2\beta}}.$$

The kinetic energy integral evaluates to

$$\mathcal{I}_K = -\int_{-\infty}^{\infty} \psi_T^* \frac{\partial^2}{\partial x^2}\psi_T dx = \beta\mathcal{I}_0,$$

whereas the potential integral becomes

$$\mathcal{I}_U = \lambda^2 \int_{0}^{\infty} \psi_T^*(x - x_0)^4\psi_T dx = \frac{3\lambda^2}{16\beta^2}\mathcal{I}_0.$$

Combining all the pieces finally yields

$$\epsilon_0(\beta) = \frac{\mathcal{I}_K + \mathcal{I}_U}{\mathcal{I}_0} = \beta + \frac{3\lambda^2}{16\beta^2}.$$

The derivative with respect to β of ϵ_0 is

$$\frac{d}{d\beta}\epsilon_0(\beta) = 1 - \frac{3\lambda^2}{8\beta^3}.$$

Setting the last expression to zero yields the optimal value of the parameter β

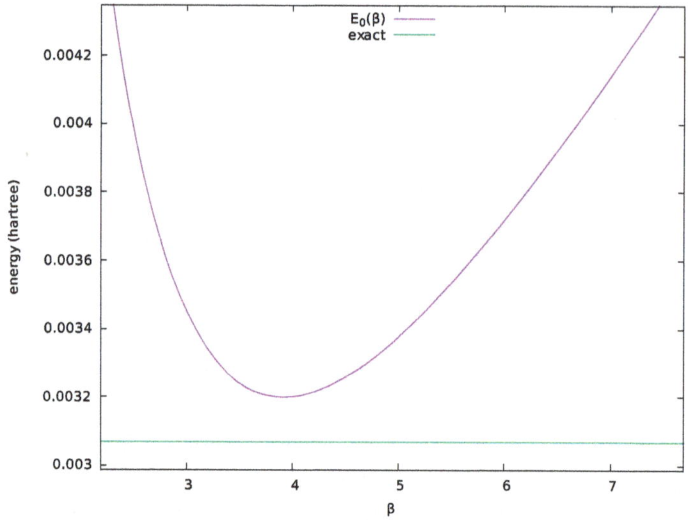

Figure 6.1: A graph of E_0 computed with equation (6.3) versus the parameter β (purple curve). The curve decreases for values of β below 3.9 and increases at larger values. The horizontal green line at $y = 0.003142$ hartree is the ground state energy estimated from the diffusion Monte Carlo algorithm.

$$\beta = \left(\frac{3\lambda^2}{8}\right)^{1/3}. \tag{6.2}$$

In Figure 6.1, we graph

$$E_0 = \frac{\hbar^2}{2m}\beta + \frac{3D}{16\beta^2} \tag{6.3}$$

at several values of β, for a particle of mass = 919.0 atomic units in a quartic potential with $D = 0.0874$ hartree bohr^{-4}. The minimum of the curve is reached at $\beta = 3.920$, in agreement with the analytical result in equation (6.2). The estimate of the ground state energy at that point is 0.003199 hartree. The estimate of the same carried out by a diffusion Monte Carlo calculation is $0.003142 \pm 6 \times 19^{-5}$ hartree, making the variational estimate slightly higher that the diffusion Monte Carlo estimate. The discrepancy in the energy estimate is actually not terrible for a one parameter trial wave function. Variational estimates with a single parameter are typically less accurate than this example. One systematically improves them by introducing more parameters and repeating the optimization process.

6.1.1 Two important results

The previous example shows how one can construct and optimize a one parameter wave function that can give us estimates of the ground state. The fact that the end result is a few percentage points off is perhaps the main reason why the variational approach is considered in many textbooks an approximation. However, for the ground state, this approach can be shown to converge rigorously to the exact result as the trial wave function is made more flexible by judiciously adding more parameters. The present goal is to demonstrate this fact, and we accomplish this in two ways. First, we show that one derives the Schrödinger equation from the variation principle. Then we prove that at any stage along the process of convergence the variational estimate of the energy is always greater or equal to the actual value of E_0.

Theorem 23. *The variational principle applied to a Hamiltonian operator and a trial wave function ψ_T is equivalent to the Schrödinger equation for any of its eigenpairs.*

Proof. Consider a trial wave function ψ_T and its variation $\delta\psi_T$. Let us insert both quantities in equation (6.1) and compare the expressions to first order in $\delta\psi_T$:

$$\langle H\rangle_a + \delta\langle H\rangle_a = \frac{\int_{\mathcal{D}}(\psi_T^* + \delta\psi_T^*)\, H\,(\psi_T + \delta\psi_T)\, dx}{\int_{\mathcal{D}}(\psi_T^* + \delta\psi_T^*)\,(\psi_T + \delta\psi_T)\, dx}.$$

This can be accomplished by expanding the right-hand side. The numerator can be manipulated algebraically,

$$\begin{aligned}\int_{\mathcal{D}}(\psi_T^* + \delta\psi_T^*)\, H\,(\psi_T + \delta\psi_T)\, dx &= \int_{\mathcal{D}} \psi_T^*\, H\, \psi_T dx + \int_{\mathcal{D}} \psi_T^*\, H\,(\delta\psi_T) dx \\ &\quad + \int_{\mathcal{D}} (\delta\psi_T^*)\, H\, \psi_T dx + \mathcal{O}(\delta^2)\end{aligned}$$

where $\mathcal{O}(\delta^2)$ means we have dropped the integral that contains the variations of both the wave function and its complex conjugate. Note that since we assume the Hamiltonian operator is Hermitian, we can write the last equation more concisely,

$$\int_{\mathcal{D}}(\psi_T^* + \delta\psi_T^*)\, H\,(\psi_T + \delta\psi_T)\, dx = \int_{\mathcal{D}} \psi_T^*\, H\, \psi_T dx + \int_{\mathcal{D}} (\delta\psi_T^*)\, H\, \psi_T dx + \text{cc}$$

where cc means the complex conjugate of the previous term and we have dropped $\mathcal{O}(\delta^2)$ for simplicity. The denominator requires the first-order estimate of $1/(1+x) = 1-x+\mathcal{O}(x^2)$ by using the Taylor series of the function on the left. Then

$$\begin{aligned}\frac{1}{\int_{\mathcal{D}}(\psi_T^* + \delta\psi_T^*)\,(\psi_T + \delta\psi_T)\, dx} &= \frac{1}{\int_{\mathcal{D}} \psi_T^*\, \psi_T\, dx + \int_{\mathcal{D}} \psi_T^*(\delta\psi_T)\, dx + \text{cc}} \\ &= \frac{1}{\int_{\mathcal{D}} \psi_T^*\, \psi_T\, dx [1 + \frac{\int_{\mathcal{D}} \psi_T^*(\delta\psi_T)\, dx}{\int_{\mathcal{D}} \psi_T^*\, \psi_T\, dx} + \text{cc}]} \\ &= \frac{1}{\int_{\mathcal{D}} \psi_T^*\, \psi_T\, dx}\left[1 - \frac{\int_{\mathcal{D}} \psi_T^*(\delta\psi_T)\, dx}{\int_{\mathcal{D}} \psi_T^*\, \psi_T\, dx} + \text{cc}\right].\end{aligned}$$

Now we can multiply the expanded numerator and denominator together:

$$\langle H\rangle_a + \delta\langle H\rangle_a = \left[\langle H\rangle_a + \frac{\int_D (\delta\psi_T^*)\, H\, \psi_T dx}{\int_D \psi_T^*\, \psi_T\, dx}\right]\left[1 - \frac{\int_D \psi_T^* (\delta\psi_T)\, dx}{\int_D \psi_T^*\, \psi_T\, dx} + \text{cc}\right].$$

Again, we only keep first-order $\delta\psi$ terms,

$$\langle H\rangle_a + \delta\langle H\rangle_a = \langle H\rangle_a - \langle H\rangle_a \frac{\int_D \psi_T^* (\delta\psi_T)\, dx}{\int_D \psi_T^*\, \psi_T\, dx} + \frac{\int_D (\delta\psi_T^*)\, H\, \psi_T dx}{\int_D \psi_T^*\, \psi_T\, dx}.$$

We drop $\langle H\rangle_a$ on both sides and slightly rearrange terms,

$$\delta\langle H\rangle_a = \frac{\int_D (\delta\psi_T^*)\, H\, \psi_T dx}{\int_D \psi_T^*\, \psi_T\, dx} - \langle H\rangle_a \frac{\int_D \psi_T^* (\delta\psi_T)\, dx}{\int_D \psi_T^*\, \psi_T\, dx} + \text{cc}.$$

Finally, note that we can collect terms into a single integral plus its complex conjugate

$$\delta\langle H\rangle_a = \frac{\int_D \delta\psi_T^*\, (H - \langle H\rangle_a)\, \psi_T dx}{\int_D \psi_T^*\, \psi_T\, dx} + \text{cc} = 0.$$

We next "set the slope to zero" $\delta\langle H\rangle_a = 0$, just as we did in our example. Note that on the right we have an integral and its complex conjugate and we want to set them equal to zero. That can hold only if the integrand in both is equal to zero for all values of x and this gives us

$$(H - \langle H\rangle_a)\, \psi_T = 0.$$

For the last equation to hold, it must be true that

$$H\psi_T = \langle H\rangle_a \psi_T.$$

This is just the Schrödinger equation where $\psi_T = \psi_n$ and $\langle H\rangle_a = E_n$ must be one of the eigenvalue eigenvector pairs of H. □

Theorem 24 (Variational upper bound). *Variational estimates of the ground state energy are always greater or equal to the exact energy.*

Proof. Let

$$\psi_T = \sum_{i=0}^{N} c_i \psi_i$$

where $\{\psi_i\}$ is a set of eigenfunctions of H, and $\sum_{i=1}^{N} c_i^* c_i = 1$, so that ψ_T is normalized. Then it is straightforward to show that

$$\int_{\mathcal{D}} \psi_T^* \, \psi_T \, dx = 1.$$

Consequently,

$$\langle H \rangle_a = \sum_{i=0}^{N} \sum_{j=0}^{N} c_i^* c_j^* \int_{\mathcal{D}} \chi_i^* \, H \chi_i \, dx = \sum_{j=0}^{N} c_j^2 E_j.$$

Since $c_j^2 = c_j^* c_j$ is strictly greater than zero, we see that if we are estimating the ground state energy $\langle H \rangle_a - E_0 \geq 0$, where the equality applies when all coefficients c_j for $j > 0$ are zero. By contrast, when any coefficient other than c_0 is nonequal to zero, a contribution of some excited energy $E_j > E_0$ increments the estimate of $\langle H \rangle_a$. □

It is important to note that when estimating an excited state energy E_j, we cannot set values of c_i for $i \in \{0, 1, \ldots, j-1\}$ to zero without knowing the states a priori. Consequently, the "trial" wave function ψ_T is always some complicated mixture of the ground and excited state and there is no guarantee that the estimate of E_j is greater than its exact value. In other words, it is futile to change parameters of an approximate excited state wave function variationally and seek a point in parameter space where the estimate of the energy goes through a minimum the same way that it happens for the ground state. This is an important point, and is the main reason why finding accurate representations of excited state wave functions is a difficult task.

6.2 The Smoluchowski operator

Let us return to the imaginary time dependent Schrödinger equation,

$$\hbar \frac{\partial \psi}{\partial \tau} = \frac{\hbar^2}{2m} \nabla^2 \psi - U\psi.$$

The idea behind importance sampling in diffusion Monte Carlo is to propagate the product $\rho = \varphi\psi$, where φ is a field that "guides" the diffusion process in areas of importance using prior knowledge, such as, e. g., known asymptotic behavior of the ground state wave function or some variationally optimized wave function. "Important" areas are places in configuration space where φ is large. We now prove this result.

Theorem 25. *If $\partial\varphi/\partial t = 0$, the differential equation satisfied by ρ is*

$$\hbar \frac{\partial \rho}{\partial \tau} = \frac{\hbar^2}{2m} \nabla^2 \rho - \frac{\hbar^2}{m} \nabla(\rho \varphi^{-1} \nabla \varphi) - E_l \rho \tag{6.4}$$

where

$$E_l = -\varphi^{-1}\frac{\hbar^2}{2m}\nabla^2\varphi + U, \tag{6.5}$$

is known as the local energy.

Proof. Inserting for ψ $\varphi^{-1}\rho$ into the imaginary time dependent Schrödinger equation gives

$$\hbar\frac{\partial}{\partial\tau}\varphi^{-1}\rho = \frac{\hbar^2}{2m}\nabla^2\varphi^{-1}\rho - U\varphi^{-1}\rho. \tag{6.6}$$

The term on the left of this equation becomes

$$\hbar\frac{\partial}{\partial\tau}\varphi^{-1}\rho = \varphi^{-1}\hbar\frac{\partial\rho}{\partial\tau},$$

since φ^{-1} is time independent by construction. We next work on the x part of the ∇^2 operator, since the y and z parts yield correspondingly identical expressions,

$$\frac{\partial^2}{\partial x^2}\varphi^{-1}\rho = \frac{\partial}{\partial x}\left(\varphi^{-1}\frac{\partial\rho}{\partial x} + \rho\frac{\partial\varphi^{-1}}{\partial x}\right).$$

Using the product rule,

$$\frac{\partial^2}{\partial x^2}\varphi^{-1}\rho = 2\left(\frac{\partial\varphi^{-1}}{\partial x}\right)\left(\frac{\partial\rho}{\partial x}\right) + \varphi^{-1}\frac{\partial^2}{\partial x^2}\rho + \rho\frac{\partial^2}{\partial x^2}\varphi^{-1}$$

followed by the chain rule for the derivatives of φ^{-1} gives

$$\frac{\partial^2}{\partial x^2}\varphi^{-1}\rho$$
$$= -2\varphi^{-2}\left(\frac{\partial\varphi}{\partial x}\right)\left(\frac{\partial\rho}{\partial x}\right) + \varphi^{-1}\frac{\partial^2}{\partial x^2}\rho + 2\rho\varphi^{-3}\left(\frac{\partial\varphi}{\partial x}\right)^2 - \rho\varphi^{-2}\frac{\partial^2\varphi}{\partial x^2}.$$

The four terms on the right can be written as follows:

$$\frac{\partial^2}{\partial x^2}\varphi^{-1}\rho = -2\varphi^{-1}\left[\frac{\partial}{\partial x}\left(\rho\varphi^{-1}\frac{\partial\varphi}{\partial x}\right) - \frac{1}{2}\varphi^{-1}\rho\frac{\partial^2\varphi}{\partial x^2}\right] + \varphi^{-1}\frac{\partial^2}{\partial x^2}\rho.$$

Adding the y and z components of ∇^2 to this result yields

$$\nabla^2\varphi^{-1}\rho = -2\varphi^{-1}\left[\nabla(\rho\varphi^{-1}\nabla\varphi) - \frac{1}{2}\varphi^{-1}\rho\nabla^2\varphi\right] + \varphi^{-1}\nabla^2\rho.$$

Inserting all the pieces back into equation (6.6) produces

$$\varphi^{-1}\hbar\frac{\partial\rho}{\partial\tau} = \varphi^{-1}\frac{\hbar^2}{2m}\nabla^2\rho - \frac{\hbar^2}{m}\varphi^{-1}\left[\nabla(\rho\varphi^{-1}\nabla\varphi) - \frac{1}{2}\varphi^{-1}\rho\nabla^2\varphi\right] - U\varphi^{-1}\rho.$$

Multiplying both sides by φ give the final result,

$$\hbar\frac{\partial\rho}{\partial\tau} = \frac{\hbar^2}{2m}\nabla^2\rho - \frac{\hbar^2}{m}\nabla(\rho\varphi^{-1}\nabla\varphi) + \frac{\hbar^2}{2m}\varphi^{-1}\rho\nabla^2\varphi - \varphi U\varphi^{-1}\rho,$$

from which equation (6.4) follows trivially. □

Equation (6.4) is the Smoluchowski differential equation in three dimension. The second term on the right is a drift force, a deterministic acceleration of the pseudo-particle. Note that equation (6.4) is exactly the diffusion-drift equation with

$$v = \frac{\hbar}{m}\varphi^{-1}\nabla\varphi.$$

6.2.1 The propagator for the diffusion-drift equation

As explained earlier, there is one last diffusion equation we need to explore that is helpful in interpreting results from nonrelativistic gauge theories, which contains generally non uniform terms proportional to the gradient of the field variables $\psi(x)$. Therefore, let us consider

$$\frac{\partial}{\partial t}\psi(x,t) = D\frac{\partial^2}{\partial x^2}\psi(x,t) - v\frac{\partial}{\partial x}\psi(x,t),$$

where v need not be constant. The usual initial condition needed to derive the Green function is

$$\psi(x,t=0) = \delta(x).$$

The Fourier transforms of the differential equation and the initial condition are

$$\frac{\partial}{\partial t}\tilde{\psi}(a,t) = -Da^2\tilde{\psi}(a,t) - iav\tilde{\psi}(a,t)$$

subject to

$$\tilde{\psi}(a,t=0) = 1.$$

The solution in the frequency domain is

$$\tilde{\psi}(a,t) = e^{-Dta^2 - iavt}.$$

Here, we are assuming that Δt is small enough for us to ignore the differences between $v(x_k)$ and $v(x_{+1})$. The inverse integral,

$$\mathcal{K}(x,0,t) = \frac{1}{2\pi}\int_{-\infty}^{\infty} da\, e^{-Dta^2 - iavt}\, e^{iax},$$

becomes a standard Gaussian integral once more by combining exponents and completing the square,

$$-Dta^2 - iavt + iax = -Dt(a+b)^2 + Dtb^2$$
$$-Dta^2 - iavt + iax = -Dta^2 - 2Dtab.$$

Solving for b,

$$b = -i\frac{(x - vt)}{2Dt}$$

and inserting into the inverse integral gives

$$\mathcal{K}(x,0,t) = \frac{1}{2\pi} e^{-(x-vt)^2/4Dt}\int_{-\infty}^{\infty} da\, e^{-Dt[a-i(x-vt)/2Dt]^2}.$$

The integral in the last expression can be transformed into a standard Gaussian integral by a u substitution on a. We arrive at the diffusion-drift propagator,

$$\mathcal{K}(x,0,t) = \frac{1}{2\sqrt{\pi Dt}} e^{-(x-vt)^2/4Dt}.$$

Therefore, the drift term is shifting the center of the Gaussian distribution of pseudo-particles over time to $-vt$, mimicking the diffusion of suspended particles in a fluid flowing with velocity v. The Green function is obtained by changing the boundary conditions to find the equivalent expression for $\mathcal{K}(x,x',t)$, where the drift v is computed at x'. To implement the propagator in a random walk, it is sufficient to modify the diffusion moves as follows:

$$x_i = x_i' + \eta + v\Delta t,$$

where η is the Gaussian random number as before. It is important to note that when importance sampling gives rise to drift terms, the estimate of the ground state energy is no longer the average U as in Chapter 5; it is E_l in equation (6.5).

6.3 Fitting φ_0 to a Fourier expansion

In this section, we consider an alternative to the traditional importance sampling approach that improves the statistical efficiency of the basic algorithm while still sampling

φ_0. To accomplish this, we first have to construct an orthogonal basis set that is appropriately parameterized to be centered about the potential minimum using the methods introduced in Chapter 2. To demonstrate the procedure, we continue to use the quartic potential as the model. Suppose we are looking for the solutions of a Schrödinger equation $H\varphi_0 = E_0\varphi_0$, where

$$H = -\frac{\hbar^2}{2m}\frac{\partial^2}{\partial x^2} + U(x) \tag{6.7}$$

with some complicated potential as in the earlier example, $U(x) = D(x - x_0)^4$, and we want to improve the efficiency of the basic diffusion Monte Carlo energy estimator. We begin by constructing an orthogonal set of functions tailored to solve a closely related Hamiltonian operator that can be computed with the methods of Chapter 2. Let $H^{(0)}$ be such approximation,

$$H^{(0)} = -\frac{\hbar^2}{2m}\frac{\partial^2}{\partial x^2} + U^{(0)}(x)$$

where, e. g., $U^{(0)} = \frac{1}{2}k(x - x_0)^2$ would be a valid choice for the quartic potential. We fit the equilibrium population of pseudo-particles, whose positions are sampled from the exact ground state wave function φ_0 to the orthogonal set created over $H^{(0)}$, following Bressanini et al. [54]. Suppose $\{\psi_i^{(0)}\}$ is the set of eigenfunctions of $H^{(0)}$. Then the task is to find coefficients $c_0, c_1, \ldots$ such that

$$\varphi_0 = \sum_{i=1}^{N} c_i\psi_i^{(0)},$$

where φ_0 is the ground state of H in equation (6.7), and φ_0 is sampled by the diffusion Monte Carlo. Here, we want to find its projection onto $\mathcal{V}^{(0)} = \operatorname{span}\{\psi_i^{(0)}\}$. One way to optimize the coefficients $c_0, c_1, \ldots, c_N$ is to minimize the integral over all space of the square of the deviations between φ_0 and ψ_T,

$$\chi^2 = \int_{\mathcal{D}} [\varphi_0 - \psi_T]^2\, dx,$$

$$\chi^2 = \int_{\mathcal{D}} \left[\varphi_0 - \sum_{i=1}^{N} c_i\psi_i^{(0)}\right]^2 dx,$$

by varying the set of coefficients $\{c_i\}$. The components of the gradient of χ^2 with respect to the coefficients are obtained by setting the parameter gradient of χ^2

$$\frac{\partial}{\partial c_k}\chi^2 = \frac{\partial}{\partial c_k}\int_{\mathcal{D}} \left[\varphi_0 - \sum_{i=1}^{N} c_i\psi_i^{(0)}\right]^2 dx,$$

to zero. Expanding the square,

$$= \frac{\partial}{\partial c_k} \int_{\mathcal{D}} \left(\varphi_0^2 - 2\varphi_0 \sum_{i=1}^{N} c_i \psi_i^{(0)} + \sum_{i=1}^{N} \sum_{j=1}^{N} c_i c_j \psi_i^{(0)} \psi_j^{(0)} \right) dx,$$

distributing the derivative operator

$$= \int_{\mathcal{D}} \left(-2\varphi_0 \sum_{i=1}^{N} \frac{\partial}{\partial c_k} c_i \psi_i^{(0)} + \sum_{i=1}^{N} \sum_{j=1}^{N} \frac{\partial}{\partial c_k} c_i c_j \psi_i^{(0)} \psi_j^{(0)} \right) dx,$$

and evaluating yield

$$= \int_{\mathcal{D}} \left(-2\varphi_0 \psi_k^{(0)} + 2 \sum_{j=1}^{N} c_j \psi_j^{(0)} \psi_k^{(0)} \right) dx,$$

$$= -2\langle \varphi_0 | \psi_k^{(0)} \rangle + 2 \sum_{j=1}^{N} c_j \langle \psi_j^{(0)} | \psi_k^{(0)} \rangle$$

where

$$\langle \varphi_0 | \psi_k^{(0)} \rangle = \int_{\mathcal{D}} \varphi_0 \psi_k^{(0)} dx,$$

and

$$\langle \psi_j^{(0)} | \psi_k^{(0)} \rangle = \int_{\mathcal{D}} \psi_j^{(0)} \psi_k^{(0)} \, dx = \delta_{jk}$$

are used. This is allowed since we have chosen an orthonormal set. Then the equation

$$\frac{\partial}{\partial c_k} \chi^2 = 0$$

translates to

$$c_k = \int_{\mathcal{D}} \varphi_0 \psi_k^{(0)} dx \approx \frac{1}{M} \sum_{i=1}^{M} \psi_k^{(0)}(x_i) \tag{6.8}$$

where on the right we use the population of M pseudo-particles to estimate the integral and the fact that $\{x_i\}$ represents φ_0 as we have shown in Chapter 5. Therefore, once the orthonormal set is constructed, one can run the diffusion Monte Carlo walk, and at every step estimate the coefficients, and compute their averages and standard deviations. Next, we use the methods introduced in Chapter 2 to tailor the set $\{\psi_i^{(0)}\}$, so that it approximates as best as possible the true solution.

6.3.1 The shifted harmonic oscillator

An approach, which should work well for many chemical physics applications, is to expand U into a Taylor series about its x_{min}. Then we simply use the curvature to estimate the parameter k of $U^{(0)} = k(x - x_{min})^2/2$. When systems of particles are considered, the eigenvalues of the Hessian of U are the force constants along the *normal modes*. While the procedure to shift the oscillator to the global minimum of U and using its curvature to estimate parameters is general, we show how the variational principle can produce a good approximation of $U^{(0)}$ as well. The example we use is the quartic model for $U = D(x - x_0)^4$, where it is clear that $x_0 = x_{min}$ but the second derivative of U at x_0 vanishes. In this case, the variational optimization of a trial wave function is critical.

Let us begin by deriving some result for a generic case in one dimension. Consider the following unnormalized wave function:

$$\varphi_0(x) = \exp(-ax^2 - bx). \tag{6.9}$$

What Schrödinger equation does φ_0 satisfy? In the $b \to 0$ limit, it is clear that we recover the solution of the harmonic oscillator with mass m and force constant k. The derivatives of φ_0 are

$$\begin{aligned} \frac{d}{dx}\varphi_0 &= -(2ax + b)\varphi_0 \\ \frac{d^2}{dx^2}\varphi_0 &= [(2ax + b)^2 - 2a]\varphi_0. \end{aligned} \tag{6.10}$$

Now we use this expression to answer the question. Namely, what is the potential energy correction and the ground state energy when φ_0 in equation (6.9),

$$-\frac{\hbar^2}{2m}\frac{d^2}{dx^2}\varphi_0 + U^{(0)}\varphi_0 = E_0\varphi_0,$$

where $U^{(0)}$ is $kx^2/2$ plus some correction term, $U^{(1)}$. Using (6.10),

$$-\frac{\hbar^2}{2m}[(2ax + b)^2 - 2a]\varphi_0 + \left(\frac{1}{2}kx^2 + U^{(1)} + C\right)\varphi_0 = E_0\varphi_0,$$

where C is some constant. Dropping φ_0,

$$-\frac{\hbar^2}{2m}[(2ax + b)^2 - 2a] + \frac{1}{2}kx^2 + U^{(1)} + C = E_0$$

and expanding yield

$$-\frac{\hbar^2}{2m}[4a^2x^2 + 4abx + b^2 - 2a] + \frac{1}{2}kx^2 + U^{(1)} + C = E_0.$$

The coefficients for every power of x must be the same on both sides. The constant terms give

$$E_0 = \frac{\hbar^2}{2m}(2a - b^2) + C. \tag{6.11}$$

The linear terms tell us that $U^{(1)}$ must also be linear,

$$U^{(1)} = 2\frac{\hbar^2 abx}{m},$$

while the quadratic coefficients yield our estimate of the force constant k,

$$k = 2\frac{\hbar^2 a^2}{m}.$$

Conversely, one can verify that equation (6.9) is a solution of

$$-\frac{\hbar^2}{2m}\frac{d^2}{dx^2}\varphi_0 + \left(\frac{1}{2}kx^2 + 2\hbar^2 abx\right)\varphi_0 = E_0\varphi_0.$$

Namely, inserting equation (6.10) and canceling one obtains equation (6.11). For our quartic potential example, we now choose $a = 3.920$ obtained from equation (6.2) for $m = 919.0$ and $D = 0.0874$ atomic units. Note that if we expand the trial wave function, we optimized variationally earlier

$$\psi_T = e^{-\beta(x-x_0)^2} = e^{-\beta x_0^2} e^{-\beta x^2 + 2\beta x_0 x},$$

with equation (6.9) we get $a = \beta$, $b = -2\beta x_0$. For x_0, we choose 1.41 bohr, the average vibrational H-H distance. Then we have the best approximation for a two-parameter guiding wave function.

In Figure 6.2, we have graphed the potentials U and $U^{(0)}$,

$$U(x) = D(x - x_0)^4,$$

and

$$U^{(0)}(x) = 2\hbar^2 a\{a(x^2 - x_0^2) + b(x - x_0)\}$$

where C is chosen here so that both functions evaluate zero at their minimum $x = x_0$. We can see that, while the optimized ψ_T may give a reasonable estimate of the ground state energy, the Fourier series for φ_0 will require a number of excited states of the shifted harmonic oscillator before convergence can be achieved. The next task is finding analytical expressions for the power integrals

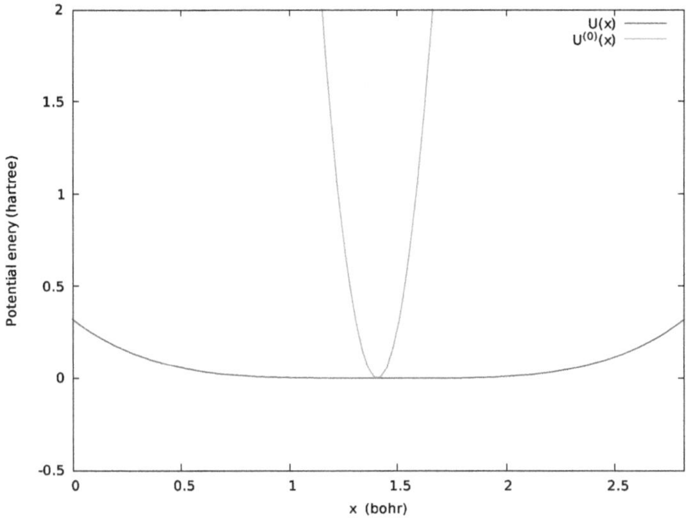

Figure 6.2: The graph of $U(x)$ (purple line), juxtaposed to the graph of $U^{(0)}(x)$ (green line). The purple line appears flat in the 1 to 2 bohr range about the minimum at 1.41 bohr, whereas $U^{(0)}(x)$ is a much narrower parabola that share the same minimum with a constant C chosen so that the potential energies coincide at the minimum point, $U^{(0)}(x_0) = U(x_0)$.

$$\mathscr{I}_n = \int_{-\infty}^{\infty} x^n \exp(-2ax^2 - 2bx)\,dx, \tag{6.12}$$

where $\exp(-2ax^2 - 2bx) = \omega(x) = \varphi_0^2$ plays the role of the weight. Note that

$$\mathscr{I}_0 = \int_{-\infty}^{\infty} \exp(-2ax^2 - 2bx)\,dx$$

is unraveled by completing the square,

$$2ax^2 + 2bx = \alpha(x - \gamma)^2 + \delta.$$

The last expression applies for all values of x. The three equations (one for each power of x) are obtained by expanding the square on the right

$$2ax^2 + 2bx = \alpha x^2 - 2\alpha\gamma x + \delta + \alpha\gamma^2$$

Therefore, $\alpha = 2a$, $\gamma = -b/(2a)$, and $\delta = -b^2/(2a)$,

$$\mathscr{I}_0 = \int_{-\infty}^{\infty} \exp(-2ax^2 - 2bx)\,dx = e^{b^2/(2a)} \int_{-\infty}^{\infty} \exp\left[-2a\left(x + \frac{b}{2a}\right)^2\right] dx,$$

$$\mathscr{I}_0 = e^{b^2/(2a)}\pi^{1/2}(2a)^{-1/2}. \tag{6.13}$$

To proceed, we need the following results.

Theorem 26. *The family of integrals in equation* (6.12) *satisfy the following derivative rule:*

$$\frac{d}{db}\mathscr{I}_n = -2\mathscr{I}_{n+1} \tag{6.14}$$

and the following two-term recursion:

$$\mathscr{I}_n = \frac{n-1}{4a}\mathscr{I}_{n-2} - \frac{b}{2a}\mathscr{I}_{n-1} \quad n \geq 2 \tag{6.15}$$

with $\mathscr{I}_0$ *as in equation* (6.13), *and* $\mathscr{I}_1 = -b\mathscr{I}_0/(2a)$.

Proof. From the definition of $\mathscr{I}_n$, it follows that

$$\frac{d}{db}\mathscr{I}_n = \int_{-\infty}^{\infty} x^n \frac{d}{db} \exp(-2ax^2 - 2bx)\, dx$$

since the integrand is bound over the entire domain, we are allowed to move the derivative operator inside the integral sign. Then

$$\frac{d}{db}\mathscr{I}_n = -2\int_{-\infty}^{\infty} x^{n+1} \exp(-2ax^2 - 2bx)\, dx$$

establishes (6.14). Moreover, by direct differentiation of equation (6.13), we find

$$\frac{d}{db}\mathscr{I}_0 = \frac{b}{a}\mathscr{I}_0$$

and, therefore, from (6.15) it follows that

$$\mathscr{I}_1 = -\frac{b}{2a}\mathscr{I}_0.$$

Evaluating derivatives with respect to b on both sides repeatedly and applying (6.14) gives

$$\begin{aligned}
\mathscr{I}_2 &= \frac{1}{4a}\mathscr{I}_0 - \frac{b}{2a}\mathscr{I}_1,\\
\mathscr{I}_3 &= \frac{1}{2a}\mathscr{I}_1 - \frac{b}{2a}\mathscr{I}_2,\\
\mathscr{I}_4 &= \frac{3}{4a}\mathscr{I}_2 - \frac{b}{2a}\mathscr{I}_3.
\end{aligned}$$

A clear pattern emerges.

Note that (6.15) is established for $2 \leq n \leq 4$. Suppose the result holds for k,

$$\mathcal{I}_k = \frac{k-1}{4a}\mathcal{I}_{k-2} - \frac{b}{2a}\mathcal{I}_{k-1}.$$

Then, taking the derivative with respect to b on both sides and using (6.14), produces

$$-2\mathcal{I}_{k+1} = -2\frac{(k-1)}{4a}\mathcal{I}_{k-1} - \frac{1}{2a}\mathcal{I}_{k-1} + \frac{b}{a}\mathcal{I}_k$$

or

$$\mathcal{I}_{k+1} = \frac{(k-1)}{4a}\mathcal{I}_{k-1} + \frac{1}{4a}\mathcal{I}_{k-1} - \frac{b}{2a}\mathcal{I}_k$$

$$\mathcal{I}_{k+1} = \frac{k}{4a}\mathcal{I}_{k-1} - \frac{b}{2a}\mathcal{I}_k. \qquad \square$$

Therefore, once the numerical value of $\mathcal{I}_0$ is known, meaning a numerical value for a and b are given, one can compute the values of the set $\{\mathcal{I}_n\}$ by the recursion and store them—coding this recursion for some given values of parameters a and b is left to the reader as an exercise. The Python function for the computation of $\psi_j^{(0)}(x) = p_j(x)e^{-ax^2-bx}$ is below:

```
def wavef(x,w,n):
    import numpy as np
    global a, b
    if (abs(x) < 1.0e-12): # test if x is not too small
        f = w[0,0]
    else:
        ex = np.exp(-a*x*x - b*x)  #  psi_0^{(0)}
        pn = 0.0
        xk = 1.0
        k = 0
        while (k < (n+1)):   # computes the polynomials p(n)
            pn += w[n,k]*xk
            xk = xk*x
            k += 1
        f = pn*ex
    return f
```

This function should be inserted before the diffusion Monte Carlo code of Chapter 5. Note that this assumes the coefficients `w[n,k]` have been computed by a function containing the code given in Chapter 2. This, too, is left as an exercise, but it should mostly be a copy and paste process with some testing. The computation of the coefficients of the Fourier expansion via equation (6.8) is carried out by the code below:

```
    sumc = np.zeros(20)  # The running estimate of c_j
    for i in range(size):
        for n in range(maxn):
            fn = wavef(x[i],w,n)
            sumc[n] += fn/float(size)
```

Note that the first line of the code should begin in column five as it is indented to be inside and at the end of the `for k in range(200):` loop, after the array reset loop that executes the `x[i] = y[i]` and `m[i] = 1.0` commands.

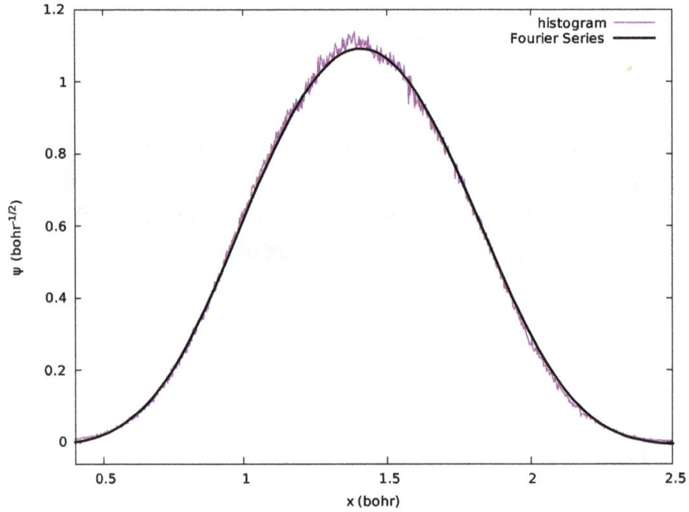

Figure 6.3: Graph of the ground state wave function of a particle of mass 919.0 atomic units in a quartic potential. The distorted bell-shaped thick black line is computed with the Fourier series, and it is juxtaposed to a plot of a 1000 bin histogram. The purple noisy line fluctuates about the Fourier series plot and seems to deviate only by a small percentage from it, near the $x = 1.41$ bohr peak.

In Figure 6.3, we compare the outcome of the fitting procedure with a histogram of φ_0 obtained from analyzing approximately 2×10^6 values of x from an equilibrated random walk. The statistically significant coefficients are $c_0 = 0.89384$, $c_2 = 0.0165$, $c_4 = -0.0355$, and $c_6 = 0.0099$. These values are averages from a single 1000 step diffusion Monte Carlo run. The values of c_k are computed with (6.8).

6.4 Exercises

1. Insert the pseudo-particle population representation of the ground state wave function

$$\varphi_0 = \frac{1}{M}\sum_{i=1}^{M}\delta(x - x_i), \tag{6.16}$$

into

$$c_k = \int_{\mathcal{D}} \varphi_0 \psi_k^{(0)} dx,$$

to derive the result in equation (6.8).

2. Show that inserting equation (6.10) into

$$-\frac{\hbar^2}{2m}\frac{d^2}{dx^2}\varphi_0 + \left(\frac{1}{2}kx^2 + \frac{\hbar^2 abx}{m} + C\right)\varphi_0 = E_0\varphi_0$$

yields equation (6.11).

3. Modify the primitive diffusion Monte Carlo codes in Chapter 5, both the initial and the restart version so that
   (a) The quartic potential $U = D(x - x_0)^4$ with $x_0 = 1.41$ bohr and $D = 0.0874$ hartree bohr^{-4} is computed with a defined function.
   (b) The branching weights are computed using the second-order convergence method.
   (c) The starting population is a Gaussian distribution about x_0 with a standard deviation equal to $(6mD)^{-1/6}$, where $m = 919.0$ atomic units and D is in part (a). This can be accomplished with
   ```
   s = (6.0*mass*0.0874)**(-1.0/6.0)
   target = 10000.0
   x = np.random.normal(1.41, s, int(2*target))
   size = int(target)
   ```
   where x is initiated with double the size to allow for growth beyond the target amount.

   Use this to estimate the ground state energy of the system by the population average of U as in the basic algorithm.

4. (a) Code equation (6.14) as well as those for $\mathscr{I}_0$ and $\mathscr{I}_1$ to compute values $\mathscr{I}_n$ for $0 \leq n \leq 19$ using $a = 3.920$, $x_0 = 1.41$, $b = -2ax_0$ atomic units. Check some selected values against those in Table 6.1.
   (b) Confidence in numerical results is earned. The previous exercise gave you some values so your code could be checked, but such luxury is not always afforded. How do we know that the result of a calculation is correct? Testing code is often more laborious than the designing and writing of it. No result should be trusted without proper testing. The following stand alone Python code estimates $\mathscr{I}_0$ using the trapezoid rule for a range of quadrature sizes:

Table 6.1: Some monomial integral values.

i	$\mathscr{I}_i$
0	3.720801×10^6
1	5.246329×10^6
2	7.634620×10^6
⋮	⋮
19	1.117098×10^{11}

```
import numpy as np
a = 3.920
x0 = 1.41
b = -2.0*a*x0
for nq in range(10,30):
    sum0 = 0.0
    x = -1.0
    dx = 5.0/float(nq)
    for j in range(nq):
        w = 1.0
        if (j == 0):
            w = 0.5
        if (j == (nq-1)):
            w = 0.5
        f = np.exp( -2.0*a*x*x - 2.0*b*x)
        sum0 += w*f*dx
        x += dx
    print(sum0)
```

Modify it so that it estimates $\mathscr{I}_3$ and compare the converged result with the analytical values obtained in the previous exercise.

5. Merge the code developed in exercises 4 part (a) with that of exercise 3 to compute the polynomials coefficients after the reindexing of the population array, but still inside the DMC step loop. Compute twenty monomial integrals and the coefficients of the first six polynomials of the shifted harmonic oscillator with the same values of a and b as in the previous exercises. Compare your coefficients with those in Table 6.2.
6. Add to the code in the previous exercise the function `wavef(x,w,n)` and the snippet that computes the coefficients of the Fourier expansion c_k in equation (6.8) and compare your values with those given in the text.
7. A Python code to read a set of values from a file containing a single column of data and compute a histogram with 1000 bins is:

Table 6.2: Some values of the coefficients of the shifted harmonic oscillator.

n	w_0^n	w_1^n	w_2^n	w_3^n	w_4^n
0	5.1842×10^{-4}				
1	-2.8945×10^{-3}	2.0528×10^{-3}			
2	1.1061×10^{-2}	-1.6209×10^{-2}	5.7479×10^{-3}		
3	-3.3292×10^{-2}	7.5862×10^{-2}	-5.5586×10^{-2}	1.3141×10^{-2}	
4	8.3360×10^{-2}	-2.6366×10^{-1}	3.0040×10^{-1}	-1.4674×10^{-1}	5.6018×10^{-2}

```
import numpy as np
x = np.loadtxt('x.dat', dtype='float')
dens, rang = np.histogram(x,bins=1000,range=None,density=True)
for i in range(1000):
print("%12.6f %12.6f " %(rang[i],dens[i]))
```

This stand alone code produces data that can be graphed using either the plotting facilities of Python with some additional lines of code, or the screen output can be piped into a file to be later graphed with any desired software. The values of x are in the first column, whereas the density (φ_0 in our case) in the second column. Create a single column of x values by adding a print statement in the last loop of the population reindexing to your restarted diffusion Monte Carlo code. A 200 step run with an average population size of 10^4 pseudo-particles will generate around 2×10^6 data points. A relatively smooth histogram results. Note that the `density=True` normalizes the wave function so that the area under its first power is one, not the area under its square. Use the histogram code (in a separate file) and modify your restarted diffusion Monte Carlo code to reproduce the graph in Figure 6.3.

8. Consider a particle of mass m subjected to a Morse potential energy [105]:

$$U(r) = D(1 - e^{-a(r-r_0)})^2 - D = D(e^{-2a(r-r_0)} - 2e^{-a(r-r_0)}).$$

This is an excellent model for a chemical bond energy. We use it here with parameters designed to mimic the vibrational states of the H_2 molecule. Modify the differential equation

$$-\frac{\partial^2}{\partial r^2}\varphi_0 + \lambda^2(e^{-2a(r-r_0)} - 2e^{-a(r-r_0)})\varphi_0 = E_0\varphi_0,$$

by rescaling the coordinate $x = ar$, $x_0 = ar_0$. Show that you get

$$-\frac{\partial^2}{\partial x^2}\varphi_0 + \lambda^2(e^{-2(x-x_0)} - 2e^{-(x-x_0)})\varphi_0 = \epsilon_0\varphi_0,$$

where $\lambda = \sqrt{2mD}/(a\hbar)$ and $\epsilon_0 = 2mE_0/(a\hbar)^2$.

9. (a) Use $\psi = e^{-\beta(x-x_0)^2}$ as the trial wave function and derive the following estimate for the ground state energy:

$$E_0 = \frac{a^2\hbar^2}{2m}\{\lambda^2[1 + e^{1/(2\beta)} - 2e^{1/(8\beta)}] + \beta\}.$$

(b) Graph E_0 from part (a) for values of β with choices for D, a, r_0, and m that mimic the vibration of a hydrogen molecule: $m = 919.0$ atomic units, $D = 0.17447$ hartree, $r_0 = 1.41$ bohr and $a = 0.708$ bohr^{-1}. Find the optimal value of β graphically, compute the best estimate of E_0 and compare its value with the exact ground state energy

$$E_0 = \left(\frac{1}{2} - \frac{1}{8\lambda}\right)\hbar\omega = 0.006829 \quad \text{hartree}$$

where $\omega = a\sqrt{2D/m} \approx 0.01379$ atomic units.

7 Excited states methods

> [It] follows that either (1) the quantum-mechanical description of reality given by the wave function is not complete or (2) when the operators corresponding to two physical quantities do not commute the two quantities cannot have simultaneous reality.
>
> A. Einstein, B. Podolsky, N. Rosen

Let us for just a moment consider the possibility that we have an accurate Fourier expansion of the ground state wave function φ_0, obtained using the fitting procedure developed in Chapter 6. Also, suppose that we have the ability to estimate the monomial integrals $\tilde{\mathscr{I}}_n$ constructed with the weight $\tilde{\omega}(x) = \varphi_0^2$, as well as the monomial integrals related to the energy $\tilde{\mathscr{Q}}_n$, $\tilde{\mathscr{P}}_n$ and $\tilde{\mathscr{U}}_n$ for a range of values of n. Then, in principle, by carrying out the same process of building excited states with the Gram–Schmidt orthogonalization procedure from the ground up, we could construct approximations for some excited states without making any assumptions about their nodal structure and without optimizing parameters for such excited states wave functions variationally. We demonstrate that for values of the mass m in the nuclear range (919 atomic units or greater) the wave functions are sufficiently accurate for the first few excited states of a system even if we know very little about it. The method in this chapter is very similar to a recently published strategy called the Ground State Probability Amplitudes (GSPA) [42], which represents the current best practice when approximating wave functions of excited states of nonidentical particles by means of orthogonal polynomials. GSPA is actually a systematic method to estimate first excitations in spaces of arbitrary dimensions, and is discussed more in Chapter 17 when that subject is taken up. In this chapter, we limit our discussion to monodimensional systems.

For demonstration purposes, we return to the quartic potential example of the last chapter. We begin in Section 7.1 by deriving and testing the algorithm to estimate the monomial integrals during a basic diffusion Monte Carlo simulation that samples φ_0. We then use these to compute the ground and estimate the first excited state energies of a test system. In Section 7.2, we develop an algorithm to estimate excited states without using fixed nodes or changing the random walk. These estimates are demonstrated to approach spectroscopic accuracy for masses in the nuclear range. Using orthogonal polynomials is not a new strategy of course. In Section 7.3, a general version of the excited VAMPIR algorithm [82], which combines variational optimization with systematic orthogonalization is developed. Though the original implementation was designed to work with vector space methods, we show how the two approaches are closely related. We conclude this chapter by introducing kinetic energy estimators based on the virial theorem of quantum mechanics.

Before we begin, however, let us note that here we introduce some new notation for the monomial integrals to avoid confusion. As in Chapter 2, we construct a vector space $\mathcal{V}$ by orthogonal polynomials once more, but this time using a weight function $\tilde{\omega}(x) = [\varphi_0]^2$ for which we have a complicated form. To distinguish the new sets of monomial integrals

https://doi.org/10.1515/9783111610207-007

from those in Chapter 2, we use a tilde symbol above them. The difference is important. Generally, $\mathscr{I}_n$, $\mathscr{Q}_n$, $\mathscr{P}_n$, and $\mathscr{U}_n$ can be expressed analytically, but their counterparts $\tilde{\mathscr{I}}_n$, $\tilde{\mathscr{Q}}_n$, $\tilde{\mathscr{P}}_n$, and $\tilde{\mathscr{U}}_n$ have to be estimated stochastically or numerically because of the complexity of φ_0 and U especially in more than one dimension. The same applies for the polynomial coefficients. In practice, both types of integrals and polynomial coefficients have to be computed during a simulation, those used to create the basis for the expansion of φ_0 using the methods of Chapter 2 from the ground state of a simpler system, e. g., the harmonic oscillator, and those built from the ground state of the actual system under study, which are identified with tilde symbols.

7.1 Stochastic estimate of the monomial integrals

We have shown in Chapter 6 that with regular diffusion Monte Carlo simulations, it is possible to sample φ_0 quite accurately, and fit φ_0 to a vector space until the desired convergence criteria is reached

$$\varphi_0 = \sum_{j=0}^{j_{\max}} c_j \chi_j, \tag{7.1}$$

where $\{\chi_i\}$ is assumed to be a set of carefully chosen orthogonal basis that satisfy the same boundary conditions as the problem under investigation and that approximate the actual solution as well as possible. We assume that the basis χ_j can be computed using the methods of Chapter 2.

Because we make no uncontrollable approximations, we can fruitfully discuss the convergence criteria of the ground state energy by our Fourier expansion. Since we generally do not have an analytical function for φ_0, the typical pointwise and uniform convergence metrics cannot be employed directly. Instead, a type of convergence in the mean has to be used. For example, one may compute the ground state energy $E_0(j_{\max})$ for several values of $j_{\max}$. When the estimate of the ground state energy agrees within statistical fluctuations with the diffusion Monte Carlo estimate, we can say we have reached convergence. Because the ground state wave function has no nodes, no uncontrolled sources of error are introduced in the process.

With the first and second derivatives of $\{\chi_i\}$ available, analytically one can estimate the four classes of monomial integrals with the following approach:

$$\tilde{\mathscr{I}}_n \approx \frac{1}{M} \sum_{i=1}^{M} x_i^n \sum_{j=0}^{j_{\max}} c_j \chi_j(x_i), \tag{7.2}$$

where we note that the second sum is simply equation (7.1) and we use the fact that

$$\varphi_0 \approx \frac{1}{M} \sum_{i=1}^{M} \delta(x - x_j) \tag{7.3}$$

is sampled by the random walk at equilibrium. The integral we seek is

$$\tilde{\mathscr{I}}_n = \int_{\mathcal{D}} x^n \varphi_0^2 \, dx.$$

Inserting (7.1) for one power of φ_0, (7.3) for the other and integrating, we arrive at equation (7.2). The same proof applies for the rest of the integrals:

$$\tilde{\mathscr{Q}}_n \approx \frac{1}{M} \sum_{i=1}^{M} x_i^n \sum_{j=0}^{j_{\max}} c_j \frac{d\chi_j}{dx}(x_i), \tag{7.4}$$

$$\tilde{\mathscr{P}}_n \approx \frac{1}{M} \sum_{i=1}^{M} x_i^n \sum_{j=0}^{j_{\max}} c_j \frac{d^2\chi_j}{dx^2}(x_i), \tag{7.5}$$

$$\tilde{\mathscr{U}}_n \approx \frac{1}{M} \sum_{i=1}^{M} x_i^n \, U(x_i) \sum_{j=0}^{j_{\max}} c_j \chi_j(x_i). \tag{7.6}$$

Therefore, it is possible in general to compute the necessary integrals once we have a fitted ground state wave function. To get the ground state energy, we first compute

$$\tilde{w}_0^{(0)} = \tilde{\mathscr{I}}_0^{-1/2}$$

followed by

$$E_0(j_{\max}) = [\tilde{w}_0^{(0)}]^2 \left(-\frac{\hbar^2}{2m} \tilde{\mathscr{P}}_0 + \tilde{\mathscr{U}}_0 \right). \tag{7.7}$$

Generally, expansions of the ground state wave function into a “good” basis set $\{\chi_i\}$ will yield more efficient (i. e., lower fluctuations) estimates of E_0 compared to the Dirac delta vector space used of the primitive algorithm, without ever using a Smoluchowski operator or a Green’s function method. Note also that we can compute other properties of the ground state wave function such as the mean position, the kinetic energy, and the momentum.

The following is the Python code for the stochastic evaluation of monomial integrals:

```
inxt = np.zeros(10)
pnxt = np.zeros(10)
qnxt = np.zeros(10)
unxt = np.zeros(10)
for i in range(size):
    psi0 = 0.0
    dpsi0 = 0.0
    d2psi0 = 0.0
    ux = pot(x[i])
```

```
for n in range(maxn):
    fn,dfn,d2fn = wavef(x[i],w,n)
    psi0 += sumc[n]*fn  # this is the unnormalized ground state
    dpsi0 += sumc[n]*dfn  # first derivative
    d2psi0 += sumc[n]*d2fn  # second derivative
xtj = 1.0
for j in range(maxn):
    inxt[j] += xtj*psi0/float(size)
    qnxt[j] += xtj*dpsi0/float(size)
    pnxt[j] += xtj*d2psi0/float(size)
    unxt[j] += xtj*ux*psi0/float(size)
    xtj *= x[i]
```

The indentation, as always, is very important. The first lines must start in column 5 and should follow the estimate of the coefficients $\{c_j\}$ in equation (7.1) inserted into the DMC code developed in the exercises of Chapter 6. Our code estimates the set $\{c_j\}$ from $j = 0$ to j_{max} at every step of the diffusion immediately after the arrays `x[i]` and `m[i]` are reindexed. The coefficient c_n is stored in `sumc[n]`. The code also assumes the necessary modifications to the subroutine `wavef(x[i],w,n)` in Chapter 6 to compute the first and second derivatives of χ_n have been made. Finally, the computation of the monomial integrals must be carried out at every step of the random walk.

The values of the monomial integrals for the particle with 919.0 atomic units of mass in the quartic potential $U(x) = D(x - x_0)^4$ with $D = 0.0874$ hartree, $x_0 = 1.41$ bohr, and j_{max} equal 6, are in Table 7.1. The ground state wavefunction φ_0 is approximated using a $j_{max} = 4$ in equation (7.1). The standard deviation in the line just below their values is computed from just ten steps of a diffusion simulation that begins from an equilibrated distribution of pseudo-particles. The row labeled "Trapezoid" represents a quadrature estimate with uncertainties far below the number of digits displayed. The diffusion Monte Carlo time step is 0.05 atomic units.

Now turning our attention to the ground state energy, estimated with equation (7.7) from a 1000 steps run using $j_{max} = 0$, namely a single term in equation (7.1) estimates the ground state energy at $0.00313744 \pm 9.2 \times 10^{-7}$ hartree. During the same run, the estimate of the ground state energy, using the population average $\langle U \rangle$ estimator of Chapter 5, yields 0.00311613 hartree $\pm 1.7 \times 10^{-5}$. The statistical fluctuations of E_0 are about 18 times smaller than those of $\langle U \rangle$ in Chapter 6. This is a remarkable result, considering we did not modify the random walk at all, the population still samples φ_0 using the basic DMC random walk. Instead of using the Smoluchowski propagator, we just use a better estimator for the energy.

The contrast is stark when the two lines fluctuating about the correct energy for the two estimators are compared. The black thick line in the center of Figure 7.1 is computed with equation (7.7) and displays considerably smaller fluctuations compared with the purple line corresponding to the $\langle U \rangle$ estimate. With $j_{max} = 0$, the simulation is only

Table 7.1: The values of some of the monomial integrals for the particle in a quartic potential.

n	0	1	2	3	4	5
$\tilde{\mathscr{I}}_n$	0.8025	1.1314	1.6489	2.476	3.821	6.045
σ	0.0023	0.0042	0.0080	0.014	0.027	0.049
Trapezoid	0.8004	1.1286	1.6455	2.472	3.820	6.049
$\tilde{\mathscr{P}}_n$	−3.0655	−4.3211	−5.9850	−8.1437	−10.5235	−12.672
σ	0.0051	0.0070	0.0089	0.0108	0.0137	0.024
Trapezoid	−3.0337	−4.2776	−5.9225	−8.0435	−10.6012	−13.217
$\tilde{\mathscr{Q}}_n$	0.0007	−0.3894	−1.1151	−2.4477	−4.9050	−9.4546
σ	0.014	0.0004	0.0010	0.0022	0.0039	0.0060
Trapezoid	0.0	−0.4002	−1.1286	−2.4682	−4.9456	−9.5491
$\tilde{\mathscr{U}}_n$	0.0008490	0.001197	0.001923	0.003405	0.0063300	0.01210
σ	0.0000086	0.000013	0.000013	0.000012	0.00027	0.00023
Trapezoid	0.0008690	0.001225	0.001976	0.003486	0.006493	0.01249

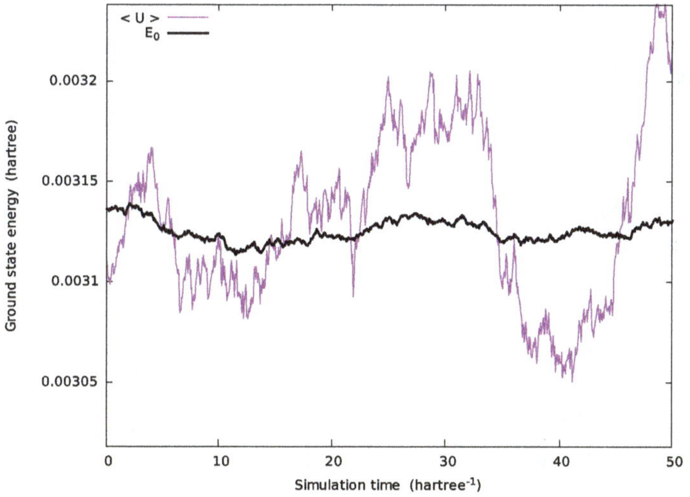

Figure 7.1: Estimates of the ground state energy for the particle in a quartic potential.

about 20 % more time consuming. Therefore, the relative efficiency of the estimator in equation (7.7) is approximately 15 times greater. It is this gain in efficiency that moved the community quickly to adapt importance sampling in diffusion Monte Carlo simulations.

7.2 Excited states of the quartic potential

At this point, the code for our diffusion Monte Carlo is ready to be enhanced to estimate excited states of the system. The approach is the same as the one used in Chapter 2. Namely, we assume that the product of a linear binomial with φ_0 is a good approximation for ψ_1. The code below, which should begin at column 5, must be appended near (but before) the end of the `for k in range(nsteps):` simulation time loop after the estimate of E_0 in the previous section.

The first through the third line estimate $\tilde{w}_0^{(0)}$, $\langle \tilde{P}_0, x\tilde{P}_0\rangle_{\tilde{\omega}}$ and $\langle x\tilde{P}_0, x\tilde{P}_0\rangle_{\tilde{\omega}}$, respectively, in terms of $\tilde{\mathscr{I}}_n$. For the $n = 1$ case, equations (2.9) through (2.12) translate to

$$\tilde{d}_1 = \left(\langle x\tilde{P}_0, x\tilde{P}_0\rangle_{\tilde{\omega}}^2 - \langle x\tilde{P}_0, \tilde{P}_0\rangle_{\tilde{\omega}}^2\right)^{1/2} \tag{7.8}$$

$$\tilde{w}_1^{(1)} = \frac{\tilde{w}_0^{(0)}}{\tilde{d}_1} \tag{7.9}$$

$$\tilde{w}_0^{(1)} = -\tilde{w}_0^{(0)} \frac{\langle x\tilde{P}_0, \tilde{P}_0\rangle_{\tilde{\omega}}}{\tilde{d}_1}. \tag{7.10}$$

The following snippet appended at the end of the previous one computes the better estimate of the ground state energy, the polynomial coefficient of the approximate first excited state and its energy estimate in equation (7.11) below:

```
e0 = (unxt[0] -pnxt[0]/(2.0*mass))/inxt[0]
tw00 = 1.0/(np.sqrt(inxt[0]))
p0xp0 = inxt[1]/inxt[0]
xp0xp0 = inxt[2]/inxt[0]
d_1 = np.sqrt(xp0xp0 - p0xp0*p0xp0)
tw10 = -p0xp0*tw00/d_1
tw11 = tw00/d_1
node = -tw10/tw11
e1t1 = - (tw10*tw11*qnxt[0] + tw11*tw11*qnxt[1])/mass
e1t2 = - (tw10*tw10*pnxt[0] + 2.0*tw10*tw11*pnxt[1] +
         tw11*tw11*pnxt[2])/(2.0*mass)
e1t3 = (tw10*tw10*unxt[0] + 2.0*tw10*tw11*unxt[1] +
       tw11*tw11*unxt[2])
e1k = e1t1 + e1t2
e1 = e1k + e1t3
```

The first excited state wave function is $\psi_1 = \tilde{P}_1\varphi_0$, or $(\tilde{w}_0^{(1)} + \tilde{w}_1^{(1)}x)\varphi_0$. Therefore, the location of the node is $x_{\text{node}} = -\tilde{w}_0^{(1)}/\tilde{w}_1^{(1)}$ and this is estimated in the `node = -tw10/tw11` line. Its location should fluctuate about the potential minimum x_0. Lines 9 through 13 use the following equation adapted from (2.17) to estimate the kinetic and total energy of the first excited state,

$$\begin{aligned} E_1 = &-\frac{\hbar^2}{m}\{\tilde{w}_0^{(1)}\tilde{w}_1^{(1)}\tilde{\mathscr{Q}}_0 + [\tilde{w}_1^{(1)}]^2\tilde{\mathscr{Q}}_1\} \\ &-\frac{\hbar^2}{2m}\{[\tilde{w}_0^{(1)}]^2\tilde{\mathscr{P}}_0 + 2\tilde{w}_0^{(1)}\tilde{w}_1^{(1)}\tilde{\mathscr{P}}_1 + [\tilde{w}_1^{(1)}]^2\tilde{\mathscr{P}}_2\} \end{aligned}$$

$$+ [\tilde{w}_0^{(1)}]^2 \tilde{\mathscr{U}}_0 + 2\tilde{w}_0^{(1)} \tilde{w}_1^{(1)} \tilde{\mathscr{U}}_1 + [\tilde{w}_1^{(1)}]^2 \tilde{\mathscr{U}}_2. \tag{7.11}$$

The energy of the first excited state of the quartic potential evaluated using ten blocks of 1000 steps is tabulated for several approximations of φ_0 computed by gradually incrementing j_{max} in equation (7.1) (see Table 7.2).

Table 7.2: Estimates of E_1 for the particle in a quartic potential.

j_{max}	0	2	4	6
E_1	0.011347	0.011366	0.0113171	0.0113091
σ	0.000012	0.000012	0.0000038	0.0000086
Δ (cm^{-1})	24.0	28.0	17.0	16.0

The row labeled with σ is the standard deviation, a reasonable estimate of the statistical fluctuations of the E_1 estimates. The row labeled Δ is computed as the difference between the diffusion Monte Carlo estimates of E_1 in the first row and the converged DVR (see the exercises in Chapter 2) energy 0.0112379 hartree. When compared with the values in row 1, we can see that the approach is producing a reasonable approximation, but the differences are significant. As j_{max} increases, the statistical noise of E_1 decreases, so does the difference between the estimate and the more accurate DVR value. We have converted the values of Δ into wave numbers (1 hartree = 219474.629 cm^{-1}) for ease of comparison. As a general rule, a computation with an error on the order of a few millihartree is considered *chemically accurate*, whereas with an error of 1–5 cm^{-1}, it is considered *spectroscopically accurate*. For $j_{max} = 6$, the precision is approaching spectroscopic quality, but it is important to uncover the main source of systematic error.

Upon inspecting the terms of equation (7.11), we note that there are eight different integrals, the average values of which are tabulated and compared with numerical quadrature values in Table 7.1. Then there are two polynomial coefficients. We can estimate these using the DVR again. This simple test does confirm that DMC is able to reproduce the DVR results within the statistical fluctuations, but only when the Fourier expansion of φ_0 has converged. The remaining source of error, comes from our assumption about ψ_1. In Figure 7.2, we graph the ratio ψ_1/ψ_0 for the system, and it is evident that $\tilde{P}_1(x)$ is not a linear function.

7.3 The VAMPIR algorithm

In Chapter 6, we have shown that optimizing parameters for the ground state wave function can always give an upper bound estimate of the ground state energy, but that the same cannot be said about excited states. That remains true, unless we deflate those

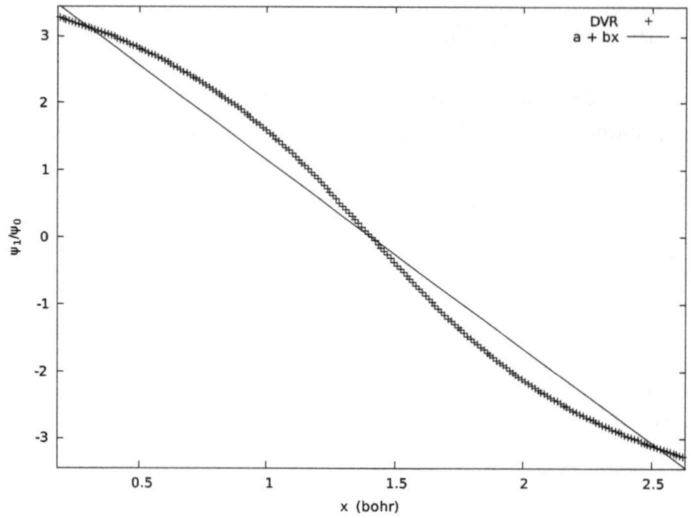

Figure 7.2: Ratio of ψ_1/ψ_0 for the particle in a quartic potential estimated by the DVR.

states that are lower in energy by the orthogonalization procedure. Therefore, it is possible to combine the optimization of some set of parameter while enforcing orthogonalization at the same time. This is not a new idea, it was first implemented by Schmid, Sheng, Grümmer, and Faessler in 1989 for the computation of electronic excited states in vector spaces comprised of Slater determinants [82]. For completeness, we preset their approach here.

The development requires additional notation. Let the set $\{\varphi_n\}$ be defined as follows:

$$\varphi_n = x^n \sum_{k=0}^{N_n} c_k^{(n)} \chi_k$$

where $\{\chi_k\}$ is the same set of function we have used for the ground state. The energy estimator for an arbitrary state n,

$$E_n = \frac{\langle \psi_n | H | \psi_n \rangle}{\langle \psi_n | \psi_n \rangle},$$

can be combined with the orthogonalization procedure by defining the overlap matrix

$$A_{ij} = \langle \psi_i | \varphi_j \rangle \quad i, j = 1, 2, \ldots, n-1,$$

where we are assuming the orthonormal set $\psi_0, \ldots, \psi_{n-1}$ have already been determined.

The *orthogonal projector* is defined as

$$P = \sum_{k=0}^{n-1} |\psi_k\rangle\langle\psi_k|. \tag{7.12}$$

Then the unnormalized orthogonal basis is given by

$$|\tilde{\psi}_n\rangle = (1 - P)|\varphi_n\rangle$$

and the energy becomes

$$\begin{aligned} E_n &= \frac{\langle\varphi_n|(1-P)H(1-P)|\varphi_n\rangle}{\langle\varphi_n|(1-P)(1-P)|\varphi_n\rangle} \\ &= \frac{\langle\varphi_n|H|\varphi_n\rangle - \langle\varphi_n|PH|\varphi_n\rangle - \langle\varphi_n|HP|\varphi_n\rangle + \langle\varphi_n|PHP|\varphi_n\rangle}{\langle\varphi_n|\varphi_n\rangle - \langle\varphi_n|P|\varphi_n\rangle}. \end{aligned}$$

Upon expanding the denominator, we encounter the matrix element $\langle\varphi_n|P^2|\varphi_n\rangle$, however, this is equal to $\langle\varphi_n|P|\varphi_n\rangle$. For P to be a projector, $P^2 = P$ must hold; it is straightforward to show that it does in our case,

$$P^2 = \sum_{k,j=0}^{n-1} |\psi_j\rangle\langle\psi_j|\psi_k\rangle\langle\psi_k| = \sum_{j=0}^{n-1} |\psi_j\rangle\langle\psi_j| = P,$$

since the set $\psi_0, \ldots, \psi_{n-1}$ is orthonormal $\langle\psi_j|\psi_k\rangle = \delta_{jk}$. Next, $\langle\varphi_n|P|\varphi_n\rangle$ unravels as follows:

$$\langle\varphi_n|P|\varphi_n\rangle = \langle\varphi_n|\left[\sum_{k=0}^{n-1} |\psi_k\rangle\langle\psi_k|\right]|\varphi_n\rangle = \sum_{k=0}^{n-1} A_{nk}^2.$$

To handle the term $\langle\varphi_n|PHP|\varphi_n\rangle$ in the numerator, we insert equation (7.12) twice:

$$\begin{aligned} \langle\varphi_n|PHP|\varphi_n\rangle &= \sum_{k,j=0}^{n-1} \langle\varphi_n|\psi_j\rangle\langle\psi_j|H|\psi_k\rangle\langle\psi_k|\varphi_n\rangle \\ &= \sum_{k,j=0}^{n-1} E_k\langle\varphi_n|\psi_j\rangle\langle\psi_j\psi_k\rangle\langle\psi_k|\varphi_n\rangle \\ &= \sum_{k=0}^{n-1} E_k A_{nk}^2, \end{aligned}$$

where $H|\psi_k\rangle = E_k|\psi_k\rangle$ is used in the second line.

Continuing along these lines, we end up with

$$E_n = \frac{H_{nn} - 2\sum_{k=0}^{n-1} A_{nk}H_{kn} + \sum_{k=0}^{n-1} E_k A_{nk}^2}{\langle\varphi_n|\varphi_n\rangle - \sum_{k=0}^{n-1} A_{nk}^2}, \tag{7.13}$$

where $H_{kn} = \langle\psi_k|H|\varphi_n\rangle$ and $E_k = \langle\psi_k|H|\psi_k\rangle$.

To further clarify the notation and relate it to the previous section, consider the case $n = 1$ and let $c_k^{(n)} = c_k$. Then equation (7.13) becomes

$$E_1 = \frac{H_{11} - 2A_{10}H_{10} + E_0 A_{10}^2}{\langle\varphi_1|\varphi_1\rangle - A_{10}^2}.$$

For example, the overlap elements are

$$A_{10} = \langle\varphi_0|\varphi_1\rangle = [w_0^{(0)}]^2 \int_{-\infty}^{\infty} x\,\varphi_0^2\,dx = [w_0^{(0)}]^2 \tilde{\mathscr{I}}_1$$

$$\langle\varphi_1|\varphi_1\rangle = [w_0^{(0)}]^2 \int_{-\infty}^{\infty} x^2\,\varphi_0^2\,dx = [w_0^{(0)}]^2 \tilde{\mathscr{I}}_2,$$

$\langle\varphi_1|\varphi_1\rangle - A_{10}^2 = \tilde{d}_1^2$ where $\tilde{d}_1$ is in equation (7.8), and E_1 becomes (7.11). Therefore, the VAMPIR method is identical to the strategy developed in Section 7.2. The strategy provides a very good starting point for the further optimization of the parameters of φ_n should one chose to refine them.

The expression for E_n in equation (7.13) contains the following integrals H_{nn}, A_{nk}, H_{kn}, and $\langle\varphi_n|\varphi_n\rangle$. For the particle in one dimension, with

$$H = -\frac{\hbar^2}{2m}\frac{d^2}{dx^2} + U,$$

assuming we have a population of M replicas that samples the ground state φ_0 projected into a vector space as in (7.1), we can approximate these four integrals as follows, e. g.,

$$\begin{aligned} H_{nn} &= \int_{-\infty}^{\infty} \varphi_n H \varphi_n\,dx \\ &= \int_{-\infty}^{\infty} \left[-\frac{\hbar^2}{2m}\varphi_n \frac{d^2\varphi_n}{dx^2} + \varphi_n U \varphi_n\right] dx \\ &\approx \frac{1}{M}\sum_{i=1}^{M}\left[\left(\frac{\varphi_n}{\varphi_0}\right)\left(-\frac{\hbar^2}{2m}\frac{d^2\varphi_n}{dx^2} + U\varphi_n\right)\right]_{x=x_i} \end{aligned}$$

where we have multiplied the integrand by $1 = \varphi_0/\varphi_0$ to account for the shape of the ground state wave function produced by the population $\{x_i\}_{i=1}^{M}$. Then the ratio φ_n/φ_0 can be evaluated with

$$\left.\frac{\varphi}{\varphi_0}\right|_{x=x_i} = \frac{x_i^n \sum_{k=1}^{k_n} c_k^{(n)} \chi_k(x_i)}{\sum_{k=1}^{k_n} c_k \chi_k(x_i)}.$$

In practice, this ratio is bound even though

$$\lim_{x\to\pm\infty} \frac{\varphi}{\varphi_0} \to \frac{x^n c_k^{(n)}}{c_k}$$

theoretically grows without bound, because the probability of sampling far into the tails of φ_0 is vanishingly small. However, having an accurate representation of φ_0 is critical for the stability of the algorithm. The remaining estimators feature the same ratio with the exception of the overlap elements A_{0k}.

Finally, it is conceivable that, for some systems, the computation of the kinetic energy may be more expansive, and more numerically unstable than the gradient of U, e. g., for electronic states. In the next section, we consider an alternative approach to estimate energies that replaces the integrals of the kinetic energy operator with those of the potential energy gradient by means of the virial theorem.

7.4 The virial theorem

The classical version of the virial theorem is stated as a time average over a period τ of the orbit

$$\frac{1}{\tau}\int_0^\tau K\,dt = \frac{1}{2\tau}\int_0^\tau \mathbf{x}\cdot\nabla U\,dt$$

where $K = \sum_i p_i^2/(2m_i)$ is the kinetic energy of n particles whose configuration is mapped with Cartesian coordinates. We prove it here for a particle in a single dimension for simplicity. The proof for the $\mathbb{R}^n$ case is not much more complicated. Let

$$\dot{x} = \frac{dx}{dt},$$

then $p = m\dot{x}$ and Newton's second law become

$$\dot{p} = -\frac{dU}{dx}.$$

Let us consider the time dependence of xp,

$$\frac{d}{dt}xp = x\dot{p} + p\dot{x}.$$

Then the time average over a period is

$$\frac{1}{\tau}\int_0^\tau \frac{d}{dt}xp\,dt = \frac{1}{\tau}\int_0^\tau (x\dot{p} + p\dot{x})\,dt.$$

Inserting Newton's equation into the first term on the right yields

$$\frac{1}{\tau}\int_0^\tau \frac{d}{dt}xp\,dt = \frac{1}{\tau}\int_0^\tau \left(p\dot{x} - x\frac{dU}{dx}\right)dt$$

but note that $p\dot{x} = p^2/m = 2K$ and now we have

$$\frac{1}{\tau} xp\Big|_0^\tau = \frac{1}{\tau} \int_0^\tau \left(2K - x\frac{dU}{dx}\right) dt. \tag{7.14}$$

If the orbit is periodic, the time average on the left-hand side vanishes leaving the one-dimensional version of the theorem. If the system is not periodic, but it is bound, letting $\tau \to \infty$ will cause the xp quantity evaluated over the two limits to vanish, because their values are finite but the denominator grows without bound.

The quantum mechanical version follows by evoking the general form of Ehrenfest theorem we derived in Chapter 4 applied to the expectations,

$$\frac{d}{dt}\langle xp\rangle = \frac{i}{\hbar}\langle [H, px]\rangle \tag{7.15}$$

where

$$[H, px] = [H, p]x - p[x, H] \tag{7.16}$$

and recall that in Chapter 4 we derive

$$[H, p] = i\hbar\frac{dU}{dx}, \quad [H, x] = -\frac{i\hbar}{m}p.$$

Inserting these into (7.16) give

$$[H, px] = i\hbar\left(x\frac{dU}{dx} - \frac{p^2}{m}\right).$$

Substitution of this result into (7.15) and integration over the period yields the virial theorem of quantum mechanics,

$$\langle K\rangle = \frac{1}{2}\left\langle x\frac{dU}{dx}\right\rangle. \tag{7.17}$$

If the gradient of U is available analytically, then rather than evaluating a number of cumbersome integrals involving derivatives of φ_0, we can use the average values of the forces to estimate the kinetic energy instead. However, trial wave functions optimized variationally as, e. g., we did in Chapter 6, do not automatically satisfy the virial theorem. Typically, more parameters are needed, and these are optimized so that the trial wave function satisfies two criteria:

- ψ_T satisfies the virial theorem
- $\langle E_n\rangle$ is minimized.

While it may be possible to impose both these constraints on the parameters of a trial wave function, the approach is only computationally advantageous if the first criteria can be satisfied analytically. It turns out that for coulombic wave functions this is possible. We sketch the procedure here.

Recall that in Chapter 6 we optimized $\psi_T = e^{-\beta(x-x_0)^2}$ to find the best variational estimate of the ground state for particle in $U = D(x-x_0)^4$. We now need to increment the parameters and let us choose $\psi_T = e^{-(ax^2+bx)}$, so that we have all the monomial integrals available from Chapter 6. The two sets of constraints that a and b must satisfy are:

- ψ_T satisfies the virial theorem

$$-\frac{\hbar^2}{2m}\mathscr{I}_0^{-1}\int_{-\infty}^{\infty}\psi_T^*\,[4a^2x^2+4abx-2a]\psi_T\,dx$$
$$=2D\mathscr{I}_0^{-1}\int_{-\infty}^{\infty}\psi_T^*\,x(x-x_0)^3\,\psi_T\,dx$$

- The ground state energy estimate

$$\langle E_0\rangle = \mathscr{I}_0^{-1}\int_{-\infty}^{\infty}\psi_T^*\left\{-\frac{\hbar^2}{2m}[4a^2x^2+4abx-2a]+D(x-x_0)^4\right\}\psi_T\,dx$$

  is minimized

where recall that $\mathscr{I}_0^{-1/2}$ is the normalization constant for ψ_T. Using the recursion in Theorem 26, we can express all the monomial integrals in terms of $\mathscr{I}_0$. Then we can evaluate the derivative with respect to b in the second equation as set it to zero. The result is a nonlinear system of two equations and two unknowns. It is algebraically messy, and rather uninteresting, but can be solved numerically by any root searching algorithm with the condition that a is strictly greater than zero. A simpler course of action is suggested in exercise 12.

7.5 Exercises

1. Derive equations (7.4) through (7.6) by following the same procedure used to derive (7.2).
2. Modify the function `wavef(x,w,n):` from Chapter 6 so that it returns not just χ_j but its first and second derivatives.
3. Derive equations (7.7) through (7.11) from equations (2.9) through (2.17).
4. Append the code for the stochastic evaluation of the monomial integrals $\tilde{\mathscr{I}}_n$, $\tilde{\mathscr{Q}}_n$, $\tilde{\mathscr{P}}_n$, $\tilde{\mathscr{U}}_n$ followed by the code that computes E_0 to the program developed in Chapter 6. Use it to reproduce the results in Table 7.1.

5. Modify the code in the previous exercise to estimate the following properties of the ground state: $\langle 0|x|0\rangle$, $\langle 0|p|0\rangle$, $\langle 0|U|0\rangle$.
6. The virial theorem of quantum mechanics, specialized for the ground state a single particle in one dimension reads

$$\langle 0|K|0\rangle = \frac{1}{2}\left\langle 0\left|x\frac{\partial U}{\partial x}\right|0\right\rangle.$$

Modify the code of exercise 4 to compute and compare both of these averages. Is the theorem satisfied? The subroutine `pot` can be modified to return both U and its first derivative, however, the call to it has to be modified from `u = pot(x[i])` to `u,du = pot(x[i])` otherwise `u` is automatically shaped as an array.
7. Modify the DVR code from Chapter 2 so that the energies of a particle of mass 919.0 atomic units in the quartic potential $U(x) = D(x - x_0)^4$ with $D = 0.0874$ hartree and $x_0 = 1.41$ bohr are computed. Be sure to check for convergence in both the lattice size as well as the cutoff error for the first few states. Repeat the process with a mass of 3645.0 atomic units.
8. Insert the code for the first excited state energy estimate into your diffusion Monte Carlo program and reproduce the results in Table 7.2. Then repeat the simulation with $m = 3645.0$ atomic units. Compare the stochastic estimate of E_1 with the deterministic one in exercise 7.
9. Fill in the steps to derive equation (7.13) by first proving that

$$\langle \varphi_n|PH|\varphi_n\rangle = \sum_{k=0}^{n-1} A_{nk}\, H_{kn}.$$

10. The following parts of this exercise pertain to the case of the excited VAMPIR algorithm when $c_k^{(n)} = c_k$:
    (a) Show that

$$H_{11} = [\tilde{w}_0^{(0)}]^2\left\{-\frac{\hbar}{2m}(2\tilde{Q}_1 + \tilde{P}_2) + \tilde{\mathcal{U}}_2\right\}.$$

    (b) Use equation (7.9) to show that

$$\frac{H_{11}}{\tilde{d}_1^2} = [\tilde{w}_1^{(1)}]^2\left\{-\frac{\hbar}{2m}(2\tilde{\mathcal{Q}}_1 + \tilde{\mathcal{P}}_2) + \tilde{\mathcal{U}}_2\right\}.$$

    (c) Show that

$$\frac{E_0 A_{10}^2}{\tilde{d}_1^2} = [\tilde{w}_0^{(1)}]^2\left\{-\frac{\hbar}{2m}\tilde{\mathcal{P}}_0 + \tilde{\mathcal{U}}_0\right\}$$

    by using equation (7.10).

(d) Derive the expression for

$$-\frac{2A_{10}H_{10}}{\tilde{d}_1^2}$$

in terms of the proper monomial integrals, and combine with the results in parts (b) and (c) to show that, when $c_k^{(n)} = c_k$,

$$E_1 = \frac{H_{11} - 2A_{01}H_{01} + E_0A_{10}^2}{\langle\varphi_1|\varphi_1\rangle - A_{10}^2}$$

becomes equation (7.11).

11. The class of homogeneous potential energy functions plays a significant role in the application of the virial theorem. A potential energy $U(x)$ is said to be homogeneous order n in x if upon rescaling $x :\rightarrow sx$ where s is some constant

$$U(sx) = s^n U(x).$$

(a) Show that the quartic potential $U(x) = D(x - x_0)^4$ is order 4 homogeneous.
(b) Is the Lennard Jones potential

$$U(r) = 4\epsilon\left[\left(\frac{\sigma}{r}\right)^{12} - \left(\frac{\sigma}{r}\right)^6\right]$$

homogeneous? The variables ϵ and σ are its parameters.
(c) Is the Coulomb potential

$$U(r) = -\frac{e^2}{4\pi\epsilon_0 r}$$

homogeneous? (Here e is the fundamental charge, ϵ_0 is the vacuum permittivity, and both are constants).
(d) Prove that if $U(x)$ is homogeneous order n, then

$$x\frac{dU}{dx} = nU.$$

Keep your proof general, i. e., do not assume any functional form for U, instead make use of the chain rule and differentiate $U(sx) = s^n U(x)$ on both sides with respect to s, and take the limit as $s \rightarrow 1$.

12. Consider a particle of mass m in a one-dimensional quartic potential,

$$-\frac{\hbar^2}{2m}\frac{d^2}{dx^2}\psi + D(x - x_0)^4\psi = E\psi.$$

(a) Change the independent variable $y = x - x_0$ use the chain rule and show that

$$-\frac{\hbar^2}{2m}\frac{d^2}{dy^2}\varphi_0 + Dy^4\varphi_0 = E\varphi_0.$$

(b) Assume $\psi = e^{-ay^2/2}$ and find an expression for the parameter a such that ψ satisfies the virial theorem. Compare your value of a with the value of β optimized in Chapter 6 for the trial wave function $\psi_T = e^{-\beta(x-x_0)^2}$.

(c) Use the solution of part (b) to estimate the ground state energy for a particle of 919.0 atomic units with $D = 0.0874$. Compare it with the variational estimate 0.003199 in Chapter 6 and 0.0031166 ± 10^{-6} hartree in Section 7.1 of this chapter and the result in exercise 7.

(d) Modify the code developed in exercise 6 to walk in the y variable. The simplest way to accomplish this is to change pot so that it computes $U = Dx^4$ and set b to zero. Use the value of a from part (b) in your diffusion Monte Carlo code, and estimate E_0 using equation (7.7). Does the wave function satisfy the virial theorem?

8 Essentials of differential geometry

Every time the subject comes up, I work on it. In fact, when I began to prepare this lecture I found myself making more analyses on the thing. Instead of worrying about the lecture, I got involved in a new problem...

Richard P. Feynman

The fusion of dynamics with geometry has become a central part of the theoretical edifice physicists have built to probe the workings of our universe. The practice of trying to explain physical phenomena by geometric arguments has a long history that started well before the celebrated theory of general relativity and gauge theory. Newtonian mechanics endowed "absolute space" with coordinates and with physical meaning. According to this view, centrifugal and coriolis are created by space, and these ideas gave a way to derive relatively simple mathematical expressions for them. Einstein's general relativity theory is based on the concept that massive objects bend spacetime. While chemistry is hardly concerned with general relativity, it turns out that "chemical theories" make frequent use of coordinates like the familiar spherical polar map of Euclidean three-dimensional space. It turns out that even the general treatment of curvilinear maps like the spherical polar one require the same geometric objects as curved manifolds, a generalized concept of space [106]. Among these objects are nonuniform metric tensors, connections, and covariant derivatives to name a few.

Another important example is the normal mode analysis for n bodies bounded together. The process involves expanding U about its minimum into a Taylor series, neglecting the constant term, and all the terms beyond the quadratic as well. At the minimum of U, by definition the gradient vanishes and one is left with a set of coupled harmonic oscillators that can be decoupled by well-established linear algebra methods. It turns out that normal mode analysis can produce uniform metrics when transforming from the Cartesian coordinates for each nucleus. In those coordinates, the Hessian of U has less physical meaning and the interpretation of the wave vector is not simple. When the Hessian of U is expressed in terms of "internal coordinates," namely bond lengths, bond angles, dihedral angles, etc, are far more meaningful. The downside is that the internal coordinate space is curved, and there are technical issues such as redundancy and the like. Moreover, rigid constraints on the configuration space of n particles create curved subspaces of $\mathbb{R}^{3n}$ and these are in many cases essential. "Changing coordinates" is one of the most frequently used tricks to simplify problems in chemical physics, and the field of differential geometry provides the appropriate mathematical structure needed for such developments. This chapter aims at introducing the essentials of differential geometry, with the focus on Riemannian manifolds. The most important result of this chapter is a powerful tool that simplifies the process of deriving the correct expression of the kinetic energy operator in an arbitrary Riemann manifold.

https://doi.org/10.1515/9783111610207-008

8.1 Riemannian geometry

We will develop the essential parts of the theory by beginning with the Euclidean three-dimensional space charted by Cartesian coordinates and following the usual process of mathematical abstraction. We start with the following definition.

Definition 27 (Gaussian surface). Consider a surface $\mathcal{M}$ in ordinary Euclidean space defined as the set of all points $\{p_i\}$ whose coordinates $x_i^{(1)}, x_i^{(2)}, x_i^{(3)}$ satisfy an analytic relation

$$f(x^{(1)}, x^{(2)}, x^{(3)}) = 0. \tag{8.1}$$

Gaussian surfaces are examples of Riemann manifolds. The choice of not using x_i, y_i, z_i is deliberate and is made with the intent to ease the path of generalization. The parenthesis are included in the superscript to avoid confusing indices with powers when natural numbers appear in our expressions. Normally, the Greek letters α, β, μ, ν, λ, σ are used as indices, but occasionally Roman letters i, j, k appear as well. In those cases, the parentheses are omitted. Because of the equation of constraint (8.1), we only have two degrees of freedom at our disposal and we may be able to use (8.1) to solve for any of the three,

$$x^{(3)} = \varphi(x^{(1)}, x^{(2)}). \tag{8.2}$$

However, equation (8.2) displays asymmetry in the three coordinates, which turns out to be too restrictive for our purposes. Gauss introduced a parametric approach to eliminate such asymmetries by using one of the central object in the study of manifolds: a chart. He choose three functions of two parameters to rewrite (8.2) in a more symmetric way,

$$x^\mu = \varphi^\mu(u^{(1)}, u^{(2)}), \quad \mu = 1, 2, 3. \tag{8.3}$$

Suppose we consider a curve on the surface in (8.3) parameterized by t, such that $u^{(1)}, u^{(2)} = u^{(1)}(t), u^{(2)}(t)$, and we want to find the length L of the curve as $t : 0 \to 1$. This is given by the following integral:

$$\begin{aligned} L = \int_0^1 ds &= \int_0^1 \left[\left(\frac{dx^{(1)}}{dt} \right)^2 + \left(\frac{dx^{(2)}}{dt} \right)^2 + \left(\frac{dx^{(3)}}{dt} \right)^3 \right]^{1/2} dt \\ &= \int_0^1 \left[\sum_{i,k=1}^{2} g_{ik} \frac{du^i}{dt} \frac{du^k}{dt} \right]^{1/2} dt. \end{aligned} \tag{8.4}$$

The object in the second line is known as the *metric tensor* of the surface $\mathcal{M}$ and it takes the following form:

$$g_{ik} = \sum_{\nu=1}^{3} \frac{\partial \varphi^{\nu}}{\partial u^{i}} \frac{\partial \varphi^{\nu}}{\partial u^{k}}, \tag{8.5}$$

which we now prove. The object,

$$\left(\frac{dx^{(1)}}{dt}\right)^2,$$

can be expressed using (8.3) and the chain rule as follows:

$$\begin{aligned}\left(\frac{dx^{(1)}}{dt}\right)^2 &= \left(\frac{\partial \varphi^{(1)}}{\partial u^{(1)}}\frac{du^{(1)}}{dt} + \frac{\partial \varphi^{(1)}}{\partial u^{(2)}}\frac{du^{(2)}}{dt}\right)^2 \\ &= \left(\frac{\partial \varphi^{(1)}}{\partial u^{(1)}}\right)^2\left(\frac{du^{(1)}}{dt}\right)^2 + \left(\frac{\partial \varphi^{(1)}}{\partial u^{(2)}}\right)^2\left(\frac{du^{(2)}}{dt}\right)^2 \\ &+ \frac{\partial \varphi^{(1)}}{\partial u^{(2)}}\frac{\partial \varphi^{(1)}}{\partial u^{(1)}}\frac{du^{(1)}}{dt}\frac{du^{(2)}}{dt} \\ &+ \frac{\partial \varphi^{(1)}}{\partial u^{(1)}}\frac{d\varphi^{(1)}}{\partial u^{(2)}}\frac{du^{(1)}}{dt}\frac{du^{(2)}}{dt}.\end{aligned} \tag{8.6}$$

Corresponding expressions can be derived for $\varphi^{(2)}$ and $\varphi^{(3)}$ by simply replacing the exponents. Now consider the i, k element of g_{ik} in equation (8.5) for the $i \neq k$ case. It follows from the third line of (8.6) when added to the equivalent lines or $\varphi^{(2)}$ and $\varphi^{(3)}$ that

$$\sum_{i,k=1}^{n} \sum_{\nu=1}^{3} \frac{\partial \varphi^{\nu}}{\partial u^{i}} \frac{\partial \varphi^{\nu}}{\partial u^{k}} \frac{du^{i}}{dt} \frac{du^{k}}{dt} = L^2$$

and when comparing this equation with the second line of equation (8.4) we arrive at (8.5).

Note that we had to separate the two cross-terms resulting from expanding the square, even though these are in fact equal. The following is an important corollary for the theory of Riemann spaces:

$$g_{\mu\nu} = g_{\nu\mu}.$$

In other words, the metric tensor is symmetric.

As an example of (8.1), and the surface $\mathcal{M}$ it constructs, consider the set of all points $\{p_i\}$ on the surface of a unit sphere $\mathbb{S}^2$,

$$x^2 + y^2 + z^2 - 1 = 0$$

where (8.2) becomes

$$z = \sqrt{1 - x^2 - y^2}.$$

This space is particularly important in the theory or linear rigid rotors and angular momentum theory. The asymmetry between z and the remaining coordinates should now be very evident—the object in equation (8.3), in fact, a map $\varphi : \mathbb{S}^2 \to \mathbb{R}^3$ for our example. The two degrees of freedom $u^{(1)}$, $u^{(2)}$ could be the angles θ and ϕ, though this is not the only way to parametrize the unit sphere:

$$x = \varphi^{(1)} = \cos\phi \sin\theta$$
$$y = \varphi^{(2)} = \sin\phi \sin\theta$$
$$z = \varphi^{(3)} = \cos\theta.$$

These are the familiar orientations of a vector in Euclidean space. We leave it as an exercise to show that the metric tensor for this case is

$$g_{\mu\nu} = \begin{pmatrix} 1 & 0 \\ 0 & \cos^2\theta \end{pmatrix}. \tag{8.7}$$

Riemann generalized the definition of the Gaussian surface by extending it to a set of points in n dimensions using the expression for the infinitesimal distance squared

$$ds^2 = \sum_{\mu,\nu=1}^{n} g_{\mu\nu} du^\mu du^\nu \tag{8.8}$$

where the metric tensor is a function of $\{u^{(1)}, u^{(2)}, \ldots, u^{(n)}\}$, the n parameters needed to transform the coordinates,

$$x^\mu = \varphi^\mu(u^{(1)}, u^{(2)}, \ldots, u^{(n)}) \quad 1 < \mu < m \tag{8.9}$$

where m need not be greater then n. These coordinate systems are defined in such way that one can go from one system to the other through one-to- one transformations

$$u^\nu = f^\nu(x^{(1)}, x^{(2)}, \ldots, x^{(n)}) \quad 1 < \nu < n. \tag{8.10}$$

The two sets of functions $\{\varphi^{(1)}, \varphi^{(2)}, \ldots, \varphi^{(m)}\}$ and $\{f^{(1)}, f^{(2)}, \ldots, f^{(n)}\}$ are assumed to be continuous and have continuous first derivatives. This implies that the Jacobian of the transformation

$$J^\nu_\mu = \frac{\partial f^\nu}{\partial \varphi^\mu} \tag{8.11}$$

and its inverse

$$J^\mu_\nu = \frac{\partial \varphi^\mu}{\partial f^\nu} \tag{8.12}$$

are continuous and never zero.

Definition 28 (Chart). A coordinate chart $\varphi : \mathcal{M} \to \mathbb{R}^m$ is the set of n one-to-one differentiable functions of m parameters in equation (8.9). Its inverse is the set of m functions of n parameters in equation (8.10). In this definition, it is assumed that $\mathbb{R}^m$ is pseudo-Euclidean, meaning it is mapped with orthogonal rectilinear coordinates.

The coordinate system defined by a chart is generally applicable only in a part of the space. To cover all points in a space, it is necessary to use more than one chart. Moreover, it is important that charts φ and φ' that map two neighboring parts of the space overlap and that a smooth one-to-one map $\varphi \to \varphi'$ is defined in the overlap area.

Definition 29 (Topology). The disjoint union of all neighborhoods of points U_i of a space that permit the constructions of coordinate charts is called a topology.

Definition 30 (Atlas). The set of all charts that cover a topology is called an atlas.

Examples of Riemann manifolds are the parameter space of the classical Lie group of transformations we encountered in Chapters 3 and 4, the familiar Euclidean three-space, the two-sphere of our previous example, the ellipsoids of inertia for a rigid body, which is also the configuration space of a three or more point particle system rigidly held in a nonlinear configuration, the Galilean space-time (a four-dimensional space with t, x, y, z with $g_{\mu\nu} = \delta_{\mu\nu}$) of Newtonian mechanics, the Lorentzian space time of special relativity endowed with metric

$$g_{\mu\nu} = \begin{pmatrix} 1 & 0 & 0 & 0 \\ 0 & -1 & 0 & 0 \\ 0 & 0 & -1 & 0 \\ 0 & 0 & 0 & -1 \end{pmatrix}$$

where the +1 so called signature is time-like and the −1 signature for the three space-like coordinates.

Definition 31 (Scalar). A scalar is a number that does not depend on the choice of reference frame. A scalar field q is a function of the points in the manifold that can be expressed in terms of the coordinates with the property that its value at a point remains is invariant under coordinate transformation.

Definition 32 (Contravariant vectors). A n-tuple of components V^μ with one index in the contravariant position (superscript). Under a coordinate change $\varphi(\{x^\mu\}) \to \varphi'(\{x^{\mu'}\})$, a contravariant vector transforms as follows:

$$V^{\mu'} = \sum_{\mu=1}^{n} \frac{\partial x^{\mu'}}{\partial x^\mu} V^\mu \tag{8.13}$$

where $\{x^\mu\} = \{x^{(1)}, \ldots, x^{(n)}\}$ etc.

Note how the coordinates x^μ are examples of contravariant vectors.

Definition 33 (Covariant vectors). A n-tuple of components F_μ with one index in the covariant position (subscript). Under a coordinate change $\varphi(\{x^\mu\}) \to \varphi'(\{x^{\mu'}\})$, a covariant vector transforms as follows:

$$F_{\mu'} = \sum_{\mu=1}^{n} \frac{\partial x^\mu}{\partial x^{\mu'}} F_\mu. \tag{8.14}$$

The forces of Newtonian mechanics are examples of covariant vectors. Covariant vectors are also called $(0,1)$ forms.

Definition 34 (Einstein summation convention). The summation symbol over coefficient from one to the dimension of the manifold is omitted whenever tensor expressions have a repeated Greek or Roman index in both a covariant and contravariant position.

This definition is just a notational convenience. Equations (8.8), (8.13), and (8.14) written with Einstein's convention become, respectively,

$$ds^2 = g_{\mu\nu} du^\mu du^\nu \tag{8.15}$$

$$V^{\mu'} = \frac{\partial x^{\mu'}}{\partial x^\mu} V^\mu \tag{8.16}$$

$$F_{\mu'} = \frac{\partial x^\mu}{\partial x^{\mu'}} F_\mu. \tag{8.17}$$

Definition 35 (Tensor). A (m,n) tensor is a quantity with m contravariant (superscripts) and n covariant (subscript) indices.

For example, the Jacobian in equations (8.11) and (8.12) are an example of $(1,1)$ tensors as they have one contravariant and one covariant index. The metric tensor we encountered earlier is a $(0,2)$ form. The quantity $T^\alpha_{\beta\kappa}$ is a $(1,2)$ tensor, etc. It is expected to transform as a 1 contravariant, 2 covariant tensor,

$$T^{\alpha'}_{\beta'\kappa'} = \frac{\partial x^{\alpha'}}{\partial x^\alpha} \frac{\partial x^\beta}{\partial x^{\beta'}} \frac{\partial x^\kappa}{\partial x^{\kappa'}} T^\alpha_{\beta\kappa}.$$

Note the use and the convenience of Einstein's summation convention. Care must be taken in the placement of the primes. Note also that indices take the prime, not the coordinates. The metric $g_{\mu\nu}$ is a $(0,2)$ tensor or a 2-form. Its transformation is very useful for our purposes,

$$g_{\mu'\nu'} = \frac{\partial x^\mu}{\partial x^{\mu'}} \frac{\partial x^\nu}{\partial x^{\nu'}} g_{\mu\nu}. \tag{8.18}$$

This expression is the generalization of (8.5) and is the tool used to derive metric tensors in manifolds.

On many occasions a repeated index may appear in a single tensor quantity.

Definition 36 (Tensor contraction). When a (n, p) tensor element has one or more repeated indices, the Einstein summation is assumed and each sum reduces the rank to the (n, p) by one in both the contravariant and covariant position.

For example,

$$\Gamma^{\mu}_{\mu\nu} = F_{\nu}$$

a $(1, 2)$ tensor becomes a $(0, 1)$ form.

Definition 37 (Raising and lowering indices). In Riemann spaces, a $(1, 0)$ vector can be transformed into a $(0, 1)$ form by means of the metric tensor.

$$V_{\mu} = g_{\mu\nu} V^{\nu} \tag{8.19}$$

where Einstein's summation has been implemented to simplify the expression.

The inverse of $g_{\mu\nu}$ is $g^{\mu\nu}$ and it can be used to raise indices, so a $(0, 1)$ form can be transformed into a $(1, 0)$ vector

$$V^{\nu} = g^{\mu\nu} V_{\mu}. \tag{8.20}$$

Note that Kronecker deltas like $\delta^{\mu\nu}$, δ^{α}_{β}, and $\delta_{\kappa\lambda}$ are not tensors; they are symbols. They are not transformed by coordinate changes. Moreover, the metric tensor or its inverse cannot be used to move their indices around. These rules help to make sense of expressions like

$$g^{\mu\lambda} g_{\nu\lambda} = \delta^{\mu}_{\nu}$$

demonstrating the fact that the inverse of $g_{\mu\nu}$ is $g^{\mu\nu}$ as explained earlier. The operation of raising and then lowering an index should leave the original quantity unchanged,

$$V_{\mu} = g_{\mu\nu} g^{\alpha\nu} V_{\alpha} = \delta^{\alpha}_{\mu} V_{\alpha} = V_{\mu}.$$

Now consider the product of a $(1, 0)$ and a $(0, 1)$ tensor,

$$V^{\nu} Q_{\nu} = B$$

a scalar quantity (no indices). This expression can be understood as the dot product or two vectors in Riemann space,

$$V^{\mu} \cdot Q^{\mu}$$

where now we insert the metric tensor to lower the index of Q^{μ},

$$V^{\mu} g_{\mu\nu} Q^{\nu} = B.$$

Therefore, we are beginning to see some of the important roles the Riemann metric plays in manifolds. For the dot product of two vectors, it plays the same role as the "weight" function does in vector spaces of functions discussed in Chapter 2. But there are more equally important roles that the metric tensor plays.

The covariant derivative

All the preliminary work and definitions are in place to derive the first very important result. The covariant derivative and the related connection coefficients are going to allow us to construct a general expression for the Laplace–Beltrami operator so that we may be able to write the Schrödinger equation in manifolds. Let us begin by showing that the partial derivative of a $(1,0)$ vector does not transform as a $(1,1)$ tensor. Take as a general case the expression,

$$\frac{\partial}{\partial x^{\mu}} V^{\nu}$$

an object that looks like a $(1,1)$ tensor, but does not transform as such. Using equations (8.16) to transform V^{ν} and (8.17) to transform the derivative, we get

$$\frac{\partial}{\partial x^{\mu'}} V^{\nu'} = \frac{\partial x^{\mu}}{\partial x^{\mu'}} \frac{\partial}{\partial x^{\mu}} \left(\frac{\partial x^{\nu'}}{\partial x^{\nu}} V^{\nu} \right)$$

and using the product rule yields

$$\frac{\partial}{\partial x^{\mu'}} V^{\nu'} = \frac{\partial x^{\mu}}{\partial x^{\mu'}} \frac{\partial x^{\nu'}}{\partial x^{\nu}} \frac{\partial}{\partial x^{\mu}} V^{\nu} + \frac{\partial x^{\mu}}{\partial x^{\mu'}} V^{\nu} \frac{\partial^2 x^{\nu'}}{\partial x^{\mu} \partial x^{\nu}}. \tag{8.21}$$

The first term to the right reads like the term we would expect under the invariance

$$\frac{\partial}{\partial x^{\mu'}} V^{\nu'} \to \frac{\partial}{\partial x^{\mu}} V^{\nu} + \frac{\partial x^{\mu}}{\partial x^{\mu'}} V^{\nu} \frac{\partial^2 x^{\nu'}}{\partial x^{\mu} \partial x^{\nu}}$$

but the second term "spoils the invariance" for us. This pattern repeats itself in gauge theory so the next few steps are critical to bridge conceptually differential geometry to it. The strategy to "restore" the transformation symmetry is one of modifying the definition of the derivative so that the transformation works as expected. The covariant derivative D_{μ} is defined as follows:

$$D_{\mu} V^{\nu} = \frac{\partial}{\partial x^{\mu}} V^{\nu} + \Gamma^{\nu}_{\mu\lambda} V^{\lambda} \tag{8.22}$$

where the object $\Gamma^{\nu}_{\mu\lambda}$ is called the connection. We now derive an equation for $\Gamma^{\nu}_{\mu\lambda}$ so that $D_{\mu} V^{\nu}$ transform properly as a $(1,1)$ tensor

$$\frac{\partial x^{\mu}}{\partial x^{\mu'}}\frac{\partial x^{\nu'}}{\partial x^{\nu}}D_{\mu}V^{\nu}=D_{\mu'}V^{\nu'}.$$

Inserting the definition of the covariant derivative (8.22) on both sides produces the following expression:

$$\frac{\partial x^{\mu}}{\partial x^{\mu'}}\frac{\partial x^{\nu'}}{\partial x^{\nu}}\frac{\partial}{\partial x^{\mu}}V^{\nu}+\frac{\partial x^{\mu}}{\partial x^{\mu'}}\frac{\partial x^{\nu'}}{\partial x^{\nu}}\Gamma^{\nu}_{\mu\lambda}V^{\lambda}$$
$$=\frac{\partial}{\partial x^{\mu'}}V^{\nu'}+\Gamma^{\nu'}_{\mu'\lambda'}V^{\lambda'}.$$

Now we use (8.21) for the first term to the right of the equal sign and (8.16) to transform $V^{\lambda'}$,

$$\frac{\partial x^{\mu}}{\partial x^{\mu'}}\frac{\partial x^{\nu'}}{\partial x^{\nu}}\frac{\partial}{\partial x^{\mu}}V^{\nu}+\frac{\partial x^{\mu}}{\partial x^{\mu'}}\frac{\partial x^{\nu'}}{\partial x^{\nu}}\Gamma^{\nu}_{\mu\lambda}V^{\lambda}$$
$$=\frac{\partial x^{\mu}}{\partial x^{\mu'}}\frac{\partial x^{\nu'}}{\partial x^{\nu}}\frac{\partial}{\partial x^{\mu}}V^{\nu}+\frac{\partial x^{\mu}}{\partial x^{\mu'}}V^{\nu}\frac{\partial^2 x^{\nu'}}{\partial x^{\mu}\partial x^{\nu}}+\Gamma^{\nu'}_{\mu'\lambda'}\frac{\partial x^{\lambda'}}{\partial x^{\lambda}}V^{\lambda}.$$

The first terms on both sides are identical and cancel, leaving

$$\frac{\partial x^{\mu}}{\partial x^{\mu'}}\frac{\partial x^{\nu'}}{\partial x^{\nu}}\Gamma^{\nu}_{\mu\lambda}V^{\lambda}.$$
$$=\frac{\partial x^{\mu}}{\partial x^{\mu'}}V^{\nu}\frac{\partial^2 x^{\nu'}}{\partial x^{\mu}\partial x^{\nu}}+\Gamma^{\nu'}_{\mu'\lambda'}\frac{\partial x^{\lambda'}}{\partial x^{\lambda}}V^{\lambda}.$$

The first term on the right can be rewritten by reindexing ν with λ. This is possible because ν is summed over and it does not matter what letter one uses for summed indices. Moreover, because V^{λ} is arbitrary it can be dropped leaving

$$\frac{\partial x^{\mu}}{\partial x^{\mu'}}\frac{\partial x^{\nu'}}{\partial x^{\nu}}\Gamma^{\nu}_{\mu\lambda}$$
$$=\frac{\partial x^{\mu}}{\partial x^{\mu'}}\frac{\partial^2 x^{\nu'}}{\partial x^{\mu}\partial x^{\lambda}}+\Gamma^{\nu'}_{\mu'\lambda'}\frac{\partial x^{\lambda'}}{\partial x^{\lambda}}.$$

Solving for $\Gamma^{\nu'}_{\mu'\lambda'}$, we finally arrive at the transformation law for the connection

$$\Gamma^{\nu'}_{\mu'\lambda'}=\frac{\partial x^{\lambda}}{\partial x^{\lambda'}}\frac{\partial x^{\mu}}{\partial x^{\mu'}}\frac{\partial x^{\nu'}}{\partial x^{\nu}}\Gamma^{\nu}_{\mu\lambda}-\frac{\partial x^{\lambda}}{\partial x^{\lambda'}}\frac{\partial x^{\mu}}{\partial x^{\mu'}}\frac{\partial^2 x^{\nu'}}{\partial x^{\mu}\partial x^{\lambda}} \tag{8.23}$$

where we have used

$$\frac{\partial x^{\lambda'}}{\partial x^{\lambda}}\frac{\partial x^{\lambda}}{\partial x^{\lambda'}}=1$$

to isolate the connection coefficient.

We should note that equation (8.23) does not provide sufficient information for us to derive a unique connection. It should be evident from this derivation that the partial derivative of a scalar should be the same as its covariant derivative,

$$D_\mu \phi = \partial_\mu \phi$$

where $\partial_\mu = \partial/\partial x^\mu$ is another useful abbreviation that save us a lot of writing. This equation tells us little about the coefficients but it becomes important when we write ϕ as the result of a scalar product $\phi = A_\lambda B^\lambda$. Then

$$(D_\mu A_\lambda)B^\lambda + A_\lambda(D_\mu B^\lambda) = (\partial_\mu A_\lambda)B^\lambda + A_\lambda(\partial_\mu B^\lambda).$$

Now let us insert (8.22) on the left-hand side,

$$\begin{aligned}(\partial_\mu A_\lambda + \tilde{\Gamma}^\lambda_{\mu\nu} A_\lambda)B^\lambda + A_\lambda(\partial_\mu B^\lambda + \Gamma^\nu_{\mu\lambda} B^\lambda)\\ = (\partial_\mu A_\lambda)B^\lambda + A_\lambda(\partial_\mu B^\lambda)\end{aligned}$$

where $\tilde{\Gamma}^\nu_{\mu\lambda}$ is the connection coefficient needed to construct the covariant derivative of a $(0,1)$ form. We do not know what this object is yet. Note that terms cancel on both sides of the last equation leaving this expression after rearranging the summed indices,

$$\tilde{\Gamma}^\lambda_{\mu\nu} A_\lambda B^\lambda + \Gamma^\lambda_{\mu\nu} A_\lambda B^\lambda = 0$$

but A_μ and B^μ are arbitrary, so we can drop them leaving us with

$$\tilde{\Gamma}^\nu_{\mu\lambda} = -\Gamma^\nu_{\mu\lambda}. \tag{8.24}$$

Using equations (8.22) and (8.24), one generalizes the covariant derivative of an arbitrary tensor as, e. g.,

$$\begin{aligned}D_\mu T^{\sigma\kappa}_{\lambda\nu\beta} = \partial_\mu T^{\sigma\kappa}_{\lambda\nu\beta} - \Gamma^\chi_{\mu\lambda} T^{\sigma\kappa}_{\chi\nu\beta} - \Gamma^\chi_{\mu\nu} T^{\sigma\kappa}_{\lambda\chi\beta} - \Gamma^\chi_{\mu\beta} T^{\sigma\kappa}_{\lambda\nu\chi}\\ + \Gamma^\sigma_{\mu\chi} T^{\chi\kappa}_{\lambda\nu\beta} + \Gamma^\kappa_{\mu\chi} T^{\sigma\chi}_{\lambda\nu\beta}.\end{aligned}$$

The pattern is now evident. A particularly useful covariant derivative formula is that of the metric tensor

$$D_\rho g_{\mu\nu} = \partial_\rho g_{\mu\nu} - \Gamma^\lambda_{\rho\mu} g_{\lambda\nu} - \Gamma^\lambda_{\rho\nu} g_{\mu\lambda}. \tag{8.25}$$

The Christoffel connection

Here, we arrive at a formula for a unique connection by requiring two properties for it

- that it is metric compatible

$$D_\rho g_{\mu\nu} = 0$$

- that it is "torsion-free," i. e., it is symmetric about the two low indices: $\Gamma^{\nu}_{\mu\lambda} = \Gamma^{\nu}_{\lambda\mu}$.

Writing equation (8.25) three times with permuted indices and setting each line to zero at the same time produces,

$$\begin{aligned}
\partial_\rho g_{\mu\nu} - \Gamma^{\lambda}_{\rho\mu} g_{\lambda\nu} - \Gamma^{\lambda}_{\rho\nu} g_{\mu\lambda} &= 0,\\
\partial_\mu g_{\nu\rho} - \Gamma^{\lambda}_{\mu\nu} g_{\lambda\rho} - \Gamma^{\lambda}_{\mu\rho} g_{\nu\lambda} &= 0,\\
\partial_\nu g_{\rho\mu} - \Gamma^{\lambda}_{\nu\rho} g_{\lambda\mu} - \Gamma^{\lambda}_{\nu\mu} g_{\rho\lambda} &= 0.
\end{aligned}$$

Subtracting the second and the third from the first yields

$$\begin{aligned}
0 = {} & \partial_\rho g_{\mu\nu} - \partial_\mu g_{\nu\rho} - \partial_\nu g_{\rho\mu}\\
& - \Gamma^{\lambda}_{\rho\nu} g_{\mu\lambda} - \Gamma^{\lambda}_{\rho\mu} g_{\nu\lambda} + \Gamma^{\lambda}_{\mu\nu} g_{\lambda\rho} + \Gamma^{\lambda}_{\mu\rho} g_{\nu\lambda} + \Gamma^{\lambda}_{\nu\rho} g_{\lambda\mu} + \Gamma^{\lambda}_{\nu\mu} g_{\rho\lambda}.
\end{aligned}$$

But using the symmetry properties of the metric tensor and the connection at the same time, we can cancel a number of terms from the last line. The reader should be able to verify that $\Gamma^{\lambda}_{\mu\nu} g_{\lambda\rho} = \Gamma^{\lambda}_{\nu\mu} g_{\rho\lambda}$ so that the third and sixth term are equal, $\Gamma^{\lambda}_{\rho\nu} g_{\mu\lambda} = \Gamma^{\lambda}_{\nu\rho} g_{\lambda\mu}$, so the first and fifth term cancel, and $\Gamma^{\lambda}_{\rho\mu} g_{\nu\lambda} = \Gamma^{\lambda}_{\mu\rho} g_{\nu\lambda}$ so the second and the fourth cancel leaving

$$0 = \partial_\rho g_{\mu\nu} - \partial_\mu g_{\nu\rho} - \partial_\nu g_{\rho\mu} + 2\Gamma^{\lambda}_{\mu\nu} g_{\lambda\rho},$$

which can be solved by moving symbol to the left and multiplying $g^{\lambda\rho}/2$ on both sides

$$\Gamma^{\lambda}_{\mu\nu} = \frac{1}{2} g^{\lambda\rho} (\partial_\mu g_{\nu\rho} + \partial_\nu g_{\rho\mu} - \partial_\rho g_{\mu\nu}). \tag{8.26}$$

This is one of the most important equations in Riemannian geometry, and according to most authors an essential part of the definition of a Riemannian manifold. The converse of the result we just derived, that is, given equation (8.26) as the definition of the Christoffel symbol $D_\rho g_{\mu\nu} = 0$ follows is known as Ricci's theorem.

Definition 38. A Riemann space (manifold) is a topology covered by an Atlas with coordinates that satisfy equation (8.8) in every chart, and a Christoffel connection.

The Laplace–Beltrami operator

The last piece of heavy lifting is to derive the covariant divergence of a vector using the Christoffel connection,

$$D_\mu V^\mu = \partial_\mu V^\mu + \Gamma^\mu_{\mu\lambda} V^\lambda,$$

a scalar quantity. Ricci's theorem states

$$\partial_\rho g_{\mu\nu} = \Gamma^\lambda_{\rho\mu} g_{\lambda\nu} + \Gamma^\lambda_{\rho\nu} g_{\mu\lambda}$$

multiplication and contraction by $g^{\mu\nu}$ leads to

$$g^{\mu\nu}\partial_\rho g_{\mu\nu} = \delta^\mu_\lambda \Gamma^\lambda_{\rho\mu} + \delta^\nu_\lambda \Gamma^\lambda_{\rho\nu},$$

which collapses the sums over λ to a single term,

$$g^{\mu\nu}\partial_\rho g_{\mu\nu} = \Gamma^\mu_{\rho\mu} + \Gamma^\nu_{\rho\nu}$$

but now switching the summed index to μ, using the symmetry of the Christoffel connection and solving gives

$$2\Gamma^\mu_{\mu\rho} = g^{\mu\nu}\partial_\rho g_{\mu\nu}. \tag{8.27}$$

From the theory of determinants reviewed in Chapter 1, we can simplify the last result by writing the inverse of the metric tensor in the following way:

$$g^{\mu\nu}\partial_\rho g_{\mu\nu} = g^{-1} M^{\mu\nu}\partial_\rho g_{\mu\nu} \tag{8.28}$$

where $M^{\mu\nu}$ is the cofactor of $g_{\mu\nu}$ and $g = \det(g_{\mu\nu})$. If we expand the determinant by row μ, then $g = g_{\mu\nu}M^{\mu\nu}$ by the cofactor expansion. From this, we can deduce that

$$M^{\mu\nu} = \frac{\partial}{\partial(g_{\mu\nu})} g \tag{8.29}$$

because $M^{\mu\nu}$ is a subdeterminant of the metric tensor that does not include $g_{\mu\nu}$. Inserting (8.29) into (8.28) and using (8.27) give

$$\begin{aligned}\Gamma^\mu_{\mu\rho} &= \frac{1}{2} g^{-1} \frac{\partial g}{\partial(g_{\mu\nu})} \partial_\rho g_{\mu\nu} = \frac{1}{2} g^{-1} \partial_\rho g \\ &= \partial_\rho \ln g^{1/2} = \frac{1}{\sqrt{g}} \partial_\rho \sqrt{g}.\end{aligned}$$

This last result allows us to write the divergence of a vector as

$$D_\mu V^\mu = \partial_\mu V^\mu + \frac{1}{\sqrt{g}} \partial_\lambda(\sqrt{g} V^\lambda). \tag{8.30}$$

Finally, we are ready to change from our Laplacian in $\mathbb{R}^n$ mapped with Cartesian coordinates

$$\nabla^2 = \frac{\partial^2}{\partial(x^{(1)})^2} + \frac{\partial^2}{\partial(x^{(2)})^2} + \cdots + \frac{\partial^2}{\partial(x^n)^2}$$

to the Laplace–Beltrami operator. We start by taking the gradient of a scalar function ψ, i. e., $\partial_\mu \psi$, we raise its index so it becomes $g^{\mu\lambda}\partial_\lambda \psi$ and the evaluate its divergence using equation (8.30),

$$\partial_\mu \partial^\mu \psi = g^{\mu\nu}\partial_\mu \partial_\mu \psi + \frac{1}{\sqrt{g}}(\partial_\lambda \sqrt{g} g^{\mu\lambda})\partial_\mu \psi. \tag{8.31}$$

This is the tool we have been looking for, a general recipe to find expressions for the kinetic energy operator in any coordinate system or any Riemannian manifold. The terms proportional to the first derivative,

$$\frac{1}{\sqrt{g}}(\partial_\lambda \sqrt{g} g^{\mu\lambda})$$

is the advection vector, a type of drift caused by the inherent curved or curvilinear nature of the coordinates.

It is high time for a familiar example. Consider the map that transforms the regular Euclidean space Cartesian coordinates into spherical polar coordinates,

$$\begin{aligned} x &= r\cos\phi\sin\theta \\ y &= r\sin\phi\sin\theta \\ z &= r\cos\theta. \end{aligned}$$

The Jacobian matrix becomes

$J =$

	x	y	z
r	$\cos\phi\sin\theta$	$\sin\phi\sin\theta$	$\cos\theta$
θ	$r\cos\phi\cos\theta$	$r\sin\phi\cos\theta$	$-r\sin\theta$
ϕ	$-r\sin\phi\sin\theta$	$r\cos\phi\sin\theta$	0

where J is a 3×3 matrix but we have labeled the columns with the dependent variables of the map and the rows with the independent variables of the map. A shorthand for the Jacobian that contains the column and row labeled as in the previous expression is for this example,

$$J = \frac{\partial(x, y, z)}{\partial(r, \theta, \phi)}.$$

The transformation in (8.18) can be written in matrix form as follows:

$$g_{\mu\nu} = JJ^T$$

where we have used the fact that in the regular Euclidean space the metric tensor is a three by three unit matrix, and the matrix J plays the role of a similarity transform. Some straightforward trigonometry and matrix multiplications should show

$g_{\mu\nu} =$

	r	θ	ϕ
r	1	0	0
θ	0	r^2	0
ϕ	0	0	$r^2 \sin^2\theta$

where the columns and rows label the degrees of freedom (the Riemannian parameters) $r, \theta, \phi = u^1, u^2, u^3$. The inverse and the determinant are respectively

$$g^{\mu\nu} = \begin{pmatrix} 1 & 0 & 0 \\ 0 & \frac{1}{r^2} & 0 \\ 0 & 0 & \frac{1}{r^2 \sin^2\theta} \end{pmatrix} \qquad g = r^4 \sin^2\theta.$$

Then the first term of (8.31) is

$$g^{\mu\nu}\partial_\mu\partial_\mu = \frac{\partial^2}{\partial r^2} + \frac{1}{r^2}\frac{\partial^2}{\partial\theta^2} + \frac{1}{r^2\sin^2\theta}\frac{\partial^2}{\partial\phi^2}$$

the second term requires a few simple calculus steps

$$\begin{aligned}\frac{1}{\sqrt{g}}(\partial_\lambda\sqrt{g}g^{\mu\lambda})\partial_\mu &= \frac{1}{r^2\sin\theta}\left[\left(\frac{\partial}{\partial r}r^2\sin\theta\right)\frac{\partial}{\partial r}\right. \\ &\quad \left.+\left(\frac{\partial}{\partial\theta}\sin\theta\right)\frac{\partial}{\partial\theta}+\left(\frac{\partial}{\partial\phi}\frac{1}{\sin\theta}\right)\frac{\partial}{\partial\phi}\right] \\ &= \frac{1}{r^2\sin\theta}\left(2r\sin\theta\frac{\partial}{\partial r}+\cos\theta\frac{\partial}{\partial\theta}\right).\end{aligned}$$

Putting all the pieces together, we end up with

$$\begin{aligned}\partial_\mu\partial^\mu &= \frac{\partial^2}{\partial r^2} + \frac{1}{r^2}\frac{\partial^2}{\partial\theta^2} + \frac{1}{r^2\sin^2\theta}\frac{\partial^2}{\partial\phi^2} \\ &\quad + \frac{2}{r}\frac{\partial}{\partial r} + \frac{\cot\theta}{r^2}\frac{\partial}{\partial\theta}.\end{aligned}$$

Those students who have derived this expression by the brute force chain rule method should have undoubtedly gained some appreciation for the power of differential geometry and the relatively small amount of tensor calculus needed to derive (8.31).

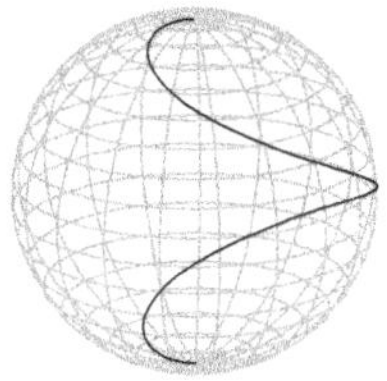

Figure 8.1: A path from the north to the south pole on a unit sphere. The top half of the curve is on the opposite side of the viewer.

8.2 Exercises

1. The path on the surface of a unit sphere drawn in Figure 8.1 is parameterized with t,

$$\theta(t) = \pi t, \quad \phi(t) = 2\pi t$$

where θ is the polar and ϕ the azimuthal angles. Use the metric tensor in equation (8.7) to compute the length of the path L defined in equation (8.4).

2. Use the polar coordinates chart over the unit sphere to derive the Jacobian

$$J^{\mu}_{\mu'} = \frac{\partial(x, y, z)}{\partial(\theta, \phi)}$$

then derive equation (8.7) two ways:
(a) by using equation (8.18);
(b) by using $g_{\mu\nu} = JJ^T$.

3. Derive expressions for the Christoffel connection coefficients for the unit sphere mapped with angles θ and ϕ, then show that

$$\Gamma^{\mu}_{\mu\rho} = \frac{1}{\sqrt{g}}\partial_{\rho}\sqrt{g}.$$

4. Derive the Laplace–Beltrami operator on the unit sphere mapped with angles θ and ϕ.

5. Consider vectors a and b in the regular three-dimensional Euclidean space. Let the unit vectors along the x, y and z directions be e_1, e_2, e_3, respectively. The three unit vectors play the same role as the familiar **i**, **j**, and **k**, but the new notation can be more easily generalized. Then a and b can be decomposed along the unit vectors,

$$a = a^{(1)}e_1 + a^{(2)}e_2 + a^{(3)}e_3$$

where a^j are the contravariant components.
(a) Write a similar decomposition for the vector b.

(b) Use the fact that the metric tensor in three-dimensional Euclidean space mapped by Cartesian coordinates is a three-by-three unit matrix to write down the expression for the dot product $a \cdot b$. Begin by expanding the expression,

$$a \cdot b = a^{\nu} b_{\nu}$$

where the vector b was turned into a $(0, 1)$ form using the metric tensor $b_{\nu} = g_{\mu\nu} b^{\mu}$. Show that the result is the same as the regular dot product.

(c) Basis vectors like e_{μ} can also be converted from $(0, 1)$ form to $(1, 0)$ tensors. Write b decomposed along the $(1, 0)$ unit vectors.

(d) Change from Cartesian to spherical polar coordinates and write the vectors a and b along the new r, θ, ϕ directions $e_{1'}, e_{2'}, e_{3'}$ in terms of the original Cartesian components $a^{(1)}, a^{(2)}, a^{(3)}$, etc.

(e) Use the metric tensor $g_{\mu'\nu'}$ that comes from part (d) to find an expression for the dot product $a^{\nu'} b_{\nu'}$ in the new coordinate system.

6. Prove in general that a scalar quantity $A_{\mu} B^{\mu}$ is invariant under a coordinate change.
7. Given any generic $(0, 2)$ form $T_{\mu\nu}$ show that

$$S_{\mu\nu} = \frac{1}{2}(T_{\mu\nu} + T_{\nu\mu})$$

is symmetric $S_{\mu\nu} = S_{\nu\mu}$, and

$$A_{\mu\nu} = \frac{1}{2}(T_{\mu\nu} - T_{\nu\mu})$$

is antisymmetric $A_{\mu\nu} = -A_{\nu\mu}$.

8. Argue that if $S^{\mu\nu} = S^{\nu\mu}$ is symmetric and $A_{\mu\nu} = -A_{\nu\mu}$ is antisymmetric, the scalar quantity

$$S^{\mu\nu} A_{\mu\nu}$$

must be equal to zero.

9. A surface defined by the constraint

$$\frac{x^2}{a^2} + \frac{y^2}{b^2} + \frac{z^2}{c^2} = 1$$

is called an ellipsoid. This surface is important when rotations of nonlinear rigid bodies are considered. A graph of an ellipsoid with $a^2 = 1.5$, $b^2 = 3.0$, and $c = 0.7$ is shown in Figure 8.2.

(a) Construct a chart for these Gaussian surfaces using a polar and an azimuthal angle like the sphere.

(b) Find an expression for the path on the ellipsoid in terms of a, b, and c if the two angles are parameterized the same ways as in exercise 1.

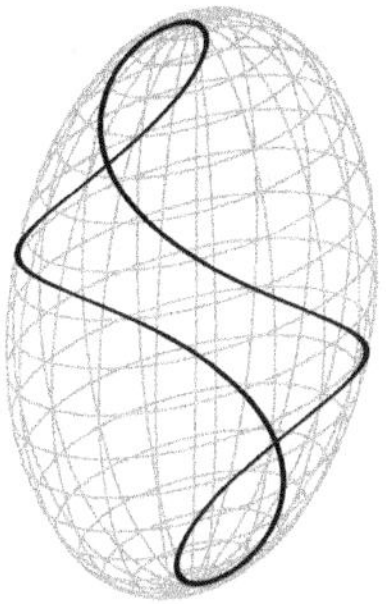

Figure 8.2: A path on an ellipsoid. Note that the path does not intersect with itself. The portion of the path that decrease from left to right is on the opposite side.

(c) Develop the metric tensor and the Christoffel connection coefficients.
(d) Work through the expression for the covariant derivative of $g_{\mu\nu}$ using the coefficients in part (c) and show that it vanishes identically.

10. There is a deep connection between the parameter spaces of Lie groups introduced in Chapter 4, their Lie algebra in Chapter 5, and differential geometry. This set of exercises is designed to provide an example.
(a) The three generators of the SO(3, R) group are the 3×3 matrices J_x, J_y, and J_z as defined in Chapter 4. It is demonstrated in Chapter 4 that

$$e^{\alpha_y J_y} = \begin{pmatrix} \cos\alpha_y & 0 & -\sin\alpha_y \\ 0 & 1 & 0 \\ \sin\alpha_y & 0 & \cos\alpha_y \end{pmatrix}.$$

Show that

$$e^{\alpha_z J_z} = \begin{pmatrix} \cos\alpha_z & \sin\alpha_z & 0 \\ -\sin\alpha_x & \cos\alpha_z & 0 \\ 0 & 0 & 1 \end{pmatrix}.$$

(b) The parameters α_x, α_z are angles and can be used to represent members of the SO(3, R) group in a number of ways. Use the result in part (a) to build the following rotation matrix:

$$R(\theta, \phi) = e^{\theta J_y} e^{\phi J_z}$$

and show that you get

$$R(\theta, \phi) = \begin{pmatrix} \cos\phi\cos\theta & \sin\phi\cos\theta & -\sin\theta \\ -\sin\phi & \cos\phi & 0 \\ \cos\phi\sin\theta & \sin\phi\sin\theta & \cos\theta \end{pmatrix}.$$

(c) Show that $R^T R = g_{\mu\nu}$ where $g_{\mu\nu}$ is the metric tensor in Euclidean three-dimensional space, a 3×3 unit matrix.

(d) An arbitrary vector of size r from the origin to point p with coordinates x, y, z is converted to a vector with components x', y', z' according to

$$\begin{pmatrix} x' \\ y' \\ z' \end{pmatrix} = R \begin{pmatrix} x \\ y \\ z \end{pmatrix}$$

and conversely we can invert the relationship to read,

$$\begin{pmatrix} x \\ y \\ z \end{pmatrix} = R^T \begin{pmatrix} x' \\ y' \\ z' \end{pmatrix} = R^T R \begin{pmatrix} x \\ y \\ z \end{pmatrix},$$

where we substituted the primed vector in the last equation with the right-hand side of the equation above it. Multiply the extreme left and the extreme right of this expression by the row vector $(x\ y\ z)$. Show that you end up with the equation that defines the sphere of radius r. This means that preserving the metric of the space the group operates on is equivalent to preserving the norm of vectors in that space.

(e) Show that

$$e^{-\phi J_z} e^{-\theta J_y} \begin{pmatrix} 0 \\ 0 \\ 1 \end{pmatrix}$$

produces the parametrization for the unit sphere introduced earlier.

9 Stereographic projections for the n-sphere

The union of relativity and quantum mechanics, needed for the description of phenomena involving simultaneously large velocities and small scales, turns out to be very difficult.

R. Shankar

In Chapter 8, we introduced the unit 2-sphere, as the set of all points in regular Euclidean space that satisfy the equation $x^2 + y^2 + z^2 - 1 = 0$, and we parameterized the surface using the polar (θ) and the azimuthal (ϕ) angles. Let us abbreviate the polar-azimuthal chart with $\Theta : \theta, \phi \to x, y, z$. In Chapter 8, we did not dive into some of the subtleties of this particular chart. In fact, most engineering and mathematics books that introduce polar coordinate transformations neglect to include a robust discussion of these because most of the issues are handled by the periodic boundary conditions that are imposed on the solutions of the differential equations one tries to solve. The fact is, however, that Θ cannot cover all the points in a sphere. Let us begin with the north pole p_N, labeled N in Figure 9.1. The point p_N is at coordinates $(0, 0, 1)$, but Θ cannot map this point. The reason is that at p_N, the polar angle is zero, but all values of ϕ can map that point. However, a chart must always be one to one from its domain to its range. Similar reasoning apply to the south pole p_S at $(0, 0, -1)$. Now imagine moving the polar angle by an infinitesimal amount from zero in the positive direction $\delta\theta$, and trace a circle using the azimuthal angle as it scans from 0 to 2π. But note that $\phi \in (0, 2\pi)$, meaning the set of points where ϕ is either equal to 0 or 2π cannot be in Θ either for the same reason, namely the chart at those points is no longer one to one. Therefore, the domain of ϕ is the open set (at both ends), not the closed set from zero to 2π. Moreover, note that the range in the polar angle is $\theta \in (0, \pi)$, also open, and does not extend all the way to 2π, since a scan from zero to π, tracing paths as we did in the immediate vicinity of the north pole and incrementing θ infinitesimally after a circular path ends, covers the entire surface, and extending θ to the remaining two quadrants would map the surface twice.

In this chapter, we introduce a different parametrization of the unit sphere that makes use of the stereographic projections of all points on a sphere. Their construction, represented in Figure 9.1, require a point on the surface at the origin of a segment that goes through the sphere (dotted lines), emerges at point P on the surface, and lands at point A on the $z = -1$ plane at coordinates u, v, Cartesian coordinates in the projection plane. Stereographic projections are used in crystallography (though they are typically defined with $z = 0$ as the projection plane) to represent a three-dimensional pattern of planes in a crystal on a two-dimensional plane, and to simplify the description of inversion axes of symmetry. Geographers, too, make use of the stereographic projection when charting the surface of the Earth. This is the origin of the technical term "atlas." We introduced stereographic projections in molecular physics [57, 59] originally to simplify the representation of a certain class of particularly efficient *Feynman path integrals*. However, we later discovered that the diffusion equation on the sphere and the ellipsoid of

https://doi.org/10.1515/9783111610207-009

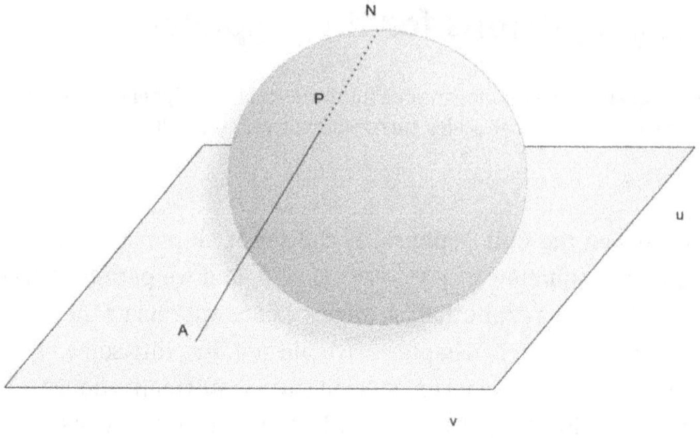

Figure 9.1: The north map of a 2-sphere.

inertia of a spherical rigid rotor (a space nearly isomorphic to a unit 3-sphere) are simplified by their use as well [60, 67]. Additionally, there is only one point that cannot be charted using the stereographic projections, p_N. Fortunately, there is a polar opposite chart projecting from the south pole that covers it. The greater symmetry of the stereographic projection charts allows us to cover every point on the surface of the sphere without using boundary conditions by seamlessly switching between the two. In practical applications, we simply switch from the north to the south chart whenever one or more projections become sufficiently large. To implement all of these geometric strategies, we need to derive the various charts, metric tensors, and advection vectors on each, and find relationships among the stereographic projections of the north chart and the corresponding ones from the south chart. This chapter is devoted to these tasks.

9.1 The charts for $\mathbb{S}^n$ spaces

An atlas $\mathscr{A}$ of charts is a set of invertible 1 to 1 maps $\Phi_\alpha : \mathcal{U}_\alpha \to \mathbb{R}^n$; $\Phi_\alpha^{-1} : \mathbb{R}^n \to \mathcal{U}_\alpha$ that can be smoothly sewn together and cover the entire manifold. The manifold $\mathcal{M}$ is the disjoint union of the domains that define its topology,

$$\mathcal{M} = \bigsqcup_{\alpha \in \mathbb{N}} \mathcal{U}_\alpha .$$

The sewing procedure of two maps Φ_α and Φ_β with adjacent domains $\mathcal{U}_\alpha$ and $\mathcal{U}_\beta$ requires the existence of an overlap between the subdomain, $\mathcal{U}_\alpha \cap \mathcal{U}_\beta \neq \emptyset$. A point $p \in \mathcal{U}_\alpha$ has a coordinate chart $\phi_\alpha^1(p), \dots, \phi_\alpha^n(p)$. If $p \in \mathcal{U}_\alpha \cap \mathcal{U}_\beta$, then p also has a coordinate chart $\phi_\beta^1(p), \dots, \phi_\beta^n(p)$. Then $\Phi_\alpha \circ \Phi_\beta^{-1} : \mathbb{R}^n \to \mathbb{R}^n$ takes the image of p from $\mathbb{R}^n$ projects it onto $\mathcal{U}_\beta$ and then from its image in $\mathcal{U}_\alpha$ projects is back to a copy of $\mathbb{R}^n$. If we include an open set of points near p, then we can say that $\Phi_\alpha \circ \Phi_\beta^{-1} : \mathbb{R}^n \to \mathbb{R}^n$ maps an open set in $\mathbb{R}^n$

back to $\mathbb{R}^n$ and, therefore, it is differentiable and integrable in the open set of $\mathcal{M}$. If the union of all open sets is a cover for $\mathcal{M}$, then the union of all charts is called an atlas.

The north chart

Consider the $\mathbb{S}^n$ manifold defined by the set of all points in $\mathbb{R}^{n+1}$ that satisfy

$$(x^1)^2 + (x^2)^2 + \cdots + (x^{n+1})^2 = 1 \tag{9.1}$$

we define n stereographic projections ϕ^μ, $\mu = 1, 2, \ldots, n$ using the north chart Φ_N,

$$\phi^\mu = 2\frac{x^\mu}{1 - x^{n+1}} \quad : \mathbb{R}^{n+1} \to \mathbb{S}^n_N. \tag{9.2}$$

However, note that the "north pole" point $p_N \in \mathbb{R}^{n+1} = (x^1, x^2, \ldots, x^{n+1}) = (0, 0, \ldots, 1)$ cannot be mapped into $\mathbb{S}^n_N$ because the denominator in equation (9.2) tends to zero. If we remove the north pole point on the n-sphere, ϕ_N is invertible: Setting

$$\sigma^2 = \sum_{k=1}^{n} (\phi^k)^2 \tag{9.3}$$

then, from equations (9.1) and (9.2) it follows that

$$1 = \left[\frac{1}{2}(1 - \zeta)\phi^{(1)}\right]^2 + \left[\frac{1}{2}(1 - \zeta)\phi^{(2)}\right]^2 + \cdots + \left[\frac{1}{2}(1 - \zeta)\phi^n\right]^2 + (\zeta)^2$$

where we have used $\zeta = x^{n+1}$. Using the definition of the quantity σ^2, we simplify the equation of constraint as follows:

$$1 = \frac{1}{4}(1 - \zeta)^2\sigma^2 + (\zeta)^2.$$

This is quadratic equation in ζ. Note that $\zeta = 1$ is a root, but this has to be rejected since it corresponds to the point p_N. The second root is

$$\zeta = \frac{\sigma^2 - 4}{\sigma^2 + 4}. \tag{9.4}$$

This expression is inserted back into equation (9.2),

$$\left(1 - \frac{\sigma^2 - 4}{\sigma^2 + 4}\right)\frac{1}{2}\phi^\mu = x^\mu,$$

and with some symbolic manipulation, we can simplify this to the main result:

$$\frac{4\phi^{\mu}}{\sigma^2+4} = x^{\mu} \quad n = 1, \ldots, n. \tag{9.5}$$

Equations (9.4) and (9.5) define $\Phi_N^{-1} : \mathbb{S}^n \rightarrow \mathbb{R}^{n+1}$, the inverse of Φ_N. The inverse map is useful to find equations for the Jacobian and the metric tensor in the projected space $\mathbb{S}_N^n = \mathbb{S}^n \backslash p_N$, where the last symbol is the domain of the north chart, i. e., the entire surface of the sphere with the north pole subtracted from it.

The Jacobian and the metric tensor of the north chart

We now follow the process outlined in Chapter 8 and derive the important properties of the sphere under the north chart:

$$\frac{\partial x^{\mu'}}{\partial \phi^{\nu}} = \frac{4}{\sigma^2+4}\delta_{\nu}^{\mu'} - \frac{8\phi^{\mu}\phi^{\nu}}{(\sigma^2+4)^2}, \tag{9.6}$$

$$\frac{\partial \zeta}{\partial \phi^{\nu}} = \frac{2\phi^{\nu}}{\sigma^2+4} - \frac{2(\sigma^2-4)\phi^{\nu}}{(\sigma^2+4)^2}. \tag{9.7}$$

At point p_N, the derivatives are not defined; therefore, neither is the metric tensor. This object has a simple general form

$$g_{\mu\nu} = \frac{\partial x^{\mu'}}{\partial \phi^{\mu}}\frac{\partial x^{\nu'}}{\partial \phi^{\nu}}\delta_{\mu'\nu'} = \frac{16}{(\sigma^2+4)^2}\delta_{\mu\nu} \quad \forall p \in \mathbb{S}_N^n, \tag{9.8}$$

i. e., the metric tensor is a unit matrix multiplied by the scalar quantity $16/(\sigma^2+4)^2\ \forall n$. This needs to be shown of course. Let us first investigate the $n = 1$ case, the unit radius ring:

$$x^2 + y^2 = 1,$$

where, for this case, $\zeta = y$. There is only one projection,

$$\phi = 2\frac{x}{1-y} \qquad : \mathbb{R}^2 \rightarrow \mathbb{R}$$

and the inverse map is

$$x = \frac{4\phi}{(\phi)^2+4}, \quad y = \frac{(\phi)^2-4}{(\phi)^2+4} \qquad : \mathbb{R} \rightarrow \mathbb{R}^2$$

the Jacobian is a 2×1 matrix,

$$\begin{pmatrix} \frac{\partial x}{\partial \phi} \\ \frac{\partial y}{\partial \phi} \end{pmatrix} = \begin{pmatrix} \frac{4}{(\phi)^2+4} - \frac{8(\phi)^2}{((\phi)^2+4)^2} \\ \frac{2\phi}{(\phi)^2+4} - \frac{2\phi((\phi)^2-4)}{((\phi)^2+4)^2} \end{pmatrix}.$$

Then equation (9.8) is a 1×1 matrix,

$$g_{11} = \begin{pmatrix} \frac{\partial x}{\partial \phi} & \frac{\partial y}{\partial \phi} \end{pmatrix} \begin{pmatrix} \frac{\partial x}{\partial \phi} \\ \frac{\partial y}{\partial \phi} \end{pmatrix}$$

$$g_{11} = \left(\frac{4}{(\phi)^2 + 4} - \frac{8(\phi)^2}{((\phi)^2 + 4)^2} \right)^2 + \left(\frac{2\phi}{(\phi)^2 + 4} - \frac{2\phi((\phi)^2 - 4)}{((\phi)^2 + 4)^2} \right)^2.$$

These equations are messy, but one can recognize a pattern after some manipulations:

$$g_{11} = \frac{16}{((\phi)^2 + 4)^2} + \frac{64(\phi)^4}{((\phi)^2 + 4)^4} - \frac{64(\phi)^2}{((\phi)^2 + 4)^3} + \frac{4(\phi)^2}{((\phi)^2 + 4)^2} + \frac{4(\phi)^2((\phi)^2 - 4)^2}{((\phi)^2 + 4)^4} - \frac{8(\phi)^2((\phi)^2 - 4)}{((\phi)^2 + 4)^3}.$$

The first term on the right is the result we want to show, which means that the remaining five terms should cancel. All five terms have a common factor,

$$g_{11} = \frac{16}{((\phi)^2 + 4)^2} + \frac{4(\phi)^2}{((\phi)^2 + 4)^2} \left[\frac{16(\phi)^2}{((\phi)^2 + 4)^2} - \frac{16}{((\phi)^2 + 4)} + 1 + \frac{((\phi)^2 - 4)^2}{((\phi)^2 + 4)^2} - \frac{2((\phi)^2 - 4)}{((\phi)^2 + 4)} \right].$$

Next, we put everything under a common denominator,

$$g_{11} = \frac{16}{((\phi)^2 + 4)^2} + \frac{4(\phi)^2}{((\phi)^2 + 4)^2} \times \frac{16(\phi)^2 - 16((\phi)^2 + 4) + ((\phi)^2 + 4)^2 + ((\phi)^2 - 4)^2 - 2((\phi)^2 - 4)((\phi)^2 + 4)}{((\phi)^2 + 4)^2}$$

canceling two terms and expanding the squares on top

$$g_{11} = \frac{16}{((\phi)^2 + 4)^2} + \frac{4(\phi)^2}{((\phi)^2 + 4)^2} \times \frac{-64 + (\phi)^4 + 8(\phi)^2 + 16 + (\phi)^4 - 8(\phi)^2 + 16 - 2(\phi)^4 + 32}{((\phi)^2 + 4)^2}$$

the top of the fraction cancels exactly, leaving only the first term:

$$g_{11} = \frac{16}{((\phi)^2 + 4)^2}.$$

Therefore, the theorem is true for the $n = 1$ case. We are now ready to prove some general results.

Theorem 27. *The diagonal elements of the metric tensor for $\mathbb{S}^n$ in the north chart are all equal to* $16/(\sigma^2 + 4)^2$.

Proof. The entries of the metric tensor are sums of $n + 1$ terms,

$$g_{\mu\mu} = \left(\frac{\partial \zeta}{\partial \phi^\mu}\right)^2 + \sum_{\mu'=1}^{n} \left(\frac{\partial x^{\mu'}}{\partial \phi^\mu}\right)^2. \tag{9.9}$$

Now we use equations (9.6) to rewrite the sum on the right-hand side as follows:

$$\begin{aligned}
\sum_{\mu=1}^{n} \left(\frac{\partial x^{\mu'}}{\partial \phi^\mu}\right)^2 &= \sum_{\mu'=1}^{n} \left(\frac{4}{\sigma^2+4}\delta_\mu^{\mu'} - \frac{8\phi^{\mu'}\phi^\mu}{(\sigma^2+4)^2}\right)^2 \\
\sum_{\mu=1}^{n} \left(\frac{\partial x^{\mu'}}{\partial \phi^\mu}\right)^2 &= \left(\frac{4}{\sigma^2+4} - \frac{8(\phi^\mu)^2}{(\sigma^2+4)^2}\right)^2 \\
&\quad + \sum_{\mu'=1,\mu'\neq\mu}^{n} \frac{64(\phi^{\mu'})^2(\phi^\mu)^2}{(\sigma^2+4)^4},
\end{aligned}$$

where we have used the sifting property of the Kronecker delta in the last step. The term $\mu' = \mu$ is missing in the last sum, but we are going to recombine it as soon as the first square is expanded,

$$\begin{aligned}
\sum_{\mu=1}^{n} \left(\frac{\partial x^{\mu'}}{\partial \phi^\mu}\right)^2 &= \frac{16}{(\sigma^2+4)^2} - \frac{64(\phi^\mu)^2}{(\sigma^2+4)^3} + \frac{64(\phi^\mu)^2(\phi^\mu)^2}{(\sigma^2+4)^4} \\
&\quad + \sum_{\mu'=1,\mu'\neq\mu}^{n} \frac{64(\phi^{\mu'})^2(\phi^\mu)^2}{(\sigma^2+4)^4} \\
\sum_{\mu=1}^{n} \left(\frac{\partial x^{\mu'}}{\partial \phi^\mu}\right)^2 &= \frac{16}{(\sigma^2+4)^2} - \frac{64(\phi^\mu)^2}{(\sigma^2+4)^3} + \frac{64\sigma^2(\phi^\mu)^2}{(\sigma^2+4)^4}.
\end{aligned}$$

Inserting this result back into equation (9.9), using equation (9.7) yields

$$\begin{aligned}
g_{\mu\mu} &= \frac{16}{(\sigma^2+4)^2} - \frac{64(\phi^\mu)^2}{(\sigma^2+4)^3} + \frac{64\sigma^2(\phi^\mu)^2}{(\sigma^2+4)^4} \\
&\quad + \left(\frac{2\phi^\mu}{\sigma^2+4} - \frac{2(\sigma^2-4)\phi^\mu}{(\sigma^2+4)^2}\right)^2, \\
g_{\mu\mu} &= \frac{16}{(\sigma^2+4)^2} - \frac{64(\phi^\mu)^2}{(\sigma^2+4)^3} + \frac{64\sigma^2(\phi^\mu)^2}{(\sigma^2+4)^4} \\
&\quad + \frac{4(\phi^\mu)^2}{(\sigma^2+4)^2} - \frac{8(\phi^\mu)^2(\sigma^2-4)}{(\sigma^2+4)^3} + \frac{4(\sigma^2-4)^2(\phi^\mu)^2}{(\sigma^2+4)^4}.
\end{aligned}$$

Following the same steps as for the $n = 1$ case, we look for the five terms on the right-hand side to cancel,

$$g_{\mu\mu} = \frac{16}{(\sigma^2+4)^2} + \frac{4(\phi^\mu)^2}{(\sigma^2+4)^2} \times \left[-\frac{16}{(\sigma^2+4)} + \frac{16\sigma^2}{(\sigma^2+4)^2} + 1 - \frac{2(\sigma^2-4)}{(\sigma^2+4)} + \frac{(\sigma^2-4)^2}{(\sigma^2+4)^2}\right]$$

under common denominator

$$g_{\mu\mu} = \frac{16}{(\sigma^2+4)^2} + \frac{4(\phi^\mu)^2}{(\sigma^2+4)^2} \times \frac{-16(\sigma^2+4) + 16\sigma^2 + (\sigma^2+4)^2 - 2(\sigma^2-4)(\sigma^2+4) + (\sigma^2-4)^2}{(\sigma^2+4)^2},$$

and expanding the squares yields a familiar result,

$$g_{\mu\mu} = \frac{16}{(\sigma^2+4)^2} + \frac{4(\phi^\mu)^2}{(\sigma^2+4)^2} \times \frac{-16\sigma^2 - 64 + 16\sigma^2 + \sigma^4 + 8\sigma^2 + 16 - 2\sigma^4 + 32 + \sigma^4 - 8\sigma^2 + 16}{(\sigma^2+4)^2}.$$

Therefore, just as in the $n = 1$ case it all cancels leaving the result we wanted to show. □

This is not sufficient to prove that equation (9.8) holds. One still needs to show that the off diagonal terms vanish for all n.

Theorem 28. *The off diagonal elements of the metric tensor vanish,*

$$g_{\mu\nu} = 0 \quad \textit{iff}\, \mu \neq \nu.$$

Proof. Starting with

$$g_{\mu\nu} = \left\{\sum_{\mu'=1}^{n}\left(\frac{\partial x^{\mu'}}{\partial\phi^\mu}\right)\left(\frac{\partial x^{\mu'}}{\partial\phi^\nu}\right)\right\} + \left(\frac{\partial\zeta}{\partial\phi^\mu}\right)\left(\frac{\partial\zeta}{\partial\phi^\nu}\right)$$

and using the general expression for the derivatives gives

$$g_{\mu\nu} = \left\{\sum_{\mu'=1}^{n}\left(\frac{4}{\sigma^2+4}\delta^{\mu'}_{\mu} - \frac{8\phi^{\mu'}\phi^\mu}{(\sigma^2+4)^2}\right)\left(\frac{4}{\sigma^2+4}\delta^{\mu'}_{\nu} - \frac{8\phi^{\mu'}\phi^\nu}{(\sigma^2+4)^2}\right)\right\} + \left(\frac{2\phi^\mu}{\sigma^2+4} - \frac{2(\sigma^2-4)\phi^\mu}{(\sigma^2+4)^2}\right)\left(\frac{2\phi^\nu}{\sigma^2+4} - \frac{2(\sigma^2-4)\phi^\nu}{(\sigma^2+4)^2}\right).$$

The sum can be written as

$$\begin{aligned}
\sum_{\mu'=1}^{n}&\left(\frac{4}{\sigma^2+4}\delta_{\mu}^{\mu'}-\frac{8\phi^{\mu'}\phi^{\mu}}{(\sigma^2+4)^2}\right)\left(\frac{4}{\sigma^2+4}\delta_{\nu}^{\mu'}-\frac{8\phi^{\mu'}\phi^{\nu}}{(\sigma^2+4)^2}\right)\\
&=\left(\frac{4}{\sigma^2+4}-\frac{8(\phi^{\mu})^2}{(\sigma^2+4)^2}\right)\left(-\frac{8\phi^{\mu}\phi^{\nu}}{(\sigma^2+4)^2}\right)\\
&+\left(\frac{4}{\sigma^2+4}-\frac{8(\phi^{\nu})^2}{(\sigma^2+4)^2}\right)\left(-\frac{8\phi^{\mu}\phi^{\nu}}{(\sigma^2+4)^2}\right)\\
&+\sum_{\mu'=1,\mu'\neq\mu,\mu'\neq\nu}^{n}\frac{64(\phi^{\mu'})^2(\phi^{\mu})^2}{(\sigma^2+4)^4}.
\end{aligned}$$

There are two terms missing in the sum, when $\mu'=\mu$ and when $\mu'=\nu$. These terms are recombined after the next steps,

$$\begin{aligned}
\sum_{\mu'=1}^{n}&\left(\frac{4}{\sigma^2+4}\delta_{\mu}^{\mu'}-\frac{8\phi^{\mu'}\phi^{\mu}}{(\sigma^2+4)^2}\right)\left(\frac{4}{\sigma^2+4}\delta_{\nu}^{\mu'}-\frac{8\phi^{\mu'}\phi^{\nu}}{(\sigma^2+4)^2}\right)\\
&=-\frac{32\phi^{\mu}\phi^{\nu}}{(\sigma^2+4)^3}+\frac{64(\phi^{\mu})^2\phi^{\mu}\phi^{\nu}}{(\sigma^2+4)^4}-\frac{32\phi^{\mu}\phi^{\nu}}{(\sigma^2+4)^3}+\frac{64(\phi^{\nu})^2\phi^{\mu}\phi^{\nu}}{(\sigma^2+4)^4}\\
&+\sum_{\mu'=1,\mu'\neq\mu,\mu'\neq\nu}^{n}\frac{64(\phi^{\mu'})^2\phi^{\mu}\phi^{\nu}}{(\sigma^2+4)^4}\\
\sum_{\mu'=1}^{n}&\left(\frac{4}{\sigma^2+4}\delta_{\mu}^{\mu'}-\frac{8\phi^{\mu'}\phi^{\mu}}{(\sigma^2+4)^2}\right)\left(\frac{4}{\sigma^2+4}\delta_{\nu}^{\mu'}-\frac{8\phi^{\mu'}\phi^{\nu}}{(\sigma^2+4)^2}\right)\\
&=-\frac{64\phi^{\mu}\phi^{\nu}}{(\sigma^2+4)^3}+\frac{64\sigma^2\phi^{\mu}\phi^{\nu}}{(\sigma^2+4)^4}.
\end{aligned}$$

Recombining gives

$$\begin{aligned}
g_{\mu\nu}&=-\frac{32\phi^{\mu}\phi^{\nu}}{(\sigma^2+4)^3}-\frac{32\phi^{\mu}\phi^{\nu}}{(\sigma^2+4)^3}+\frac{64\sigma^2(\phi^{\mu})^2}{(\sigma^2+4)^4}\\
&+\left(\frac{2\phi^{\mu}}{\sigma^2+4}-\frac{2(\sigma^2-4)\phi^{\mu}}{(\sigma^2+4)^2}\right)\left(\frac{2\phi^{\nu}}{\sigma^2+4}-\frac{2(\sigma^2-4)\phi^{\nu}}{(\sigma^2+4)^2}\right)\\
g_{\mu\nu}&=-\frac{64\phi^{\mu}\phi^{\nu}}{(\sigma^2+4)^3}+\frac{64\sigma^2\phi^{\mu}\phi^{\nu}}{(\sigma^2+4)^4}\\
&+\frac{4\phi^{\mu}\phi^{\nu}}{(\sigma^2+4)^2}-\frac{8(\sigma^2-4)\phi^{\mu}\phi^{\nu}}{(\sigma^2+4)^3}+\frac{4(\sigma^2-4)^2\phi^{\mu}\phi^{\nu}}{(\sigma^2+4)^4}.
\end{aligned}$$

Factoring a common term,

$$g_{\mu\nu}=\frac{4\phi^{\mu}\phi^{\nu}}{(\sigma^2+4)^2}\left[1-\frac{16}{(\sigma^2+4)}+\frac{16\sigma^2}{(\sigma^2+4)^2}-\frac{2(\sigma^2-4)}{(\sigma^2+4)}+\frac{(\sigma^2-4)^2}{(\sigma^2+4)^2}\right]$$

putting all fractions inside the bracket under common denominator,

$$g_{\mu\nu} = \frac{4\phi^\mu\phi^\nu}{(\sigma^2+4)^2} \times \frac{(\sigma^2+4)^2 - 16(\sigma^2+4) - 16\sigma^2 - 2(\sigma^2-4)(\sigma^2+4) + (\sigma^2-4)^2}{(\sigma^2+4)^2}$$

and expanding the squares finally yields the result:

$$g_{\mu\nu} = \frac{4\phi^\mu\phi^\nu}{(\sigma^2+4)^2} \times \frac{\sigma^4 + 8\sigma^2 + 16 - 16\sigma^2 - 64 + 16\sigma^2 - 2\sigma^4 + 32 + \sigma^4 - 8\sigma^2 + 16}{(\sigma^2+4)^2} = 0.$$

□

We now need to repeat the process for the south chart. Since some expressions mirror those developed for the north chart, the tasks of finding the Jacobian and metric tensors are simpler.

The south chart

We define n stereographic projections φ^μ, $\mu = 1, 2, \ldots, n$ using the south chart Φ_S,

$$\varphi^\mu = 2\frac{x^\mu}{1+\zeta} \qquad : \mathbb{R}^{n+1} \to \mathbb{S}^n_N. \tag{9.10}$$

The "south pole" point $p_S \in \mathbb{R}^{n+1} = (x^1, x^2, \ldots, \zeta) = (0, 0, \ldots, -1)$ cannot be mapped into $\mathbb{S}^n_N$ because the denominator in equation (9.10) tends to zero. Setting

$$\varsigma^2 = \sum_{k=1}^{n} (\varphi^k)^2 \tag{9.11}$$

then, from equations (9.1) and (9.10), it follows that

$$1 = \left[\frac{1}{2}(1+\zeta)\varphi^1\right]^2 + \left[\frac{1}{2}(1+\zeta)\varphi^2\right]^2 + \cdots + \left[\frac{1}{2}(1+\zeta)\varphi^n\right]^2 + (\zeta)^2$$

where again, $\zeta = x^{n+1}$,

$$1 = \frac{1}{4}(1+\zeta)^2\varsigma^2 + (\zeta)^2$$

a quadratic equation in ζ similar to that in the north map. The two roots are

$$\zeta = \frac{4-\varsigma^2}{\varsigma^2+4}, \qquad \zeta = -1. \tag{9.12}$$

The $\zeta = -1$ root is discarded since it corresponds to the south pole. The first root is inserted back into equation (9.10),

$$\begin{aligned}\left(1+\frac{4-\varsigma^2}{\varsigma^2+4}\right)\frac{1}{2}\varphi^\mu &= x^\mu \\ \frac{4\varphi^\mu}{\varsigma^2+4} &= x^\mu \quad n=1,\dots,n. \end{aligned} \tag{9.13}$$

Now note that

$$\mathbb{S}^n_N = \mathbb{S}^n \backslash p_N \quad \mathbb{S}^n_S = \mathbb{S}^n \backslash p_S,$$

i. e., the domain of Φ_N excludes the point p_N, and the domain of Φ_S excludes p_S. Moreover,

$$\mathbb{S}^n_N \cap \mathbb{S}^n_S = \mathbb{S}^n \backslash \{p_N, p_S\},$$

i. e., the intersection of the two domains excludes both points. The map $\Phi_S \circ \Phi_N^{-1} : \mathbb{S}^n_N \to \mathbb{S}^n_S$ and its inverse $\Phi_N \circ \Phi_S^{-1} : \mathbb{S}^n_S \to \mathbb{S}^n_N$ are powerful instruments. For the former, we seek expressions for φ^μ in terms of ϕ^μ and these can be derived by directly inserting equations (9.5) and (9.4) into (9.10) as follows:

$$\begin{aligned}\varphi^\mu &= 2\frac{x^\mu}{1+\zeta} = \frac{8\phi^\mu}{\sigma^2+4}\left(1+\frac{\sigma^2-4}{\sigma^2+4}\right)^{-1} \\ \varphi^\mu &= \frac{8\phi^\mu}{\sigma^2+4}\left(\frac{2\sigma^2}{\sigma^2+4}\right)^{-1} = 4\frac{\phi^\mu}{\sigma^2},\end{aligned}$$

whereas $\Phi_N \circ \Phi_S^{-1}$ is derived by inserting equations (9.13) and (9.12) into (9.2),

$$\begin{aligned}\frac{4\varphi^\mu}{\varsigma^2+4} = x^\mu \quad x^{n+1} &= \frac{4-\varsigma^2}{\varsigma^2+4} \\ \phi^\mu &= 2\frac{x^\mu}{1-\zeta} = \frac{8\varphi^\mu}{\varsigma^2+4}\left(1-\frac{4-\varsigma^2}{\varsigma^2+4}\right)^{-1} \\ \phi^\mu &= \frac{8\varphi^\mu}{\varsigma^2+4}\left(\frac{2\varsigma^2}{\varsigma^2+4}\right)^{-1} = 4\frac{\varphi^\mu}{\varsigma^2}.\end{aligned}$$

The Jacobian and the metric tensor of the south chart

Here are the general expressions for the derivatives:

$$\begin{aligned}\frac{\partial x^{\mu'}}{\partial \varphi^\nu} &= \frac{4}{\sigma^2+4}\delta^{\mu'}_\nu - \frac{8\varphi^\mu\varphi^\nu}{(\sigma^2+4)^2}, \\ \frac{\partial \zeta}{\partial \varphi^\nu} &= -\frac{2\varphi^\nu}{\varsigma^2+4} - \frac{2(4-\varsigma^2)\varphi^\nu}{(\varsigma^2+4)^2}.\end{aligned}$$

The metric tensor for the south chart is

$$g_{\mu\nu} = \frac{\partial x^{\mu'}}{\partial \varphi^{\mu}} \frac{\partial x^{\nu'}}{\partial \varphi^{\nu}} \delta_{\mu'\nu'} = \frac{16}{(\varsigma^2+4)^2} \delta_{\mu\nu} \quad \forall p \in \mathbb{S}^n_S. \tag{9.14}$$

That is, the matrix is a unit matrix multiplied by $16/(\varsigma^2+4)^2$. Because the partial derivative of ζ is slightly different, this result is not immediately obvious. However, only few steps are necessary here.

Theorem 29. *The expression for all diagonal elements of the metric tensor is the same:*

$$g_{\mu\mu} = \frac{16}{(\varsigma^2+4)^2} \quad \forall n,$$

i. e., the diagonal elements of the metric tensor are all equal to $16/(\varsigma^2+4)^2$.

Proof. The entries of the metric tensor are sums of $n+1$ terms,

$$g_{\mu\mu} = \left(\frac{\partial \zeta}{\partial \varphi^{\mu}}\right)^2 + \sum_{\mu'=1}^{n} \left(\frac{\partial x^{\mu'}}{\partial \varphi^{\mu}}\right)^2$$

the derivation of this intermediate result is unchanged since the derivatives involved have exactly the same form,

$$\sum_{\mu=1}^{n} \left(\frac{\partial x^{\mu'}}{\partial \varphi^{\mu}}\right)^2 = \frac{16}{(\varsigma^2+4)^2} - \frac{64(\varphi^{\mu})^2}{(\varsigma^2+4)^3} + \frac{64\varsigma^2(\varphi^{\mu})^2}{(\varsigma^2+4)^4}.$$

The derivatives of ζ are

$$\left(-\frac{2\varphi^{\nu}}{\varsigma^2+4} - \frac{2(4-\varsigma^2)\varphi^{\nu}}{(\varsigma^2+4)^2}\right)^2.$$

The quantity inside the parenthesis can be rearranged trivially to read

$$\left(-\frac{2\varphi^{\nu}}{\varsigma^2+4} + \frac{2(\varsigma^2-4)\varphi^{\nu}}{(\varsigma^2+4)^2}\right)^2,$$

which is the additive inverse of

$$\left(\frac{2\phi^{\mu}}{\sigma^2+4} - \frac{2(\sigma^2-4)\phi^{\mu}}{(\sigma^2+4)^2}\right)^2$$

derived earlier on the north map. Because the quantity is squared, the change in sign is inconsequential and the remaining steps are identical. □

For the off diagonal elements to vanish, we proceed as before.

Theorem 30. *The metric tensor is diagonal*

$$g_{\mu\nu} = 0 \quad \textit{iff}\, \mu \neq \nu.$$

Proof. Once more we start with

$$g_{\mu\nu} = \left(\frac{\partial \zeta}{\partial \varphi^{\mu}}\right)\left(\frac{\partial \zeta}{\partial \varphi^{\nu}}\right) + \sum_{\mu'=1}^{n}\left(\frac{\partial x^{\mu'}}{\partial \varphi^{\mu}}\right)\left(\frac{\partial x^{\mu'}}{\partial \varphi^{\nu}}\right).$$

The sum is isomorphic to its north map counterpart; therefore, the first product yields

$$\left(\frac{\partial \zeta}{\partial \varphi^{\mu}}\right)\left(\frac{\partial \zeta}{\partial \varphi^{\nu}}\right) = \left(-\frac{2\varphi^{\mu}}{\varsigma^2+4} - \frac{2(4-\varsigma^2)\varphi^{\mu}}{(\varsigma^2+4)^2}\right)\left(-\frac{2\varphi^{\nu}}{\varsigma^2+4} - \frac{2(4-\varsigma^2)\varphi^{\nu}}{(\varsigma^2+4)^2}\right),$$

this rearranges trivially to

$$\left(\frac{\partial \zeta}{\partial \varphi^{\mu}}\right)\left(\frac{\partial \zeta}{\partial \varphi^{\nu}}\right) = \left(\frac{2\varphi^{\mu}}{\varsigma^2+4} - \frac{2(\varsigma^2-4)\varphi^{\mu}}{(\varsigma^2+4)^2}\right)\left(\frac{2\varphi^{\nu}}{\varsigma^2+4} - \frac{2(\varsigma^2-4)\varphi^{\nu}}{(\varsigma^2+4)^2}\right),$$

which is isomorphic to the counterpart from the north map:

$$\left(\frac{2\phi^{\mu}}{\sigma^2+4} - \frac{2(\sigma^2-4)\phi^{\mu}}{(\sigma^2+4)^2}\right)\left(\frac{2\phi^{\nu}}{\sigma^2+4} - \frac{2(\sigma^2-4)\phi^{\nu}}{(\sigma^2+4)^2}\right).$$ □

9.2 The Laplace–Beltrami operator on the n-sphere

We are ready to derive the expression for $\partial_\mu \partial^\mu$ for the general case. We can do this with either chart since the metric tensors were shown to be isomorphic. Then

$$\partial_\mu \partial^\mu = g^{\mu\nu}\partial_\mu \partial_\nu + \mathcal{G}^\mu \partial_\mu$$

where

$$\mathcal{G}^\mu = g^{-1/2}\partial_\lambda g^{1/2} g^{\mu\nu},$$

defines the advection vector,

$$g^{\mu\nu} = \frac{(\sigma^2+4)^2}{16}\delta^{\mu\nu}$$

is the inverse of the metric tensor and the square root of the determinant is

$$g^{\pm 1/2} = \left(\frac{4}{\sigma^2 + 4}\right)^{\pm n}.$$

The advection vector becomes after some straightforward steps,

$$\begin{aligned} \mathcal{G}^\mu &= \frac{1}{16}(2-n)(\sigma^2+4)\delta^{\mu\lambda}\partial_\lambda \sigma^2 \\ &= \frac{1}{8}(2-n)(\sigma^2+4)\phi^\mu \end{aligned} \tag{9.15}$$

on the north map. Putting all the pieces together, we end up with

$$\partial_\mu \partial^\mu = \sum_{\mu=1}^{n}\left(\frac{(\sigma^2+4)^2}{16}\frac{\partial^2}{\partial(\phi^\mu)^2} + \frac{1}{8}(2-n)(\sigma^2+4)\phi^\mu \partial_\mu\right). \tag{9.16}$$

Since $g_{\mu\nu}$ has exactly the same algebraic expression on the south chart and can be obtained from the equivalent on the north chart by replacing $\phi^\mu \to \varphi^\mu$ and $\sigma \to \varsigma$, the same exchange can be used to derive the Laplace–Beltrami operator in the south chart. It is interesting to note the special feature of the 2-sphere, its Laplace–Beltrami operator is advection-free. The three most important n-spheres are the $n = 1$ case, namely the set of all point in a ring of radius 1, the sphere (also known as the 2-sphere), and the 3-sphere. Equation (9.16) for the first case becomes

$$\partial_\mu \partial^\mu = \frac{(\xi+4)^2}{16}\frac{\partial^2}{\partial \xi^2} + \frac{1}{8}(\xi^2+4)\xi\frac{\partial}{\partial \xi}, \tag{9.17}$$

whereas for $n = 2$ we get

$$\partial_\mu \partial^\mu = \frac{(u^2+v^2+4)^2}{16}\left(\frac{\partial^2}{\partial u^2} + \frac{\partial^2}{\partial v^2}\right) \tag{9.18}$$

with the replacements $\xi \to \mathring{\xi}$ and $u, v \to \mathring{u}, \mathring{v}$,

$$\mathring{\xi} \leftarrow \frac{4}{\xi}, \quad \mathring{u}, \mathring{v} \leftarrow \frac{4u}{u^2+v^2}, \frac{4v}{u^2+v^2}$$

to go from the north to the south chart, and

$$\xi \leftarrow \frac{4}{\mathring{\xi}}, \quad u, v \leftarrow \frac{4\mathring{u}}{\mathring{u}^2+\mathring{v}^2}, \frac{4\mathring{v}}{\mathring{u}^2+\mathring{v}^2}$$

for the reversed direction. These expressions allow us to smoothly switch between the two charts whenever a point becomes too close to the point of projection, and we have used this method to prevent simulations to be abruptly interrupted by floating point exceptions. Switching between charts during a simulation, either via DMC or molecular dynamics require some additional considerations that are straightforward to address [49].

9.3 Exercises

1. In Chapter 8, we derived the Laplace–Beltrami operator in spherical polar coordinates,

$$\partial_\mu \partial^\mu = \frac{\partial^2}{\partial r^2} + \frac{1}{r^2}\frac{\partial^2}{\partial \theta^2} + \frac{1}{r^2 \sin^2\theta}\frac{\partial^2}{\partial \phi^2}\frac{2}{r}\frac{\partial}{\partial r} + \frac{\cot\theta}{r^2}\frac{\partial}{\partial \theta}.$$

The same object in Cartesian coordinates reads

$$\nabla^2 = \frac{\partial^2}{\partial x^2} + \frac{\partial^2}{\partial y^2} + \frac{\partial^2}{\partial z^2}$$

and the Cartesian coordinates can have values in $(-\infty, \infty)$. In particular, the point at the origin $(x, y, z) = (0, 0, 0)$ is fine for the regular Laplacian operator but seems to create a problem for the Laplace– Beltrami operator in spherical polar coordinates, since the radius r approaches zero, and all its terms become singular in that limit. Explain, using concepts from this chapter and Chapter 8, why from the theoretical perspective the origin is not a problem for either operator.

2. Insert the definition of ϕ^μ in equation (9.2), into equation (9.1), use $\zeta = x^{n+1}$ and (9.3) to derive

$$1 = \frac{1}{4}(1-\zeta)^2\sigma^2 + (\zeta)^2.$$

Then solve for ζ and show that

$$\zeta = 1, \quad \frac{\sigma^2 - 4}{\sigma^2 + 4}.$$

3. Consider the ring $\mathbb{S}^1$ mapped with the stereographic projection ξ of the north chart

$$x = \frac{2\xi}{\xi^2 + 4}, \quad y = \frac{\xi^2 - 4}{\xi^2 + 4}.$$

(a) Use these two equations to show that

$$x^2 + y^2 = 1.$$

(b) Derive the metric tensor (a 1×1 matrix) of the north chart, by finding the Jacobian and transforming the 2×2 unit matrix, the metric of the Euclidean plane, i. e.,

$$g(\xi) = \left(\frac{\partial x}{\partial \xi}\right)^2 + \left(\frac{\partial y}{\partial \xi}\right)^2.$$

(c) Derive the metric tensor (a 1×1 matrix) $g(\phi)$ using the regular polar chart,

$$x = \cos\phi, \quad y = \sin\phi$$

by following the same procedure of part (b).

(d) Derive the following result:

$$\phi = \tan^{-1}\left(\frac{\xi^2 - 4}{4\xi}\right).$$

(e) Evaluate the derivative

$$\frac{\partial\phi}{\partial\xi}$$

and use it to transform the metric tensor in part (c) $g(\phi)$ into the metric tensor in part (b) $g(\xi)$, i. e.,

$$g(\xi) = \left(\frac{\partial\phi}{\partial\xi}\right)^2 g(\phi).$$

4. Derive equation (9.15).
5. The Schrödinger equation for a particle in a ring of radius r and reduced mass μ in terms of the stereographic projection ξ,

$$-\frac{\hbar^2}{2\mu r^2}\left[\frac{(\xi+4)^2}{16}\frac{\partial^2}{\partial\xi^2} + \frac{1}{8}(\xi^2+4)\xi\frac{\partial}{\partial\xi}\right]\psi + U\psi = E\psi, \tag{9.19}$$

where we have taken the Laplace–Beltrami operator derived on a unit sphere, and we have divided the kinetic energy operator by the moment of inertia μr^2. The reduced mass is obviously placed in the correct location, however, we need to verify that deriving the kinetic energy operator with a unit sphere and simply dividing by the square of the radius is a legitimate strategy. To do so, derive the stereographic projection for a ring of radius r by following the directions below.

(a) Inspect the circle of radius r drawn in Figure 9.2. Its center is located at point O. The segments $\overline{NA}$ and $\overline{AB}$ are perpendicular. Point P on the ring has coordinates x, y. Its projection from the north point N is at point B; therefore, $\eta = \overline{AB}$ is the stereographic projection coordinate of point P. By using the similarity of two triangles find a relationship between $\overline{NA}$ and $\overline{AB}$ that involves r and y.

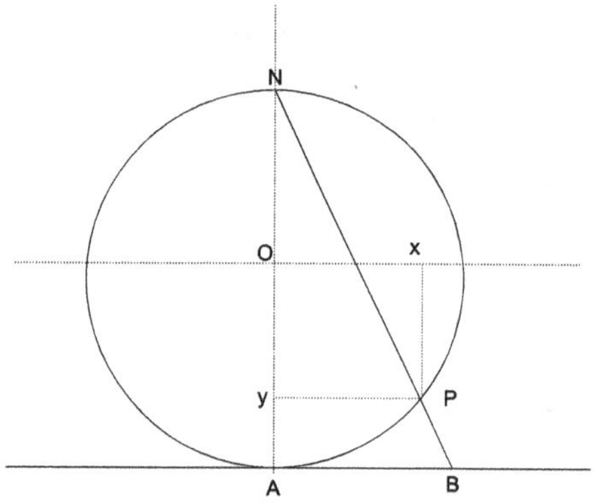

Figure 9.2: A ring of radius r, defining the stereographic projection η.

(b) Use the result in part (a) together with $x^2 + y^2 = r^2$ to derive

$$x = \frac{4r^2\eta}{\eta^2 + 4r^2}, \quad y = r\frac{\eta^2 - 4r^2}{\eta^2 + 4r^2}.$$

(c) Derive the only component of the metric tensor,

$$g = \frac{16r^4}{(\eta^2 + 4r^2)^2}.$$

(d) Show that the advection vector becomes

$$\mathcal{G}^\mu = \frac{\eta^2 + 4r^2}{8r^4}\eta.$$

(e) Use the inverse of the metric tensor and the advection vector to write the Schrödinger equation

$$-\frac{\hbar^2}{2\mu}(g^{\mu\nu}\partial_\mu\partial_\nu + \mathcal{G}^\mu\partial_\mu)\psi + U\psi = E\psi$$

in terms of η and r. Then perform a change of independent variable $\xi = \eta/r$ and show that the result is (9.19). This procedure, meaning using the unit sphere to derive the derivative terms and then divide by μr^2 to find the kinetic energy operator, can be generalized to the n-sphere, but cannot be generalized for non-linear rigid rotors as the next exercise demonstrates.

6. The configuration space of the asymmetric rigid rotor is an inertia "weighted" ellipsoid. It can be mapped using three angles θ, ψ, ϕ defined in equation (3.12). It can be shown that the metric tensor in the ellipsoid of inertia takes the following general form:

$$g_{\mu\nu} = \begin{pmatrix} I_1 \cos^2\psi + I_1 \sin^2\psi & (I_1 - I_2)\sin\theta\cos\psi\sin\psi & 0 \\ (I_1 - I_2)\sin\theta\cos\psi\sin\psi & I_1\sin^2\theta\sin^2\psi + I_2\sin^2\theta\cos^2\psi + I_3\cos^2\theta & I_3\cos\theta \\ 0 & I_3\cos\theta & I_3 \end{pmatrix} \tag{9.20}$$

where I_1, I_2, I_3 are the moments of inertia of the rigid body along the principle axis of inertia. For a rigid n point particle system with masses m_i, the inertia tensor is computed with

$$I_{\mu\nu} = \begin{pmatrix} \sum_{k=1}^{n} m_k(y_k^2 + z_k^2) & -\sum_{k=1}^{n} m_k x_k y_k & -\sum_{k=1}^{n} m_k x_k z_k \\ -\sum_{k=1}^{n} m_k x_k y_k & \sum_{k=1}^{n} m_k(x_k^2 + z_k^2) & -\sum_{k=1}^{n} m_k y_k z_k \\ -\sum_{k=1}^{n} m_k x_k z_k & -\sum_{k=1}^{n} m_k y_k z_k & \sum_{k=1}^{n} m_k(x_k^2 + y_k^2) \end{pmatrix}$$

where x_k, y_k, z_k are the coordinates of particle k relative to the center of mass of the system. The inertia tensor can be diagonalized by a rotation that aligns the new x, y, z axis along the principal axis of inertia. In that case, $I_{\mu\nu} = \mathrm{diag}(I_1, I_2, I_3)$.

(a) Derive the form of the metric tensor for a spherical rigid rotor ($I_1 = I_2 = I_3$) from equation (9.20).

(b) The configuration space of the spherical rigid rotor is related to the 3-sphere as follows. One defines the unit quaternions $q^{(1)}, q^{(2)}, q^{(3)}, q^{(4)}$ with the following expressions:

$$q^{(1)} = \cos\frac{\theta}{2}\cos\left(\frac{\psi+\phi}{2}\right)$$
$$q^{(2)} = \sin\frac{\theta}{2}\cos\left(\frac{\psi-\phi}{2}\right)$$
$$q^{(3)} = \sin\frac{\theta}{2}\sin\left(\frac{\psi-\phi}{2}\right)$$
$$q^{(4)} = \cos\frac{\theta}{2}\sin\left(\frac{\psi+\phi}{2}\right).$$

Show that $(q^{(1)})^2 + (q^{(2)})^2 + (q^{(3)})^2 + (q^{(4)})^2 = 1$

(c) Note how the result in part (b) defines a 3-sphere. One derives the north and south stereographic projections charts as in the general case, however, the metric tensor is obtained by transforming the metric obtained in part (a), not a 4×4 unit matrix. The results for the n-spheres derived earlier in the chapter assume that $\mathbb{R}^{n+1}$ is endowed with an Euclidean metric, i. e., a unit $(n+1)\times(n+1)$ matrix. The result for the spherical rigid rotor is closely related to the metric tensor of the 3-sphere,

$$g_{\mu\nu} = \frac{64I_1}{(\sigma^2+4)^2}\begin{pmatrix} 1 & 0 & 0 \\ 0 & 1 & 0 \\ 0 & 0 & 1 \end{pmatrix}.$$

Use this expression to write the Schrödinger equation for the spherical rigid rotor in the north map.

7. The Cartesian coordinates of a methane molecule (Point groups T_d), in the center of mass, oriented along the principal axis are

	x	*y*	*z*
C	0	0	0
H	*a*	*a*	*a*
H	−*a*	−*a*	*a*
H	−*a*	*a*	−*a*
H	*a*	−*a*	−*a*

where $a = 1.20299$ bohr. Use the definition of the inertia tensor in exercise 6 to show that for methane the inertia tensor is

$$I_{\mu\nu} = 8m_H a^2 \begin{pmatrix} 1 & 0 & 0 \\ 0 & 1 & 0 \\ 0 & 0 & 1 \end{pmatrix},$$

where m_H is the mass of a hydrogen atom. Therefore, in its equilibrium position methane is a spherical top.

10 Coordinates for systems of point particles

Great interest is to be attributed to this atom-model; for, as Rutherford has shown, the assumption of the existence of nuclei, as those in question, seems to be necessary in order to account for the results of the experiments on large angle scattering of the α rays.

N. Bohr

Thanks to the gold foil experiment that Rutherford, Geiger, and Marsden carried out more than a century ago, we know today that nuclei occupy approximately point-like volumes at the atomic and molecular scale. Therefore, to apply all the ideas we explored in the previous chapters to chemical physics, we need to consider how such concepts translate to the theory of systems of point particles. Assuming an accurate model for U as a function of the nuclei coordinates is available, how do we extract information in the form of measurable chemical and physical properties from it?

Suppose, for instance, we are interested in correcting the binding energy of two molecules with their zero point energy of the intermolecular interactions. This is a purely quantum effect, and we could be tempted to simply implement a multidimensional version of the diffusion Monte Carlo method to find the ground state. The set of n Cartesian 3-vectors of Euclidean space would not work well for such purpose. The binding energy of each molecule is typically thousands of time greater than the intramolecular one, and the random fluctuations of the total energy would drown out completely the effect we are looking for. Therefore, unless one uses rigid constraints and the tools developed in the last two chapters, the task is untractable. Moreover, excited states of systems of point particles need some classification. If there are no external force fields, then the ground state of the system is always in the $J = 0$ rotation state. However, excited states are always a combination of rotation and internal vibration modes. Neither of those dynamics are well represented by our set of n Cartesian 3-vectors.

In Chapter 9, we have seen that rotations can be better represented with angular or projection-like coordinates. Historically, internal modes of vibrations have been represented by a linear transformation of the mass-weighted n Cartesian 3-vectors. The transformation is carried out by the orthogonal matrix that diagonalizes the Hessian matrix of the potential U evaluated at its global minimum, where the Hessian matrix contains the second partial derivatives of U with respect to the configuration space coordinates. The resulting coordinates are known as *normal modes*. When expressed in normal modes, the Hessian matrix of U is diagonal, and its eigenvalues are precisely the force constants k_i from which the natural frequency of oscillation of mode i, $\omega_i = (k_i)^{1/2}$ can be computed. The procedure we just outlined is in essence the process of decoupling harmonic oscillators. Then the ground state of the n particle system is simply $\sum_{i=1}^{3n-6} \hbar\omega_i/2$ and any excited state energy along a particular normal mode k can be computed by simply adding $\hbar\omega_k$ to the ground state. After analyzing the eigenvectors using group theory,

https://doi.org/10.1515/9783111610207-010

the proper selection rules are determined, and one can use frequencies of allowed transitions to reproduce the fundamentals of vibrational spectroscopy, both infrared as well as Raman. Six frequencies are zero when evaluated at the minimum of U. Three correspond to the translation of the whole system of n point particles, and three others (for nonlinear configurations) correspond to the rotations.

Beyond the point-like particle nature of the nuclei and the accuracy of the Born–Oppenheimer approximations, normal mode analysis adds three sources of systematic errors in the estimate of the ground and excited state energies. The sources of one type of such errors are the cubic and higher order terms of the expansion of U that are neglected. These can be reintroduced by representing the tensor like objects such as, e. g., $\partial_\mu \partial_\nu \partial_\sigma U$ in the normal modes space, and correct the energy levels perturbatively. Features in the vibrational spectrum such as overtones, combinations, and Fermi resonances can be reproduced by such approach. The second source of error occurs when the system is floppy. The nuclei can find themselves quite far from the minimum of U, and may in fact visit other local minima of U. In those cases, the harmonic frequencies can be hundreds of wave numbers off even for the fundamental transitions and the power series expansion of U while perhaps still convergent does so inefficiently from the computational point of view. Finally the third and, arguably, most difficult source of error to correct is the inherent coupling between the rotations and the vibrations degrees of freedom of the system. The Watson Hamiltonian [75] is arguably the best way to correct for the last source of error.

A decisively easier approach to handle all of the problems of normal mode analysis at once is to use classical *molecular dynamics*, i. e., by numerically integrating Newton's second law. Of course, the problem with molecular dynamics is that it strictly treats nuclei as classical particles, when they really are not. Consequently, overtones, combinations, and Fermi resonance cannot be extracted from the transformation of the time autocorrelation functions computed with molecular dynamics [1]. Only recently has progress been made along these lines. A powerful method called ring polymer molecular dynamics has been formulated to systematically correct for the classical nature of the nuclei. The approach is based on the Feynman path integral and it is a useful instrument for us; therefore, it is discussed in a later chapter.

In this chapter, we propose a transformation of independent variables for a n point particle system that allows us to find ground and excited states by diffusion Monte Carlo discussed in Chapter 7. In this chapter, we propose a number of linear transformations of coordinates, which like the normal mode analysis, attempts to separate rotation degrees of freedom from the internal ones without introducing uncontrollable approximations. Unlike the normal mode transformation, it is not based on a power series expansion of U, and more importantly does not neglect the rotation-vibration coupling (the so-called *coriolis* forces), but instead provides the minimum coupling between these degrees of freedom when the system is in the vicinity of its *reference frame*, a concept that we carefully define in this chapter.

Finding a representation that minimizes the coupling between rotations and vibrations becomes particularly important when interpreting the approximate forms for the excited state wave functions, as it gives the investigator a way of partitioning the first excitations. Because the transformation is linear, the resulting metric tensor is uniform, and the Laplace–Beltrami operator is advection-free. However, the metric tensor is not diagonal, which requires us to implement correlated sampling procedures in order to diffuse the replicas properly. We deal with correlated sampling in a later chapter. Finally, one can easily estimate the systematic error that neglecting coriolis effects produces on the desired physical properties. One simply suppresses those degrees of freedom that are rotation like in the course of the simulation and compares the properties with those from an unconstrained simulation.

10.1 Preliminary concepts

Here are some fundamental definitions that are helpful in developing the theory in this chapter. To avoid unnecessary repetitions, we refer to a set of n point particles, simply as “system” in all that follows in this chapter.

Definition 39 (Configuration space). Configuration space is the set of all possible positions of a system. A “configuration” is a single point in configuration space.

Definition 40 (State space). The state space of a system is the Cartesian product $\mathbb{R}^{3n} \otimes \mathbb{R}^{3n}$ of all possible positions and velocities of the system.

For example, for a single point particle the state space is the set of all values of x, y, z, $\dot{x}, \dot{y}, \dot{z}$, where $\dot{x}$ represents the derivative of x with respect to time t. In many differential geometry textbooks, the velocity space is called the *tangent bundle.*

Definition 41 (Classical state). The classical state of a system is a point in state space at time t.

We use the words “classical state” here so as to not confuse this concept with the state idea discussed in previous chapters. The laws of quantum mechanics prevents us from measuring the classical state of a system with infinite precision, but the concept is very useful nonetheless.

Definition 42 (Law of inertia). A body at rest, or moving at constant speed in a straight line will remain at rest, or move at constant speed in a straight line, unless acted upon by a force.

Definition 43 (Reference frame). A set of three orthogonal vectors oriented arbitrarily and a particular value of time t.

The configuration space mapped with coordinates relative to the axis of the reference frame is called the *reference configuration.* In the chemical physics literature, the reference frame is often called "the laboratory frame." The two terms are synonymous.

The frame is a key concept that was developed to derive the laws of special relativity. Here, we make use of Galilean frames, where time is an absolute parameter. The large majority of experiments that probe the dynamics of nuclei involve energies that are far too small to require taking into account relativistic effects. Therefore, Galilean frames are more than adequate for the present tasks. Associated with frames of reference is another key principle, the principle of invariance, one of the pillars of gauge theories.

Definition 44 (Inertial frames). An inertial frame is a frame of reference in which the law of inertia and other laws of physics hold.

Since Galilean frames are inertial, it is impossible for an observer in a frame to know if his or her frame is moving. This is the basic principle of relativity: Absolute classical states cannot be determined, only relative ones. Therefore, when dealing with frames, it is always necessary to compare the properties in one frame with another. However, the laws of physics should be the same no matter how one chooses to set up the frame.

Definition 45 (Principle of invariance). All nonrelativistic laws of physics hold under Galilean transformation.

For example, Newton's laws should hold irrespective of where the observer decides to set up its origin of coordinates, its axis orientation, what distance scale is used, and even how fast the axis move in a straight line, as long as the velocity is constant. The same applies to the Schrödinger equation. These are far reaching ideas that will help us to navigate the material in later chapters after Hamilton's variation principle is introduced. There, instead of talking about inertial frames, we speak of invariant theories. Then the invariance principle becomes a test to rule out theoretical hypothesis. The principle of invariance is another powerful instrument for the theoretician. In this chapter, we need these definitions to interpret the process of deriving the various systems of coordinates, and we have taken the opportunity to lay the ground for things to come as well.

It is important to note that Galilean transformations are carried out by members of groups (see Chapter 3). The group of all Galilean transformations is called, not surprisingly, the Galilean group, which include translations, rotations, and Galilean boosts to go from a frame at rest to one that moves at constant speed in a straight line. It should not come as a surprise that there is a corresponding Galilean algebra since frame transformations are continuous and smooth. Finally, accelerated frames (those that do not move in a straight line, as well as those that do not move at constant speed) are by definition noninertial. Such changes can only be brought about by a force. Since all frames we consider here are Galilean, we omit the adjective heretofore and simply use the word "frame," which should be understood to mean "Galilean frame."

Probably the best example to explain how rotating frames accelerate the system was offered by Newton's spinning bucket of water. A bucket of water is suspended by a rope that has been twisted a number of times. An observer inside the bucket would be spinning at the same angular speed as the rope unwinds. As long as the rotation is not accelerated, the observer in the rotating frame cannot measure the spinning rate since its axes are rotating at exactly the same rate. Consequently, the observer would assume the water is a rest. However, he or she would notice that the surface of the water curves as the result of the centrifugal force. Therefore, the observation contradicts the observer hypothesis. For our systems, these ideas translate to the coupling between the rotation and the internal degrees of freedom mentioned earlier. Our analysis in this chapter shows that the ro-vibrational coupling are closely related to off diagonal elements of the metric tensor in a particular set of coordinates we derive in the next two sections.

10.2 Relative coordinates

The change of coordinates in this chapter takes place in two stages. The transformation in this section produces what we call "relative" coordinates, and these alone can be particularly advantageous for centrosymmetric molecules of the type XY_n, e. g., CH_4, and have also been called Radau coordinates [69] or baricentric coordinates. The first order of business is to define a vector from the origin of the lab frame to a point in configuration space. Let

$$x^{\mu} = (x_1, x_2, \ldots, z_n), \quad 1 \leq \mu \leq 3n \tag{10.1}$$

represent the Cartesian like coordinate vector in $\mathbb{R}^{3n}$ for the system, where x_i is the x component of the Euclidean 3-vector r_i for particle i, etc. Ordering the vector with all x components first, followed by y and z ones is a deliberate choice that simplifies the structure of the matrices we derive from the transformations.

To find coordinate systems that minimize the rotation-vibration coupling, we need to consider two separate steps. We first remove the translation degrees of freedom of the center of mass. In other words, we transform from a rest frame to a frame with the center of mass at the origin and that moves at the same speed and direction as the center of mass. First, we find the coordinates of the center of mass x_c, y_c, z_c,

$$x_c = \frac{1}{m_t} \sum_{i=1}^{n} m_i x_i \tag{10.2}$$

with corresponding equations for y_c and z_c. Here, m_i is the mass of particle i, and $m_t = \sum_{i=1}^{n} m_i$ is the total mass of the system. To transform into the center of mass frame, it is not sufficient to simply use $x_i' = x_i - x_c$ and corresponding (y, z) equations, because that would leave us with $3n$ degrees of freedom when in reality we have $3n - 3$ independent ones. Equation (10.2) and corresponding components written in the primed system reads

$$\frac{1}{m_t}\sum_{i=1}^{n} m_i x_i' = 0, \tag{10.3}$$

i. e., three equations of constraint are created. The reader should recall that equations of constrains on coordinates create subspaces that have to be carefully handled, as they could be curved, have a different metric tensor, etc. In other words, points in a frame must *always* be mapped by a set of independent parameters. Because the equations of constraint are linear, the subspace is also linear. Before we derive a map for the subspace, let us prove (10.3) by inserting $x_i' = x_i - x_c$ into the sum on right-hand side,

$$\frac{1}{m_t}\sum_{i=1}^{n} m_i x_i' = \frac{1}{m_t}\sum_{i=1}^{n} m_i(x_i - x_c),$$

and separating into two sums, we obtain

$$= \frac{1}{m_t}\sum_{i=1}^{n} m_i x_i - x_c\frac{1}{m_t}\sum_{i=1}^{n} m_i = x_c - x_c = 0,$$

where we have distributed the summation sign to the two terms, taken x_c out of the second sum, used the definition of m_t to cancel the second sum, and used (10.2) to substitute the first. To map the subspace, we define a set of center and relative coordinates,

$$q^\mu = (x_c, y_c, z_c, d_x^{(1)}, d_x^{(2)}, \dots, d_z^{n-1}) \quad 1 \le \mu \le 3n \tag{10.4}$$

where $d_x^{(1)} = x_1 - x_n$, $d_x^{(2)} = x_2 - x_n$, etc. This defines a map $\Phi : x^\mu \to q^\mu$ and its inverse $\Phi^{-1} : q^\mu \to x^\mu$. It is straightforward to show that the map Φ is a bijection, meaning is one to one in the forward and reverse directions. The equations for Φ^{-1} are obtained by solving a system of $3n$ linear equations, n for each of the three dimensions,

$$d_x^{(1)} = x_1 - x_n, \quad \dots, \quad d_z^{n-1} = z_{n-1} - z_n. \tag{10.5}$$

Theorem 31. *The expressions for Φ^{-1} obtained by solving a system of $3n$ equations are*

$$x_i = d_x^i + x_c - \frac{1}{m_t}\sum_{j=1}^{n-1} m_j d_x^j \quad (1 \le i \le n-1), \tag{10.6}$$

$$x_n = x_c - \frac{1}{m_t}\sum_{i=1}^{n-1} m_i d_x^i. \tag{10.7}$$

with equivalent ones for the y and z coordinates.

We prove (10.6) and leave the proof of equation (10.7) as an exercise since it can be carried out with exactly the same technique.

Proof. We rewrite (10.6) by using (10.2) to eliminate x_c in favor of the sum, and (10.5) to eliminate d_x^j,

$$x_i = x_i - x_n + \frac{1}{m_t}\sum_{j=1}^{n} m_j x_j - \frac{1}{m_t}\sum_{j=1}^{n-1} m_j (x_j - x_n).$$

Next, we distribute the last sum to the two terms,

$$x_i = x_i - x_n + \frac{1}{m_t}\sum_{j=1}^{n} m_j x_j - \frac{1}{m_t}\sum_{j=1}^{n-1} m_j x_j + x_n \frac{1}{m_t}\sum_{j=1}^{n-1} m_j,$$

where x_n is a constant that was factored out of the sum. The remaining sum over the masses is just $m_t - m_n$; therefore,

$$x_i = x_i - x_n + \frac{1}{m_t}\sum_{j=1}^{n} m_j x_j - \frac{1}{m_t}\sum_{j=1}^{n-1} m_j x_j + x_n \frac{m_t}{m_t} - x_n \frac{m_n}{m_t}$$

but now we can move the $-m_n x_n / m_t$ term on the extreme right back into the second summation sign changing its top limit from $n-1$ to n. Then the two sums become identical with opposite signs and cancel out leaving,

$$x_i = x_i - x_n + x_n \frac{m_t}{m_t}.$$

The identity tells us that equation (10.6) holds. □

The last proof demonstrates that Φ is a bijection. The vector q^μ includes both the subspace of relative coordinates as well as the center of mass coordinates. The latter ones are not part of the frame, however, they are included here so that equations (10.5) through (10.7) can be written as linear transformations of vectors using matrices:

$$q^\mu = D^\mu_\nu x^\nu$$

for Φ and

$$x^\nu = J^\nu_\mu q^\mu$$

where $J = D^{-1}$ is the Jacobian. Note that D and its inverse are members of $\mathrm{Gl}(\mathbb{R}, 3n)$ with constant coefficients since D is invertible and both J and D are $3n \times 3n$ matrices. If we write q^μ and x^ν as column vectors with $3n$ components and we use (10.2) and (10.5), we can read off directly the entries of the $3n \times 3n$ matrix D. Its superstructure is

$$D = \begin{bmatrix} 0 & M_{3\times 3n} & 0 \\ D_x & 0 & 0 \\ 0 & D_y & 0 \\ 0 & 0 & D_z \end{bmatrix}$$

where

$$D_x = D_x = D_z = \begin{pmatrix} 1 & 0 & \cdots & 0 & -1 \\ 0 & 1 & \cdots & 0 & -1 \\ \vdots & \vdots & \cdots & \vdots & \vdots \\ 0 & 0 & \cdots & 1 & -1 \end{pmatrix}$$

are $(n-1)\times n$ matrices, and the $3\times 3n$ matrix contains the mass ratios needed to construct the sum in equation (10.2):

$$M_{3\times 3n} = \frac{1}{m_t}\begin{pmatrix} m_1 & \cdots & m_n & 0 & \cdots & 0 & 0 & \cdots & 0 \\ 0 & \cdots & 0 & m_1 & \cdots & m_n & 0 & \cdots & 0 \\ 0 & \cdots & 0 & 0 & \cdots & 0 & m_1 & \cdots & m_n \end{pmatrix}.$$

Using the same strategy, we can read the components of the matrix J^T, where

$$J = \frac{\partial(x_1, \ldots, z_n)}{\partial(x_c, y_c, z_c, d_x^{(1)}, \ldots, d_z^{n-1})}.$$

Its superstructure is

$$J^T = \begin{bmatrix} 0 & B & 0 & 0 \\ I_{3n\times 3} & 0 & B & 0 \\ 0 & 0 & 0 & B \end{bmatrix}, \tag{10.8}$$

where

$$B = \frac{1}{m_t}\begin{pmatrix} m_t - m_1 & -m_2 & \cdots & -m_{n-1} \\ -m_1 & m_t - m_2 & \cdots & -m_{n-1} \\ \vdots & \vdots & \cdots & \vdots \\ -m_1 & -m_2 & \cdots & m_t - m_{n-1} \end{pmatrix}$$

is shaped as a $n \times (n-1)$ array, and

$$I_{3n\times 3} = \begin{pmatrix} 1 & 0 & 0 \\ \vdots & \vdots & \vdots \\ 0 & 1 & 0 \\ \vdots & \vdots & \vdots \\ 0 & 0 & 1 \\ \vdots & \vdots & \vdots \end{pmatrix},$$

i. e., the first n rows have entries 1 0 0, etc.

10.2.1 The Jacobian

The relative coordinates of the $n = 2$ case are special because they transform the metric tensor into a diagonal object. This is not the case for $n > 2$. Manipulating the transformation matrices by hand is a tedious exercise for the $n = 3$ case, but coding the general structure is straightforward if we look at specific equations derived by evaluating the derivatives of equations (10.6) and (10.7) rewritten here in terms of the coordinates q^μ defined in equation (10.4). Note that these are the elements of the Jacobian matrix whose structure we have just inspected. The following are explicit equations for its elements, and these are essential for coding purposes. One could code D and numerically invert it instead. However, the effort of coding D is almost the same, but its numerical inversion adds execution time and carries unnecessary rounding errors:

$$x^\mu = q^1 + q^{\mu+3}(1 - \delta_{\mu n}) - \frac{1}{m_t}\sum_{i=1}^{n-1} m_i q^{i+3} \quad 1 \le \mu \le n \tag{10.9}$$

for the x_i block of equations (10.6) and (10.7),

$$x^\mu = q^2 + q^{\mu+2}(1 - \delta_{\mu(2n)}) - \frac{1}{m_t}\sum_{i=1}^{n-1} m_i q^{i+n+2} \quad n+1 \le \mu \le 2n \tag{10.10}$$

for the y_i block of the vector, and finally,

$$x^\mu = q^3 + q^{\mu+1}(1 - \delta_{\mu(3n)}) - \frac{1}{m_t}\sum_{i=1}^{n-1} m_i q^{i+2n+1} \quad 2n+1 \le \mu \le 3n \tag{10.11}$$

for the z_i block.

Then the equations to construct the Jacobian are ($1 \le \nu \le n-1$ for all these)

$$\frac{\partial x^\mu}{\partial q^1} = 1 \quad 1 \le \mu \le n, \tag{10.12}$$

$$\frac{\partial x^\mu}{\partial q^{\nu+3}} = \delta_{\mu\nu} - \frac{m_\nu}{m_t} \quad 1 \le \mu \le n, \tag{10.13}$$

for the x_i block,

$$\frac{\partial x^\mu}{\partial q^2} = 1 \quad n+1 \le \mu \le 2n,$$
$$\frac{\partial x^\mu}{\partial q^{\nu+n+2}} = \delta_{\mu(n+\nu)} - \frac{m_\nu}{m_t} \quad n+1 \le \mu \le 2n,$$

for the y_i block, and finally,

$$\frac{\partial x^\mu}{\partial q^3} = 1 \quad 2n+1 \le \mu \le 3n,$$
$$\frac{\partial x^\mu}{\partial q^{\nu+2n+1}} = \delta_{\mu(2n+\nu)} - \frac{m_\nu}{m_t} \quad 2n+1 \le \mu \le 3n,$$

for the z_i block. The rest of the Jacobian matrix elements are zero.

10.2.2 The metric tensor

In Chapter 9, we derive transformation of coordinates for rotation degrees of freedom. Then, by working through the differential geometry analysis in Chapter 8, we derive equations for the metric of the space. The metric tensors in the last two chapters are purely geometrical objects. In order to apply them to physical systems, we had to use our intuition and place the effective inertial mass in the correct place. However, there is a rigorous procedure to derive the correct moment of inertia, and more generally the effective mass associated with the parameters of the points in a chart. To do so, we have to transform a modified version of the Euclidean metric (i. e., the unit matrix used in Chapters 8 and 9) for our system. The correct effective masses in any coordinate system are the elements of the Hessian metric we define here.

Definition 46 (Hessian metric). The Hessian metric is the effective inertial mass of the system.

For a system of n particles whose configuration space is mapped with Cartesian coordinates, the Hessian metric tensor is a diagonal $3n \times 3n$ matrix with mass m_i corresponding to the coordinates x_i, y_i, z_i for particle i. The structure of the vector x^μ as, e. g., in equation (10.1) informs how the masses should be arranged along the main diagonal

$$g_{\mu\nu} = \mathrm{diag}\{m_1, \dots, m_n, m_1, \dots, m_n, m_1, \dots, m_n\}. \tag{10.14}$$

Under a change of coordinates, the metric tensor transforms as a 2-form in the usual way,

$$g_{\mu'\nu'} = \frac{\partial x^{\mu}}{\partial q^{\mu'}}\frac{\partial x^{\nu}}{\partial q^{\nu'}} g_{\mu\nu} = J\,G J^{T}. \tag{10.15}$$

Then, when writing the kinetic energy operator, one simply uses

$$-\frac{\hbar^2}{2}(g^{\mu\nu}\partial_{\mu}\partial_{\nu} + \mathcal{G}^{\mu}\partial_{\mu}).$$

The inverse of the Hessian tensor in the last equation properly places the effective mass in the correct place. Similar statements apply for the advection vector.

For the present case, matrix multiplication could be implemented directly to find the metric tensor. However, it is possible to derive a general algebraic form for it. In general, its structure is block diagonal,

$$g_{\mu'\nu'} = \begin{pmatrix} m_t & 0 & 0 & 0 \\ 0 & g^{(xx)} & 0 & 0 \\ 0 & 0 & g^{(yy)} & 0 \\ 0 & 0 & 0 & g^{(zz)} \end{pmatrix}, \tag{10.16}$$

where the top block is a 3×3 diagonal matrix with m_t for all three eigenvalues. The remaining three blocks are of size $(n-1) \times (n-1)$ and have the following entries:

$$g^{(xx)}_{\mu\nu} = g^{(xx)}_{\mu\nu} = g^{(xx)}_{\mu\nu} = m_{\mu}\left(\delta_{\mu\nu} - \frac{m_{\nu}}{m_t}\right). \tag{10.17}$$

For example, for the $n = 3$ system, consider the 1, 1 element of g,

$$g_{11} = \sum_{\mu=1}^{3n}\left(\frac{\partial x^{\mu}}{\partial q^{1}}\right)^2 m_{\mu}.$$

By inspecting the superstructure of J, as well as equations (10.12), (10.13) with those corresponding to the y and z components, one can show that there are only three surviving terms:

$$g_{11} = \left(\frac{\partial x^{1}}{\partial q^{1}}\right)^2 m_1 + \left(\frac{\partial x^{2}}{\partial q^{1}}\right)^2 m_2 + \left(\frac{\partial x^{3}}{\partial q^{1}}\right)^2 m_3$$

$$g_{11} = \left(\frac{\partial x_{1}}{\partial x_{c}}\right)^2 m_1 + \left(\frac{\partial x_{2}}{\partial x_{c}}\right)^2 m_2 + \left(\frac{\partial x_{3}}{\partial x_{c}}\right)^2 m_3 = m_t$$

since all the partial derivatives are equal to 1. Next, consider the 1, 2 element

$$g_{12} = \sum_{\mu=1}^{3n}\left(\frac{\partial x^{\mu}}{\partial q^{1}}\right)\left(\frac{\partial x^{\mu}}{\partial q^{2}}\right) m_{\mu} = 0$$

since all derivatives like $(\partial x_1/\partial y_c)$ are equal to zero. In fact, in writing these two cases the reader may have realized that to implement equation (10.15) one could multiply column μ by the respective masses in (10.14), and evaluate the dot product of the resulting column with column ν of J. For the $n = 3$ case, as an example,

$$g^{(xx)} = g^{(yy)} = g^{(zz)} = \begin{bmatrix} m_1(1-\frac{m_1}{m_t}) & -\frac{m_1 m_2}{m_t} \\ -\frac{m_1 m_2}{m_t} & m_2(1-\frac{m_2}{m_t}) \end{bmatrix}.$$

Because the metric tensor in this case is not diagonal the relative coordinates are not orthogonal to each other. Since $g_{\mu\nu}$ is uniform, we could diagonalize it once for all configurations. However, the next transformation to arrive at our definition of Eckart coordinates would still create off diagonal terms, so this additional step is unnecessary.

In this section, we have transformed the Cartesian coordinate chart for our general system into a relative coordinate chart, and we produce analytical equations for the Hessian metric tensor. As previously noted, the linear transformation of coordinates yields a uniform metric. The purpose of this transformation is to change the frame of reference from a laboratory fixed one to a inertial frame moving with the center of mass. In the absence of external forces on the system, as we show in a later chapter, the center of mass in not accelerated and the three degrees of freedom can be removed from the analysis while remaining in the same frame. To do so formally, we now reshape and rename the relative coordinate vector

$$d^{\mu} = (d_x^{(1)}, \ldots, d_z^{n-1}) \tag{10.18}$$

now of size $3n - 3$. Moreover, because there are no off diagonal terms that couple the center of mass coordinates with the relative ones in the metric tensor we derive, we can safely remove the three leftmost columns and the top three rows of $g_{\mu\nu}$ giving us a $(3n - 1) \times (3n - 1)$ block diagonal matrix. The equations for the elements $(n - 1) \times (n - 1)$ are in equation (10.17). It is this reduced metric tensor that we need to transform in the next section.

10.3 Eckart coordinates systems

In this section, we explore a coordinate transformation inspired by the Eckart condition named after Carl H. Eckart. The condition was developed in an attempt to separate the rotation degrees of freedom of a system (considered external) from the internal ones. Since we are always able to remove the center of mass degrees of freedom as long as there are no external forces acting on our system, it makes sense to try to further reduce the size of our system by removing the rotations as well. In the absence of external forces, space is uniform, meaning after a translation space looks exactly the same. Under the same conditions, we can rotate our axis, and space still looks the same; the technical term is *isotropic*. However, as explained earlier, a rotating frame (which is what is needed to "remove" rotations), in noninertial. Consequently, even if space is isotropic, our system is affected by rotational motion. However, the Eckart condition is still quite useful as we shall see, because it allows us to finds a system of coordinates (a special rotating frame) that minimizes the coupling between the vibrations and the rotations of the system.

To get started, we introduce a definition.

Definition 47 (Reference configuration). A point in the configuration space of the system measured in a reference frame.

For the reference configuration of a system, we use the following notation. The vector $r_i^{(0)}$ is the 3-vectors Euclidean space measured in the reference frame. There are two conditions for a system to be in an Eckart frame [55, 76, 77].

Definition 48 (Eckart conditions).
- The system center of mass is at the origin.
- The configuration has zero angular momentum relative to the reference configuration.

10.3.1 Relative-Eckart frames

In the previous section, we proved that the n 3-vectors relative to the center of mass (we denoted it as the primed coordinates) satisfies the first Eckart condition. Here, we want to show that the set of relative coordinates does not change the properties of the prime coordinates. The result should be obvious since we are simply translating the location from the nonprimed n 3-vectors to the primed once, and we should be able to translate the relative 3-vectors the same way. However, it is instructive to see how this plays out, as the result helps with the second Eckart condition when it is expressed in terms of the relative coordinates.

First, we need a preliminary result.

Theorem 32. *Let us begin by proving the following result,*

$$\sum_{i=1}^{n-1} m_i d^i = m_t(r_n - r_c). \tag{10.19}$$

Proof. We just need to consider the x axis portion of the theorem as the expressions for y and z can be obtained by simple substitutions $x \to y$ and $x \to z$. On the left hand side we have

$$\sum_{i=1}^{n-1} m_i d_x^i = \sum_{i=1}^{n-1} m_i (x_i - x_n),$$

where we inserted the definition of d_x^i. It followis immediately that,

$$\sum_{i=1}^{n-1} m_i d_x^i = \sum_{i=1}^{n-1} m_i x_i - x_n \sum_{i=1}^{n-1} m_i.$$

After distributing the sum, we arrive at the desired result:

$$\begin{aligned}\sum_{i=1}^{n-1} m_i d_x^i &= m_t x_c - m_n x_n - x_n (m_t - m_n) \\ &= m_t (x_c - x_n).\end{aligned}$$ □

Now let us construct the following sum:

$$\frac{1}{m_t} \sum_{i=1}^{n} m_i x_i'$$

where the primed coordinates were defined by subtracting the center of mass,

$$\begin{aligned}&= \frac{1}{m_t} \sum_{i=1}^{n} m_i (x_i - x_c) = \frac{1}{m_t} \left[m_n (x_n - x_c) + \sum_{i=1}^{n-1} m_i (d_x^i + x_n - x_c) \right] \\ &= \frac{m_n}{m_t} (x_n - x_c) + \frac{1}{m_t} \sum_{i=1}^{n-1} m_i d_x^i + (x_n - x_c) \frac{(m_t - m_n)}{m_t}.\end{aligned}$$

We insert the x component of equation (10.19), and we obtain

$$\begin{aligned}&\frac{1}{m_t} \sum_{i=1}^{n} m_i x_i' \\ &\quad = \frac{m_n}{m_t} (x_n - x_c) + x_c - x_n + (x_n - x_c) \frac{(m_t - m_n)}{m_t}.\end{aligned}$$

Everithing in the last line cancels. In practice, this result means that, given a configuration represented with the n Euclidean 3-vectors, we first compute the center of mass of the system, if the center of mass is not zero, we simply shift all 3-vectors to the prime system. Then we just redefine $d_i = r_i' - r_n'$ and the resulting relative coordinates automatically satisfy the first Eckart condition.

The next theorem is an important result, since it tells us that imposing Eckart's second condition can be done using the Euclidean n 3-vectors as well as by using the $n - 1$ relative 3-vectors.

Theorem 33. *The following identity holds,*

$$\sum_{i=1}^{n-1} m_i r_i^{(0)} \times d_i = \sum_{i=1}^{n} m_i r_i^{(0)} \times r_i. \tag{10.20}$$

Proof. Using the definition of the relative 3-vectors, we write

$$\sum_{i=1}^{n-1} m_i r_i^{(0)} \times d_i = \sum_{i=1}^{n-1} m_i r_i^{(0)} \times (r_i - r_n).$$

The upper limit of the sum on the right can be extend to n, because we just add zero in doing so:

$$\sum_{i=1}^{n-1} m_i r_i^{(0)} \times d_i = \sum_{i=1}^{n} m_i r_i^{(0)} \times (r_i - r_n).$$

Then

$$\begin{aligned}&\sum_{i=1}^{n-1} m_i r_i^{(0)} \times d_i\\ &\quad = \sum_{i=1}^{n} m_i r_i^{(0)} \times r_i - \left(\sum_{i=1}^{n} m_i r_i^{(0)}\right) \times r_n\end{aligned}$$

where in the second sum r_n is a constant factor and in the second line we use the distributive property of the cross product. However, the sum inside the parenthesis is equal to zero since we choose the reference configuration to be in the center of mass. □

The last result allows us to reformulate the second Eckart condition:

$$\sum_{i=1}^{n-1} m_i r_i^{(0)} \times d_i = 0. \tag{10.21}$$

This last expression is simply another constraint for the relative coordinates. That is, we want the net angular momentum of the system relative to the reference configuration to be a constant equal to zero. To see how that works, recall that the reference frame is a set of constant parameters, i. e., they do not depend on time. Then, if we take the derivative with respect to time on the right-hand side of (10.21), we get

$$\sum_{i=1}^{n} m_i r_i^{(0)} \times \dot{r}_i$$

but note that $m_i r_i^{(0)} \times \dot{r}_i$ can be rewritten $r_i^{(0)} \times m_i \dot{r}_i$, i. e., in terms of the momentum p_i, the summand is the angular momentum of particle i about the origin $r_i^{(0)} \times p_i$ and the sum becomes the total angular momentum in the limit as r_i approaches the reference frame. In the vicinity of the reference frame, the second Eckart condition cannot prevent the system from rotating, but it provides the minimal coupling between vibrations and rotations possible. To turn this framework into a coordinate transformation, we look to a powerful linear algebra tool.

10.3.2 The singular value decomposition

Equation (10.21) can be expressed as yet another matrix, multiplying a vector and yielding a null vector as a result, $Ad = 0$. The solution of the problem becomes one of finding the null space of the matrix A. Then those linear combinations of the relative vectors that are associated with zero eigenvalues of A represent the null space of A. It is beginning to sound like a diagonalization problem, however, the regular eigenanalysis of chapter one does not work here because the matrix A is neither symmetric nor is it square. It is a $3 \times 3(n-1)$, i. e., the three rows and $3(n-1)$ columns we extract from equation (10.21). We rewrite this as

$$A^{\nu}_{\mu} d^{\mu} = 0, \tag{10.22}$$

where the $3(n-1)$ vector d^{μ} is defined in equation (10.18). The first observation that can be made from equation (10.22), regards the components of A only being dependent on the reference frame parameters, and these are constant. Therefore, A is constant. After writing equation (10.21) component-by-component and comparing the result with equation (10.22), one arrives at the following expression for the components of A:

$$A = \begin{pmatrix} 0 & \cdots & -m_1 z_1^{(0)} & \cdots & m_1 y_1^{(0)} & \cdots \\ m_1 z_1^{(0)} & \cdots & 0 & \cdots & -m_1 x_1^{(0)} & \cdots \\ -m_1 y_1^{(0)} & \cdots & m_1 x_1^{(0)} & \cdots & 0 & \cdots \end{pmatrix}, \tag{10.23}$$

where only the first entry ($i = 1$) of the three $n-1$ row vectors strung together is shown, and the dots to the right of each element in the row imply the same pattern for the next $i = 1, 2, \ldots, n-1$ entries. Therefore, the first row has $n-1$ zeros, followed by $-m_1 z_2^{(0)}$, $z_2^{(0)}$ up to $m_{n-1} z_{n-1}^{(0)}$. Similar statements apply for the next $n-1$ elements of row one and the remaining two rows.

To see how one derives equation (10.23), we demonstrate the process for the three-body problem. There are two terms in the sum on the left of equation (10.20),

$$m_1 r_1^{(0)} \times d_1 + m_2 r_2^{(0)} \times d_2 = 0,$$

expanding the two cross-products in terms of components gives the following three lines in the **i**, **j**, **k** basis. The component along **i** is

$$m_1 \left(y_1^{(0)} d_z^{(1)} - z_1^{(0)} d_y^{(1)}\right) + m_2 \left(y_2^{(0)} d_z^{(2)} - z_2^{(0)} d_y^{(2)}\right),$$

whereas

$$m_1\,(z_1^{(0)}d_x^{(1)} - x_1^{(0)}d_z^{(1)}) + m_2\,(z_2^{(0)}d_x^{(2)} - x_2^{(0)}d_z^{(2)}),$$

for the **j** component, and

$$m_1\,(x_1^{(0)}d_y^{(1)} - y_1^{(0)}d_x^{(1)}) + m_2\,(x_2^{(0)}d_y^{(2)} - y_2^{(0)}d_x^{(2)})$$

is the **k** component. Since A multiplies $d^\mu = (d_x^{(1)}, d_x^{(2)}, d_y^{(1)}, d_y^{(2)}, d_z^{(1)}, d_z^{(2)})^T$, the elements of A have to be

$$A = \begin{pmatrix} 0 & 0 & -m_1z_1^{(0)} & -m_2z_2^{(0)} & m_1y_1^{(0)} & m_2y_2^{(0)} \\ m_2z_2^{(0)} & m_1z_1^{(0)} & 0 & 0 & -m_1x_1^{(0)} & -m_2x_2^{(0)} \\ -m_1y_1^{(0)} & -m_2y_2^{(0)} & m_1x_1^{(0)} & m_2x_2^{(0)} & 0 & 0 \end{pmatrix}.$$

Therefore, A is a rectangular matrix with 3 rows and $3(n-1)$ columns. Its rank is three.

The singular value decomposition process produces three matrices from A that are as follows:

$$A = P\Sigma V^T$$

where both P and V are orthogonal,

$$P^TP = PP^T = I_{3\times3}, \quad V^TV = VV^T = I_{3(n-1)\times3(n-1)},$$

and the $I_{k\times k}$ in the last equation are unit matrices. Σ is a $3(n-1)\times3(n-1)$ diagonal matrix with three nonzero eigenvalues, and a number of zero (singular) eigenvalues as the size of the null space of A, i. e., $3n-6$. These are associated with the corresponding columns of V. These are the "internal degrees of freedom" vectors we seek. V is of the same shape as Σ.

The matrix P is 3×3 and it contains the eigenvectors of AA^T, as we now show. Using the decomposition, and the fact that $A^T = V\Sigma P^T$,

$$AA^T = P\Sigma V^T V\Sigma P^T = P\Sigma^2P^T$$

whereas the columns of V matrix are the eigenvectors of A^TA,

$$A^TA = V\Sigma P^TP\Sigma V^T = V\Sigma^2V^T.$$

The nonzero values of Σ^2 are the eigenvalues of AA^T, which is a type of moment of inertia tensor similar to the one defined in Chapter 9,

$$AA^T = \begin{pmatrix} \sum_{i=1}^{n-1} m_i^2(y_i^2 + z_i^2) & -\sum_{i=1}^{n-1} m_i^2 x_i y_i & -\sum_{i=1}^{n-1} m_i^2 x_i z_i \\ -\sum_{i=1}^{n-1} m_i^2 x_i y_i & \sum_{i=1}^{n-1} m_i^2(x_i^2 + z_i^2) & -\sum_{i=1}^{n-1} m_i^2 y_i z_i \\ -\sum_{i=1}^{n-1} m_i^2(x + iy_i) & -\sum_{i=1}^{n-1} m_i^2 y_i z_i & \sum_{i=1}^{n-1} m_i^2(x_i^2 + y_i^2) \end{pmatrix}. \tag{10.24}$$

We are now ready for the definition of a new set of coordinates.

Definition 49 (Relative-Eckart coordinates). The vector ζ^μ

$$\zeta^\mu = (V^T)^\mu_\nu d^\nu \tag{10.25}$$

with d^ν as in equation (10.18) expressed as a column vector, defines the relative-Eckart vector.

To see how one can "satisfy" both Eckart conditions with these coordinates without producing an embedding, consider the following procedure:

- Obtain a reference frame $r_i^{(0)}$, and compute its center of mass vector, $r_c^{(0)} = (x_c^{(0)}, y_c^{(0)}, z_c^{(0)})$.
- Translate the reference frame $r_i^{(0)}$ to the center of mass

$$r_i^{(0)}{}' = r_i^{(0)} - r_c^{(0)}.$$

- Use equation (10.25) and x^μ built from the primed vectors in the previous step to find the vector $(\zeta^{(0)})^\mu$ that correspond to the reference frame.

This produces an initial point in the configuration space of the system mapped by the ζ^μ coordinates. The simulation takes place by updating the ζ^μ vector from this position. Equation (10.25) is never used in the course of a simulation. Typically, the inverse of equation (10.25) is needed to compute the potential energy. To further clarify how all $3n$ Eckart coordinates are linearly independent, let us recall that the Eckart conditions are only fully satisfied at one point in the configuration space spanned by ζ^μ, i. e., $(\zeta^{(0)})^\mu$. At any other point, the net zero angular momentum condition applies only to make minimal deviations from a zero "rotation," as it does not restrict the motion, whereas placing the reference frame in the center of mass is sufficient to insure that all points in the configuration space spanned by ζ^μ transforms into a center of mass frame.

10.3.3 A three particle D_{3h} example

It is possible to derive closed forms for the elements of V^T for $n \geq 2$. However, even for the $n = 2$ case the equations are quite messy and not very informative. To illustrate the concept of the rovibrational coupling, we compute the metric tensor in the ζ^μ basis for a $n = 3$, uniform system with unit masses. The reference frame is arranged as an equilateral triangle on the $x - y$ plane with the following coordinates and masses:

$$\begin{array}{c|cccc} & x & y & z & m \\ \hline r_1^{(0)} & \frac{\sqrt{3}}{2} & -\frac{1}{2} & 0 & 1 \\ r_2^{(0)} & 0 & 1 & 0 & 1 \\ r_3^{(0)} & -\frac{\sqrt{3}}{2} & -\frac{1}{2} & 0 & 1 \end{array} \tag{10.26}$$

With these coordinates, the matrix A in equation (10.23) is a three row, 6 column rectangular object,

$$A = \begin{pmatrix} 0 & 0 & 0 & 0 & -\frac{1}{2} & 1 \\ 0 & 0 & 0 & 0 & -\frac{\sqrt{3}}{2} & 0 \\ \frac{1}{2} & -1 & \frac{\sqrt{3}}{2} & 0 & 0 & 0 \end{pmatrix}. \tag{10.27}$$

The singular value decomposition is carried out with the following (minimal) Python code:

```
import numpy as np
sq32 = np.sqrt(3.0)/2.0
a = np.array([[0,0,0,0,-0.5,1.0],                    # Part 1 code A
              [0,0,0,0,-sq32,0],
              [0.5,-1.0,sq32,0,0,0]])
U, S, Vh = np.linalg.svd(a, full_matrices=True)    # Part 2 SDV of A
tt = 2.0/3.0                                   # Part 3 transform the metric
ot = 1.0/3.0
g = np.array([[tt,-ot,0,0,0,0],
              [-ot,tt,0,0,0,0],
              [0,0,tt,-ot,0,0],
              [0,0,-ot,tt,0,0],
              [0,0,0,0,tt,-ot],
              [0,0,0,0,-ot,tt]])
g1 = np.matmul(g,Vh)
gp = np.matmul(V,g1)
```

The metric tensor is obtained by transforming the matrix G in equation (10.16) using $G' = VGV^T$. The numerical output is a 6×6 array:

$$G' = \begin{pmatrix} 0.94439 & 0.14171 & 0.00329 & 0.00000 & -0.08333 & 0.08333 \\ 0.14171 & 0.58894 & -0.17156 & 0.00000 & 0.16667 & -0.16667 \\ 0.00329 & -0.17156 & 0.46667 & 0.00000 & -0.14434 & 0.14434 \\ 0.00000 & 0.00000 & 0.00000 & 0.66667 & 0.23570 & 0.23570 \\ -0.08333 & 0.16667 & -0.14434 & 0.23570 & 0.66667 & 0.00000 \\ 0.08333 & -0.16667 & 0.14434 & 0.23570 & 0.00000 & 0.66667 \end{pmatrix}.$$

There are several observations we can make about this result. First, we note that the resulting coordinates are not orthogonal. Recall that the metric tensor governs how the unit vectors ϵ_μ associated with the ζ^μ components behave. Clearly, $e_1 \cdot e_2 = 0.14171 \neq 0$ in our system of coordinates. Second, the elements along the main diagonal are all larger compared with those off the main diagonal. Third, the metric tensor is clearly symmetric. The fourth observation regards the top left and bottom right 3×3 blocks. These are separated by a number of zeros in row four and column four. In the top left block the elements along the main diagonal are all different, whereas in the right bottom block they are all numerically equivalent. The top left 3×3 block is associated with the rotation degrees of freedom along with the nonzero singular values of the matrix Σ. Python returns a 3-array with elements sorted in descending order. The algorithm also sorts the rows and columns of Q and V. The nonzero values are 1.41421, 1.22474, 0.70711. The rest of the elements of Σ are zero. The bottom right block is associated with the null space of A, since the corresponding eigenvalues in Σ are zero. The remaining two 3×3 blocks are the ro-vibrational coupling terms.

In conclusion, we have presented several linear transformations of coordinates for systems of n nonidentical point particles that have found practical applications in chemical physics. The relative coordinates, generally only remove three degrees of freedom and by themselves do not seem useful for three or more particles. The relative coordinates we have introduced have found applications for molecular systems of the type XY_n, where X is a heavier atom. It is important to note that the procedure to generate Eckart frame used here is not unique. There are other ways that have been explored in the literature to construct Eckart coordinates. The reader is guided through a procedure that transforms the x^μ vector directly without using relative coordinates at all. (See exercise 9.) These are better suited for uniform systems of particles of the X_n type, but the null space of A contains more than just internal vibrations.

In the exercises, we also introduce mass weighted Cartesian coordinates, and we briefly discussed normal modes. In regards to the latter set, we should clarify that aside from neglecting ro-vibrational couplings, the coordinates themselves are not the sources of inaccuracies. For anharmonic systems, the energy estimates by the harmonic oscillator approximation is incorrect, but the normal mode coordinates need not be a poor representation of the internal modes of vibration. Moreover, though we did not consider it here in detail, it is possible to transform into an Eckart frame, perhaps at the point $(\eta^{(0)})^\mu$, which is also the global minimum of U. Then use the Hessian of U transformed into the null space of A spanned by η^μ,

$$\frac{\partial}{\partial \eta^\mu}\frac{\partial}{\partial \eta^\nu} U = \frac{\partial x^{\mu'}}{\partial \eta^\mu}\frac{\partial x^{\nu'}}{\partial \eta^\nu}\partial_{\mu'}\partial_{\nu'} U,$$

diagonalize the resulting Hessian matrix, and use the eigenvectors to form a new set of coordinates that (a) minimize the rovibrational coupling, and (b) decouple the harmonic terms in the potential. To the best of our knowledge, no such transformation has ever been proposed. However, we are certain there are chemical physics problems where such set of coordinates would be quite useful. Though the process sounds laborious, it is still just a set of two linear transformations that produces a uniform metric tensor; therefore, it only needs to be perform as part of a set up before the simulations begin, adding little cost if any.

10.4 Exercises

1. Prove equation (10.7).
2. Show that the cross-product is distributive. This was critical for the proof of Theorem 33. The best way to accomplish this is to compare the **i** component of $a \times (b + c)$ with those of $a \times b + a \times c$.
3. This exercise is an exploration of the relative coordinate transformation steps for a two-particle system. It takes the reader through the various steps outlined in the chapter, and the end result should be a familiar concept from introductory physics: The reduced mass of a two-body system.
   (a) Write equation (10.2) and equation (10.5) for the $n = 2$ case.
   (b) Insert the expressions in part (a) into equations (10.6) and (10.7), which for the $n = 2$ case read

$$x_1 = d_x + x_c - \frac{1}{m_1 + m_2} m_1 d_x \tag{10.28}$$

$$x_2 = x_c - \frac{1}{m_1 + m_2} m_1 d_x \tag{10.29}$$

   and show that the equalities are satisfied.
   (c) Use equations (10.28) and (10.29) along with the corresponding one for y and z components to evaluate the Jacobian matrix,

$$J = \frac{\partial(x_1, x_2, y_1, y_2, z_1, z_2)}{\partial(x_c, y_c, z_c, d_x, d_y, d_z)}.$$

   (d) Write the transpose of J in part (c) and compare it with equation (10.8) specialized for the $n = 2$ case.
   (e) Write an expression for the Hessian metric tensor in the $(x_1, x_2, y_1, y_2, z_1, z_2)$ coordinate system.
   (f) Transform the Hessian metric tensor in part (e) into the $(x_c, y_c, z_c, d_x, d_y, d_z)$ coordinate system, then compare your expression with equation (10.16) specialized for the $n = 2$ case. Recall that the metric tensor transforms as follows:

$$g_{\mu'\nu'} = \frac{\partial x^\mu}{\partial x^{\mu'}} \frac{\partial x^\nu}{\partial x^{\nu'}} g_{\mu\nu}$$

where in this case $g_{\mu\nu}$ is the diagonal matrix of part (e) and the terms $\frac{\partial x^{\mu}}{\partial x^{\mu'}}$ are elements of J in part (c). The order of the primed system as given in part (f) must be carefully maintained. The fact that $g_{\mu\nu}$ is diagonal turns the double sum in the transformation law into a single sum, and the block structure of J further simplifies the operation. For instance, the g_{11} associated with the "x_c, x_c" element of $g_{\mu'\nu'}$ reads as follows:

$$g_{11} = \left(\frac{\partial x_1}{\partial x_c}\right)^2 m_1 + \left(\frac{\partial x_2}{\partial x_c}\right)^2 m_2 + \left(\frac{\partial y_1}{\partial x_c}\right)^2 m_1 + \left(\frac{\partial y_2}{\partial x_c}\right)^2 m_2 + \left(\frac{\partial z_1}{\partial x_c}\right)^2 m_1 + \left(\frac{\partial z_2}{\partial x_c}\right)^2 m_2$$

but a number of these partial derivatives are zero, leaving only

$$g_{11} = \left(\frac{\partial x_1}{\partial x_c}\right)^2 m_1 + \left(\frac{\partial x_2}{\partial x_c}\right)^2 m_2.$$

(g) Equation (10.17) applied to the $n = 2$ case should yield only one by one blocks for $g^{(xx)}$, $g^{(yy)}$ and $g^{(zz)}$. Use equation (10.17) to show that for the $n = 2$ case the three one-by-one blocks are the reduced mass of the system.

$$\mu = \frac{m_1 m_2}{m_1 + m_2}.$$

(h) Consider equation (10.21) specialized for the $n = 2$ case. Express in your own words what type of rotation is involved in it.

(i) Write expression for the matrix A in equation (10.23) or the $n = 2$ case.

4. Derive equations (10.28) and (10.29) by solving a system of two equations, two unknowns (x_1, x_2) given x_c and d_x, and their definitions.
5. Work through the 2×2 $g^{(xx)}$ block of $g_{\mu'\nu'}$ for the $n = 3$ case, where by the primed system we mean the relative coordinates defined in equations (10.4) and (10.5). You do not need to evaluate all the derivatives of the nine by nine Jacobian to accomplish this task. Begin by writing the single sum for each element of $g^{(xx)}$ and recall that partials of x-like components with respect to the y and z components of the relative coordinates vanish, etc. Compare your result with equation (10.17).
6. In this chapter, we introduced the Hessian metric tensor to automatically include the masses in the kinetic energy operator, i. e., we can now simply write

$$K = -\frac{\hbar^2}{2}(g^{\mu\nu}\partial_\mu\partial_\nu + \mathcal{G}^{\mu}\partial_\mu),$$

where

$$\frac{1}{\sqrt{g}}\partial_\lambda \sqrt{g} g^{\mu\lambda} = \mathcal{G}^{\mu}$$

in any coordinate system. Write the expression for K using the Hessian metric tensor for the $n = 2$ case in the following coordinate systems:
- Cartesian $(x_1, x_2, y_1, y_2, z_1, z_2)$.
- Relative (d_x, d_y, d_z).

7. When deriving normal mode coordinates it is convenient to introduce a mass-weighted system of Cartesian coordinates using the following trivial transformation:

$$x_i' = m_i^{1/2} x_i \tag{10.30}$$

with corresponding equations for the y and z components. The following applies to the two particle system.
   (a) Find an expression for $g_{\mu'\nu'}$. Remember to transform the Hessian metric.
   (b) Write an expression for the kinetic energy operator K (see the previous exercise) in the primed system defined in equation (10.30). Explain why the masses do not appear in your expression and why it is okay.
   (c) Write equations for the center of mass using the primed coordinates.
   (d) Define a doubly primed system using $x^{\mu''} = (x_c, y_c, z_c, d_x', d_y', d_z')$ where, e. g., $d_x' = x_1' - x_2'$ and derive the equivalent of equations (10.28) and (10.29), i. e., the equations to map $x^{\mu''} \to x^{\mu'}$. (See exercise 4.)
   (e) Find the expressions for

$$J = \frac{\partial(x_1', x_2', y_1', y_2', z_1', z_2')}{\partial(x_c, y_c, z_c, d_x', d_y', d_z')}$$

   then use it to find $g_{\mu''\nu''}$. To expedite the process, it helps to define the following two quantities:

$$\alpha_m = \frac{m_1^{1/2}}{m_1^{1/2} + m_2^{1/2}}, \quad \beta_m = \frac{m_1 + m_2}{m_1^{1/2} + m_2^{1/2}}.$$

   You may leave your answer in terms of α_m and β_m since your expressions do not simplify further.
   (f) Write an expression for the matrix A in equation (10.23) in the mass weighted primed system.

8. Use the result in part (f) of the last exercise to obtain expressions for AA^T, and compare it with the inertia tensor in exercise 9.6.
9. Run the Python code for the D_{3h} example and verify the result in the chapter. You may need to add proper printing statements. Then perform the following task by adding appropriate code:
   (a) Check that $P^T P = PP^T$ is a unit 3×3 matrix, and that $V^T V = VV^T$ is a 6×6 unit matrix.

(b) Check that the product $P\Sigma V^T$ reconstructs A. You may need to pad with zeros a new matrix Σ, using the vector `S` nonzero entries to make a 6×6 matrix, else the routine `matmul` will get confused. For instance, for the P matrix we had to use

```
up = np.array([[U[0,0] ,U[0,1],U[0,2],0,0,0],
          [U[1,0] ,U[1,1],U[1,2],0,0,0],
          [U[2,0] ,U[2,1],U[2,2],0,0,0],
          [0,0,0,0,0,0],
          [0,0,0,0,0,0],
          [0,0,0,0,0,0]])
```

to turn it into the correct shape. The reconstructed A matrix will be a 6×6 array with the bottom three rows computed as zeros. This step is critical to verify that the null space is identified properly with the correct columns of V.

10. Consider the three particle system as in the previous exercise. This exercise guides you to construct the most general Eckart coordinates using the second Eckart condition and transforming directly from the 9 Cartesian coordinates.

(a) Write the matrix A that corresponds to general Eckart condition

$$\sum_{i=1}^{n} m_i r_i^{(0)} \times r_i = 0$$

using the vector x^μ ordered according to the definition in (10.1). This simply adds properly placed columns to A in the previous exercise. The coordinates and masses to be used are in equation (10.26). A should be a 3×9 rectangular array and its null space is six-dimensional.

(b) Modify the Python code from the previous exercise to compute the new A matrix and the new metric tensor. Note that because we are skipping over the relative coordinate transformation step, the proper Hessian metric tensor to be transformed is in equation (10.14) adapted to the present case. What happened to the ro-vibrational coupling terms?

(c) Repeat the exercise in part (b) but change the mass of particle 3 to 10 and carefully inspect the metric tensor. Note, however, that the reference frame in equation (10.26) is no longer in the center of mass.

(d) What conclusions can be drawn from the exercises in part (a) and (b)?

11 Diffusion Monte Carlo in manifolds

Hard study on the great models has ever brought out the strong; and of such must be our new scientific generation if it is to be worthy of the era to which it is born and of the struggles to which it is destined.

Eugenio Beltrami

Generalizing the diffusion Monte Carlo and other simulation strategies to more than one dimension is relatively straightforward as long as we continue to use Cartesian coordinates. However, as we have argued in Chapter 10, these are not always desirable. In the previous chapters, we have encountered three important objects that naturally occur whenever we change coordinates. These are (a) the Laplace–Beltrami operator, (b) the metric tensor, and (c) atlases. The most general type of diffusion equation we want to be able to solve is

$$\hbar\partial_t\psi = \frac{\hbar^2}{2}(g^{\mu\nu}\partial_\mu\partial_\nu + \mathcal{G}^\mu\partial_\mu)\psi - U\psi, \tag{11.1}$$

where the kinetic energy term is the Laplace–Beltrami operator in equation (8.31) multiplied by $\hbar^2/2$ and $g^{\mu\nu}$ is the system effective inverse mass. Equation (11.1) can be obtained from the time dependent Schrödinger equation by transforming time as we did in Chapter 5. In the most general case, the metric tensor is not uniform, and is not diagonal. Moreover, a single chart may cause singularities in the kinetic energy operator creating numerical overflow in the course of simulations. The question we ask in this chapter is how does one handle all of the added complexities when simulating a system with non-Cartesian coordinates. We break down the process of generalizing the basic algorithm introduced in Chapter 5 into several parts. First, we begin investigating the case of a constant metric tensor in a system of nonorthogonal coordinates like, e. g., in an Eckart frame of Chapter 10. Simulating the diffusion process requires drawing random numbers from a *correlated multivariate* Gaussian distribution. To build intuition, we begin with a two-dimensional manifold endowed with a constant metric tensor.

Second, we have already derived a "drift-like" move that advection vectors produce in Chapter 5 when the implementation of the Smoluchowski operator is discussed. The monodimensional case treated in Chapter 5 is sufficient to explain the process. We show that the update to the population configuration is deterministic, meaning it does not use random variables. Of course, it is possible to combine both update terms and extend them to many dimensions for nonuniform metrics and advection vectors $\mathcal{G}^\mu$ all at the same time. To derive the propagators for nonuniform metrics and advections, we assume these objects to be approximately uniform provided that a sufficiently small step in time is taken. The resulting propagator when implemented to update the coordinates is called the Euler–Maruyama method. This approximation is the equivalent of neglecting linear terms in Δt in the power series expansion of the Green function. Therefore,

https://doi.org/10.1515/9783111610207-011

for the most general case the values of physical properties are expected converge only as $\Delta t^{1/2}$.

This convergence behavior is worse than the basic algorithm in Cartesian coordinates, which is linear in Δt, and quadratic when the second-order branching term is introduced. However, the $\Delta t^{1/2}$ behavior can only be proved using the tools of stochastic calculus. We do discuss briefly this modern field of mathematics and present recently developed strategies to achieve second-order convergence (meaning excluding $\mathcal{O}(\Delta t^3)$ terms) for diffusion Monte Carlo simulations when combined with second-order branching. The method is called the Milshtejn propagator [81]. We simply lift it from the literature and present it as useful information. Having said that, our experience with the Euler–Maruyama method when combined with the second-order branching is that the convergence is quadratic when masses are in the nuclear range, and the reason for the observed behavior is clear when inspecting the denominators of the additional terms the Milshtejn propagator introduces. We conclude this chapter with a numerical example. Together with some of the exercises at the end of the chapter the example should confirm the computational advantage stereographic projection coordinates, introduced in Chapter 9, have to offer when dealing with angular momentum states.

11.1 Covariant multivariate sampling

Suppose we have the following diffusion equation to solve:

$$\hbar \partial_t \psi = \frac{\hbar^2}{2}\left(a\frac{\partial^2}{\partial x^2} + c\frac{\partial^2}{\partial y^2} + 2b\frac{\partial^2}{\partial x \partial y} \right)\psi \tag{11.2}$$

where a, b, and c are constant real numbers that satisfy

$$(a+c)^2 - 4(ac - b^2) > 0.$$

One way to derive the update for the population coordinates is to evoke the Fourier transform as we did in Chapter 5. Its generalization in $\mathbb{R}^2$ is straightforward,

$$\mathscr{F}[f(x,y,t)] = \tilde{f}(\alpha_x, \alpha_y, t) = \int_{-\infty}^{\infty} \left[\int_{-\infty}^{\infty} f(x,y,t)\, e^{-i(\alpha_x x + \alpha_y y)}\, dy \right] dx,$$

To proceed, we need the following results.

Theorem 34. *The following four results apply,*
(a)

$$\mathscr{F}\left[\frac{\partial}{\partial t} f(x,y,t)\right] = \frac{\partial}{\partial t}\tilde{f}(\alpha_x, \alpha_y, t).$$

(b)

$$\mathcal{F}\left[\frac{\partial}{\partial x} f(x, y, t)\right] = i\alpha_x \tilde{f}(\alpha_x, \alpha_y, t).$$

(c)

$$\mathcal{F}\left[\frac{\partial^2}{\partial x^2} f(x, y, t)\right] = -\alpha_x^2 \tilde{f}(\alpha_x, \alpha_y, t)$$

with a corresponding equations for the partials with respect to y,

(d)

$$\mathcal{F}\left[\frac{\partial^2}{\partial y \partial x} f(x, y, t)\right] = -\alpha_x \alpha_y \tilde{f}(\alpha_x, \alpha_y, t).$$

We prove parts (b) and (d) in detail and leave the remaining parts as an exercise.

Proof. Evaluating the derivative with respect to x of the definition of the Fourier transform

$$\frac{\partial}{\partial x} \mathcal{F}[f(x, y, t)] = \frac{\partial}{\partial x} \tilde{f}(\alpha_x, \alpha_y, t) = 0$$

since $\tilde{f}(\alpha_x, \alpha_y, t)$ does not depend on x. It follows from the definition that

$$\frac{\partial}{\partial x} \int_{-\infty}^{\infty} \left[\int_{-\infty}^{\infty} f(x, y, t)\, e^{-i(\alpha_x x + \alpha_y y)}\, dy \right] dx = 0$$

and assuming the integrand is bound over the domain, we can exchange the order of operation,

$$\int_{-\infty}^{\infty} \left[\int_{-\infty}^{\infty} \frac{\partial}{\partial x} f(x, y, t)\, e^{-i(\alpha_x x + \alpha_y y)}\, dy \right] dx = 0$$

$$\int_{-\infty}^{\infty} \left[\int_{-\infty}^{\infty} \left(\frac{\partial}{\partial x} f(x, y, t) \right) e^{-i(\alpha_x x + \alpha_y y)} - i\alpha_x f(x, y, t)\, e^{-i(\alpha_x x + \alpha_y y)}\, dy \right] dx = 0,$$

where we have applied the product rule in the last step. Using the linearity of the integral operator, we can rearrange this expression to read

$$\begin{aligned} &\int_{-\infty}^{\infty} \left[\int_{-\infty}^{\infty} \left(\frac{\partial}{\partial x} f(x, y, t) \right) e^{-i(\alpha_x x + \alpha_y y)}\, dy \right] dx \\ &= i\alpha_x \int_{-\infty}^{\infty} \left[\int_{-\infty}^{\infty} f(x, y, t)\, e^{-i(\alpha_x x + \alpha_y y)}\, dy \right] dx. \end{aligned}$$

But by using the definitions of $\mathcal{F}[f(x,y,t)]$ and $\mathcal{F}[\frac{\partial}{\partial x}f(x,y,t)]$, we arrive at the desired result.

To prove part (d), we also note that

$$\frac{\partial}{\partial y}\mathcal{F}\left[\frac{\partial}{\partial x}f(x,y,t)\right] = 0$$

since $\tilde{f}(\alpha_x,\alpha_y,t)$ does not depend on either x or y. Therefore,

$$\int_{-\infty}^{\infty}\left[\int_{-\infty}^{\infty}\frac{\partial}{\partial y}\left(\frac{\partial}{\partial x}f(x,y,t)\right)e^{-i(\alpha_x x+\alpha_y y)}\,dy\right]dx = 0.$$

The derivative of the integrand unravels as follows:

$$\begin{aligned}
&\frac{\partial}{\partial y}\left(\frac{\partial}{\partial x}f(x,y,t)\right)e^{-i(\alpha_x x+\alpha_y y)}\\
&\quad = \left(\frac{\partial^2}{\partial y\partial x}f(x,y,t)\right)e^{-i(\alpha_x x+\alpha_y y)} - i\alpha_y\left(\frac{\partial}{\partial x}f(x,y,t)\right)e^{-i(\alpha_x x+\alpha_y y)}.
\end{aligned}$$

Inserting these two terms back into the double integral, we arrive at

$$\begin{aligned}
&\int_{-\infty}^{\infty}\left[\int_{-\infty}^{\infty}\left(\frac{\partial^2}{\partial y\partial x}f(x,y,t)\right)e^{-i(\alpha_x x+\alpha_y y)}\,dy\right]dx\\
&\quad = i\alpha_y\int_{-\infty}^{\infty}\left[\int_{-\infty}^{\infty}\left(\frac{\partial}{\partial x}f(x,y,t)\right)e^{-i(\alpha_x x+\alpha_y y)}\,dy\right]dx,
\end{aligned}$$

using the definition of the Fourier transform, this reads

$$\mathcal{F}\left[\frac{\partial^2}{\partial y\partial x}f(x,y,t)\right] = i\alpha_y\mathcal{F}\left[\frac{\partial}{\partial x}f(x,y,t)\right]$$

replacing the right-hand side of the last expression with that of part 2 yields the desired result. □

We now transform equation (11.2),

$$\frac{\partial}{\partial t}\tilde{\psi} = -\frac{\hbar}{2}(a\alpha_x^2 + c\alpha_x^2 + 2b\alpha_x\alpha_y)\tilde{\psi},$$

which is a first-order ordinary differential equation. This can be integrated to yield

$$\tilde{\psi}(\alpha_x,\alpha_y,t) = \exp\left[-\frac{\hbar}{2}(a\alpha_x^2 + c\alpha_x^2 + 2b\alpha_x\alpha_y)t\right].$$

The task now is to invert this solution in the frequency domain back to the space domain,

$$\psi(x,y,t) = \int_{-\infty}^{\infty}\left[\int_{-\infty}^{\infty} \exp\left[-\frac{\hbar}{2}(aa_x^2 + ca_x^2 + 2ba_x a_y)t\right] e^{i(a_x x + a_y y)}\, da_y\right] da_x$$

$$= \int_{-\infty}^{\infty}\left[\int_{-\infty}^{\infty} \exp\left[-\frac{\hbar}{2}(aa_x^2 + ca_x^2 + 2ba_x a_y)t + ia_x x + ia_y y\right] da_y\right] da_x.$$

The integrand is a generalized Gaussian form [104]. The following result, the proof of which is left as an exercise, helps us to evaluate it in closed form

$$\int d^n x \exp\left(-\frac{1}{2}\sum_{i,j=1}^{n} \mathcal{A}_{ij}x_i x_j + \sum_{i=1}^{n} \mathcal{B}_i x_i\right)$$
$$= (2\pi)^{n/2}\left[\det(\mathcal{A})\right]^{-1/2} \exp\left(\frac{1}{2}\sum_{i,j=1}^{n} \mathcal{A}_{ij}^{-1}\mathcal{B}_i\mathcal{B}_j\right), \tag{11.3}$$

by making the following identifications: $n = 2$, $(x_1, x_2) = (a_x, a_y)$, $(\mathcal{B}_1, \mathcal{B}_2) = (ix, iy)$, and

$$\mathcal{A} = \begin{pmatrix} a & b \\ b & c \end{pmatrix} t.$$

Then

$$\mathcal{A}^{-1} = \frac{1}{\det \mathcal{A}}\begin{pmatrix} c & -b \\ -b & a \end{pmatrix},$$

and ψ becomes

$$\psi(x,y,t) = (2\pi)\left[\det(\mathcal{A})\right]^{-1/2} \exp\left[-\frac{1}{2}\,(x \quad y)\,\mathcal{A}^{-1}\begin{pmatrix} x \\ y \end{pmatrix}\right].$$

This expression may not seem very helpful, however, there are well-known procedures to sample correlated multivariate Gaussian random variables like these. Let us recall that $\mathcal{A}$ plays the role of $g^{\mu\nu}$, a constant inverse metric tensor, a positive definite matrix. In order to relate this to the diffusion steps performed in Chapter 5, $x(t+\Delta t) = x(t)+\eta$ recall that in $\mathbb{R}^1$ the solution of the free diffusion equation

$$\frac{\partial}{\partial t}\psi(x,t) = D\frac{\partial^2}{\partial x^2}\psi(x,t)$$

is

$$\psi(x,t) = \frac{1}{2\sqrt{\pi D t}}\, e^{-x^2/(4Dt)},$$

where $D = \hbar^2/(2m)$. The form of ψ tells us that x should be updated $x :\longrightarrow x + \eta$ where η is a Gaussian variable with zero mean and variance $\sigma^2 = 2Dt$ from the distribution,

$$p(\eta) = e^{-\eta^2/(2\sigma^2)}.$$

Such sampling is accomplished by drawing a Gaussian random number with zero mean and unit variance and multiplying it by the correct σ,

$$\sigma = \sqrt{2D\Delta t} = \hbar\sqrt{\frac{\Delta t}{m}}.$$

We now need to extend this procedure to our $\mathbb{R}^2$ case. Multivariate correlated Gaussian sampling occurs in many areas of mathematics, including statistics, where $\mathcal{A}$ is interpreted as a variance matrix Σ. The off-diagonal terms of the variance matrix give rise to correlated variance or covariance. One approach to sample such distribution involves diagonalizing Σ,

$$\begin{pmatrix} \sigma_1 & 0 \\ 0 & \sigma_2 \end{pmatrix}^2 = P^T \Sigma P,$$

then we can write the quadratic form in the exponent of $\psi(x, y, t)$ as follows:

$$\begin{aligned}(x \quad y)\, \mathcal{A}^{-1} \begin{pmatrix} x \\ y \end{pmatrix} &= (x \quad y)\, P \begin{pmatrix} \sigma_1 & 0 \\ 0 & \sigma_2 \end{pmatrix}^{-2} P^T \begin{pmatrix} x \\ y \end{pmatrix} \\ &= (x \quad y)\, P \begin{pmatrix} \sigma_1^{-2} & 0 \\ 0 & \sigma_2^{-2} \end{pmatrix} P^T \begin{pmatrix} x \\ y \end{pmatrix} = (u \quad v) \begin{pmatrix} \sigma_1^{-2} & 0 \\ 0 & \sigma_2^{-2} \end{pmatrix} \begin{pmatrix} u \\ v \end{pmatrix}\end{aligned}$$

where

$$\begin{pmatrix} u \\ v \end{pmatrix} = P^T \begin{pmatrix} x \\ y \end{pmatrix}.$$

Therefore, one procedure to draw the correct random numbers for our diffusion simulation is to

- Select $(\eta_1, \eta_2)^T$ a vector of 2 Gaussian random numbers with zero mean and unit variance.
- Transform $(\eta_1, \eta_2)^T :\to (\eta_u, \eta_v)^T$

$$\eta_u = \eta_1 \sigma_1^{-1} \sqrt{\Delta t}, \quad \eta_v = \eta_2 \sigma_2^{-1} \sqrt{\Delta t}.$$

Transform $(\eta_u, \eta_v)^T :\to (\eta_x, \eta_y)^T$

$$\begin{pmatrix} \eta_x \\ \eta_y \end{pmatrix} = P \begin{pmatrix} \eta_u \\ \eta_v \end{pmatrix}.$$

Alternatively, we can split $\mathcal{A}^{-1}$ into its square root and perform a single transformation. There are a number of ways to do this. For example, the Cholesky decomposition finds a lower triangular matrix L such that $\mathcal{A}^{-1} = LL^T$. However, we have found that a more symmetric decomposition always produces uniform convergence in Δt for all degrees of freedom, whereas occasionally the Cholesky decomposition does not. To achieve a symmetric decomposition, we write

$$(x \quad y)\, \mathcal{A}^{-1} \begin{pmatrix} x \\ y \end{pmatrix} = (x \quad y)\, \mathcal{D}^2 \begin{pmatrix} x \\ y \end{pmatrix}$$

where

$$\mathcal{D} = P \begin{pmatrix} \sigma_1^{-1} & 0 \\ 0 & \sigma_2^{-1} \end{pmatrix} P^T.$$

Then

$$\begin{pmatrix} \eta_x \\ \eta_y \end{pmatrix} = \mathcal{D} \begin{pmatrix} \eta_1 \\ \eta_2 \end{pmatrix}$$

where $(\eta_1, \eta_2)^T$ is as above. Therefore, the general approach is to diagonalize $g_{\mu\nu}$ and use its eigenvalues and eigenvectors to split the inverse $g^{\mu\nu}$ into its "square root" matrix

$$g^{\mu\nu} = \sum_{\lambda=1}^{n} \sigma^{\mu\lambda} \sigma^{\lambda\nu},$$

and use $\sigma^{\mu\lambda}$ to convert Gaussian random numbers with zero mean and variance $(\Delta t)^2$ into the properly covariant ones. There is a frequently used symbol for the set of Gaussian random numbers with mean μ and variance s^2: $\mathcal{N}(\mu, s^2)$. We use this symbol as a convenient shorthand for the rest of this chapter.

11.2 The Fourier transform of the diffusion equation

We begin with the Fourier transform of a multidimensional function $\psi : \mathcal{M}, t \to \mathbb{R}$, a function of n coordinates for the system's configuration and time:

$$\mathcal{F}(\psi) = \int_{\mathcal{M}} e^{-i\alpha_\mu x^\mu} \psi \, d^n x = \widetilde{\psi}$$

where $\widetilde{\psi}$ is a function of $n + 1$ variables, namely α_μ and t. Note that the definition of the Fourier transform integral does not require a Jacobian. Technically, we are using a fundamental property of manifolds that allow us to set up a system of local Cartesian-like coordinates in the neighborhood of a point. If Δt is sufficiently small, then we can treat $g^{\mu\nu}$ and $\mathcal{G}^\mu$ as constant objects. Moreover, because we are in a tangent space set up at a point in $\mathcal{M}$, we can extend the tangent space rectangular coordinates to $\pm\infty$, which allows us to use the theorem in equation (11.3). With these caveats in mind, the three results we need are as follows.

Theorem 35. *The generalized Fourier transform of the time derivative satifies the following identity*

$$\mathcal{F}(\partial_t \psi) = \partial_t \tilde{\psi}.$$

Proof. By using the definition,

$$\mathcal{F}(\partial_t \psi) = \int_{\mathcal{M}} e^{-ia_\mu x^\mu} \partial_t \psi \, d^n x$$

and permuting the integration with the time derivative we arrive at the desired expression. □

Theorem 36. *The generalized Fourier transform of the gradient satifies the following identity*

$$\mathcal{F}(\partial_\mu \psi) = ia_\mu \tilde{\psi}.$$

Proof. By using the definition,

$$\mathcal{F}(\partial_\mu \psi) = \int_{\mathcal{M}} e^{-ia_\nu x^\nu} \partial_\mu \psi \, d^n x$$

using integration by parts, and assuming ψ vanishes asymptotically, we get

$$\mathcal{F}(\partial_\mu \psi) = \int_{\mathcal{M}} e^{-ia_\nu x^\nu} ia\psi \, d^n x = ia \int_{\mathcal{M}} e^{-ia_\nu x^\nu} \psi \, d^n x.$$ □

Theorem 37. *If $g^{\mu\nu}$ is uniform,*

$$\mathcal{F}(g^{\mu\nu} \partial_\mu \partial_\mu \psi) = -g^{\mu\nu} a_\mu a_\nu \tilde{\psi}.$$

Proof. From definition of the Fourier transform, and using the uniformity of the metric, we write

$$\mathcal{F}(g^{\mu\nu} \partial_\mu \partial_\mu \psi) = g^{\mu\nu} \int_{\mathcal{M}} e^{-ia_\sigma x^\sigma} \partial_\mu \partial_\nu \psi \, d^m x.$$

Using Theorem 36 twice, first for the ∂_ν integration and then for the other partial derivative, this becomes

$$\mathcal{F}(g^{\mu\nu}\partial_\mu\partial_\mu\psi) = -g^{\mu\nu}\alpha_\nu\alpha_\mu \int_{\mathcal{M}} e^{-i\alpha_\sigma x^\sigma}\psi\, d^m x,$$

from which the theorem follows trivially. □

Equation (11.1) transforms into

$$\hbar\partial_t\widetilde{\psi} = -\frac{\hbar^2}{2}(g^{\mu\nu}\alpha_\mu\alpha_\nu\widetilde{\psi} - i\mathcal{G}^\mu\alpha_\mu)\widetilde{\psi} - U\widetilde{\psi}.$$

Solving the ordinary DE in time with the usual initial conditions produces

$$\widetilde{\psi}(\alpha, t) = \widetilde{\psi}(\alpha, 0)\, e^{-Dg^{\mu\nu}\alpha_\mu\alpha_\mu t + iD\mathcal{G}^\mu\alpha_\mu t - Ut}$$

and $D = \hbar/2$. The inverse transform integral in n dimensions becomes

$$\mathcal{K}(x, 0, t) = \frac{1}{(2\pi)^n} e^{-Ut} \int_{-\infty}^{+\infty} e^{-Dg^{\mu\nu}\alpha_\mu\alpha_\mu t + iDt\mathcal{G}^\mu\alpha_\mu + ix^\mu\alpha_\mu}\, d^m\alpha.$$

The argument of the exponential function is a general quadratic form with a $n \times n$ coupling matrix

$$\mathcal{A}^{\mu\nu} = 2Dtg^{\mu\nu}$$

and n vector

$$\mathcal{B}^\mu = i(x^\mu + Dt\mathcal{G}^\mu).$$

Using equation (11.3) once more, we arrive at a general for the propagator:

$$\mathcal{K}(x, 0, t) = g^{1/2}\sqrt{\frac{1}{(4\pi Dt)^n}}\, e^{-Ut}\, e^{-g_{\mu\nu}(x^\mu + Dt\mathcal{G}^\mu)(x^\nu + D\mathcal{G}^\mu)/(4Dt)}. \tag{11.4}$$

This is the most general form we can derive using the Fourier transform approach. The translation from the propagator to the update rule for the position of replicas is a combination of the covariant multivariate sampling along with the deterministic drifts proportional to $\mathcal{G}^\mu$. The resulting strategy may or may not converge to the correct ground state the same way that the basic first-order diffusion Monte Carlo method in Cartesian coordinate does. Convergence behavior is highly dependent on the coordinates chosen for the charts, and the subtleties of the topology of the manifold. Careful testing is required when developing methods in this area. Assuming we have dealt with possible numerical singularities and nonuniform convergence issues, the approach should converge linearly with Δt. In the next section, we discuss more efficient methods.

11.3 Stochastic differential equations

Equation (11.1) can be interpreted as a type of stochastic differential equation whose solution can be written in terms of a new type of integral over a random variable. The configuration space update for a finite time interval in the language of Itô calculus is written as

$$x^{\mu}_{n+1} = x^{\mu}_{n} + \int_{t_n}^{t_{n+1}} \sigma^{\mu}_{\nu}\, dW^{\nu}_{s} + D \int_{t_n}^{t_{n+1}} \mathcal{G}^{\mu} ds$$

where $D = \hbar/2$. The second integral on the right is a regular Riemann type in one (time-like) dimension, whereas the first one defines a new type of operation known as the Itô integral. The symbol W^{μ}_{n} represents a set of d random variables drawn at time t_n of the $\mathcal{N}(0, \Delta t^2)$ type, i. e.,

$$\langle W^{\nu}_{t} \rangle = 0,$$
$$\langle (W^{\nu}_{t})^2 - \langle W^{\nu}_{t} \rangle^2 \rangle = (\Delta t)^2.$$

The Itô integral is a sum over a set of random trajectories in configuration space distributed over time according to W_n and these are used to build the expectations in the last two equations.

If σ^{μ}_{ν} is uniform, then the Itô integral simply evaluates to $\sigma^{\mu}_{\nu} W^{\nu}_{n}$. The resulting algorithm is known as the Euler–Maruyama method,

$$x^{\mu}_{n+1} = x^{\mu}_{n} + \sigma^{\mu}_{\nu} W^{\nu}_{n} + D\mathcal{G}^{\mu} \Delta t,$$

and this update strategy is the same as the one derived from the Green propagator in equation (11.4). However, if σ^{μ}_{ν} is not uniform, then the Euler–Maruyama method converges to first order in Δt. To achieve second-order convergence with a new term, proportional to the gradient of σ^{μ}_{ν} has to be included,

$$x^{\mu}_{n+1} = x^{\mu}_{n} + \sigma^{\mu}_{\nu} W^{\nu}_{n} + D\mathcal{G}^{\mu} \Delta t + \frac{1}{2}(\sigma^{\lambda}_{\kappa} \partial_{\lambda} \sigma^{\mu}_{\nu})(W^{\kappa}_{n} W^{\nu}_{n} + \Omega^{\kappa\nu}_{n})$$

where the random numbers $\Omega^{\kappa\nu}_{n}$ are drawn from a uniform distribution as follows:

1. Draw a uniform random number $z \in (0, 1)$.
2. If $z < 1/2$ and $\kappa < \nu$, then $\Omega^{\kappa\nu}_{n} = -\Delta t$. If $z > 1/2$ and $\kappa < \nu$, then $\Omega^{\kappa\nu}_{n} = -\Delta t$.
3. Finally, $\Omega^{\nu\kappa}_{n} = -\Omega^{\kappa\nu}_{n}$.

The name of the quadratically convergent algorithm is the Milshtejn method.

We just took a very brief journey in a relatively new field of mathematics called stochastic calculus. Our objective is to find the right terms to improve the efficiency of algorithms that aim to estimate physical properties statistically. Stochastic calculus, however, represents a paradigm shift for physics. For a good part of the last century, theoreticians attempted to solve the quantum problem using the classical mathematics of the nineteenth century. There is glaring logical inconsistency in the processes of seeking a solution to purely random events described by some distribution using deterministic means. Stochastic calculus provides the proper tools to build mathematical models for purely random events without requiring deterministic terms. The theoretical physics community has made the shift to stochastic theories for quantum mechanics on few occasions. Perhaps the most famous example is the Feynman path integral, an integral over random trajectories not too different from the Itô integral we have just encountered briefly.

Stochastic calculus is an important frontier in theoretical physics, and consequently, theoretical chemistry. As another well-known example where stochastic methods can improve theoretical models, consider the following recent advance. Perhaps the reader is familiar with the collapse of the wave packet conundrum. As a brief reminder, prior to an experiment aiming to measure a physical property such as, e. g., momentum, the state of a system is thought to be in a linear superposition of pure momentum states. The act of measuring momentum collapses the linear superposition into a pure momentum state with a probability that can be computed using the rules of quantum mechanics. These probabilities are then confirmed by repeating the experiment a sufficient number of times. The difficulty for the deterministic time evolution of a wave packet emerging from the Schrödinger formalism is that there is no built-in mathematical process able to reproduce such instantaneous (or more likely very fast) collapse. However, one can think of the interaction between the system and the external "apparatus," as some type of second-order stochastic correction to the Schrödinger equation, much the same way we had to evoke stochastic corrections to capture second-order convergence in this section. In later chapters, the reader will see how from a first-order expansion of the Feynman path integral one derives the determinist Schrödinger equation, suggesting that the stochastic view of quantum systems may indeed produce more accurate models and perhaps solve some of the vexing problem plaguing the theoretical communities.

11.4 Numerical examples

An implementation of the basic diffusion Monte Carlo method to simulate the ground state of a particle of mass m on a sphere of radius r using stereographic projections u, v is presented in this section. The u and v coordinates on the projection plane from the north pole of the sphere have been introduced in Chapter 9. A minimal simulation code follows in this section. The code is tested with a sphere of unit radius, a mass of 919.0 atomic units, and a potential energy corresponding to the interaction of a unit charge with an electric dipole,

$$U = U_0 \frac{\sigma^2 - 4}{\sigma^2 + 4}$$

where $\sigma^2 = (u)^2 + (v)^2$. This corresponds to $U = U_0 \cos\theta$ when polar angles are used. The minimum of the potential is a $u, v = (0, 0)$, or in terms of xyz coordinates on the sphere, at $z = -1$.

```
import numpy as np
import random as rnd
rnd.seed()    # seed with system time
mass,radius,mu,deltat,eref = 919.0,1.0,0.0,0.1,0.0
m12 = np.sqrt(mass)
x = np.zeros((2,20000))   # x[0,i] = theta_i, x[1,i] = phi_i
y = np.zeros((2,20000))
m = np.zeros(20000)
sigma = np.sqrt(deltat)
target = 10000.0
size = int(target)
for k in range(1000): # perform some time steps
    sume = 0.0
    isum = 0.0
    for i in range(size):
        g = metric(x[0,i],x[1,i])
        for j in range(2):
            x[j,i] += rnd.gauss(mu,sigma)/(g*m12)
        u = pot(x[0,i],x[1,i])         # first order branching
        weight = np.exp(-(u - eref)*deltat)
        r = rnd.random()
        iw = int(weight+r)
        m[i] = min(iw,2)
        sume += m[i]*u
        isum += m[i]
    ae = sume/isum
    eref = ae - np.log(isum/target)
    print("%5.4f %10.5f %6d %10.5f" %((k+1)*deltat,ae,size,eref))
    index = 0  # reset the arrays
    for i in range(size):
        j = 1
        while j <= m[i]:
            j += 1
            for l in range(2):
                y[l,index] = x[l,i]
```

```
            index += 1
    size = int(isum)
    for i in range(size):
        for l in range(2):
            x[l,i] = y[l,i]
        m[i] = 1.0
```

The two-dimensional arrays x and y are initiated with zeros; therefore, the initial state of the population is a Dirac delta at the potential minimum. The code demonstrates how relatively simple it can be to update the positions of replicas using this set of coordinates. Because the update is symmetric among the coordinates, we are able to loop through the two updates,

$$u_{n+1} = u_n + \frac{W_n^1}{\sqrt{mg}},$$

with a corresponding equation for v_{n+1} using a second $\mathcal{N}(0, \Delta t^2)$ random number W_n^2. The function `metric` returns $4/(\sigma^2 - 4)$, the square root of the diagonal element of $g_{\mu\nu}$. The rest of the code is very similar to the basic algorithm introduced in Chapter 5. The branching weights are the first- order type since we have implemented the Euler–Maruyama method for the present numerical test. The graph in Figure 11.1 displays the evolution of the ground state energy estimate as a function of simulation time. The horizontal black line represents the exact ground state energy, computed using angular momentum theory. Its value is −0.9672856 hartree.

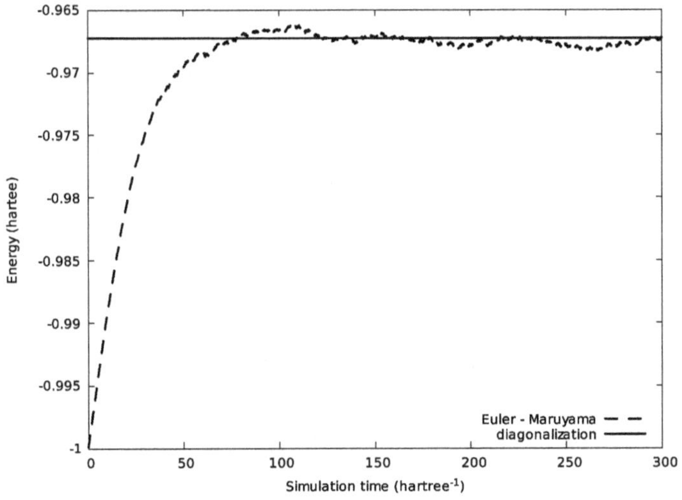

Figure 11.1: The ground state energy estimate of the particle on the surface of a sphere mapped using stereographic projections.

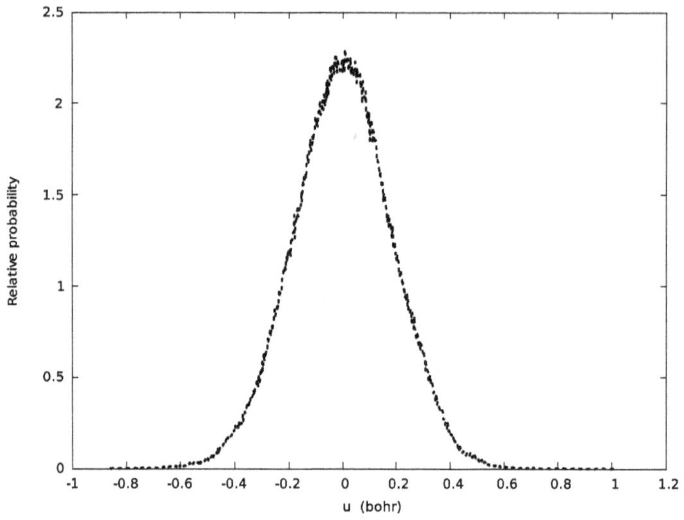

Figure 11.2: Distribution of the projection coordinate “u.”

For this particular example, there is no need to employ the south chart. The graph in Figure 11.2 best explains why this is so. The two projection coordinates from the north are equally and symmetrically distributed near zero. To produce this graph, we process into a 1000 bin histogram approximately a million values of the coordinate u from 100 consecutive diffusion steps collected after the random walk has equilibrated. The v distribution is statistically indistinguishable, and the probability of reaching absolute values greater than one for either projection coordinate is minuscule. Therefore, the near totality of the wave function for the particle on the sphere resides in the southern hemisphere with a peak at the geographic “south pole.” The south chart would be relevant if there were significant chances for a replica to land on the northern hemisphere where values of u, or v would be greater than 2.

Therefore, this example demonstrates that solving the diffusion equation on the surface of a unit sphere presents no challenges other than, perhaps a switch from one chart to another.

The same system is significantly more challenging to simulate with angular coordinates. The Laplace–Beltrami operator is the well-known Legendrian,

$$\Delta_{\text{LB}} = I^{-1}\left\{\frac{\partial^2}{\partial\theta^2} + \frac{1}{\sin^2\theta}\frac{\partial^2}{\partial\phi^2} + \frac{\cos\theta}{\sin\theta}\frac{\partial}{\partial\theta}\right\}. \tag{11.5}$$

In terms of the stochastic calculus, the updates for the two angles are

$$\theta_{n+1} = \theta_n + \frac{1}{I^{1/2}}\int_{t_n}^{t_{n+1}} dW_s + \frac{1}{2I^{1/2}}\int_{t_n}^{t_{n+1}} \frac{\cos\theta}{\sin\theta}\,ds, \tag{11.6}$$

$$\phi_{n+1} = \phi_n + \frac{1}{I^{1/2}} \int_{t_n}^{t_{n+1}} \frac{1}{\sin\theta}\, dW_s, \tag{11.7}$$

where $I = mr^2$ is the moment of inertia. Therefore, the Euler–Maruyama method applied to these yields

$$\theta_{n+1} = \theta_n + \frac{W_n^1}{I^{1/2}} + \frac{\cos\theta_n}{2I\,\sin\theta_n}\Delta t\,, \tag{11.8}$$

$$\phi_{n+1} = \phi_n + \frac{1}{I^{1/2}\,\sin\theta_n} W_n^2, \tag{11.9}$$

where $W_n^{1,2}$ number are $\mathcal{N}(0, \Delta t^2)$. However, it should be noted that the minimum of the potential U happens to be outside of the conventional $\theta\phi$ chart for reasons explained in Chapter 9. Computationally, the random walk will concentrate the population in the immediate neighborhood of the singularity. To avoid numerical overflows when θ_n becomes sufficiently close to either 0 or π, one needs to change the updates. After some experimenting [30], we have settled to the following criteria to identify a potentially problematic region $\theta_n < \Delta t I^{-1}\pi^{-1}$ and for $(\theta_n - \pi) < \Delta t I^{-1}\pi^{-1}$. In this region, the updates are changed to $\theta_{n+1} = \theta_n + I^{-1/2} W_n^1$, and $\phi_{n+1} = \phi_n$. However, even when singularities are at the periphery of the region of importance, defined as the region where the ground state wave function is relatively large, these "fixes" degrade significantly the convergence of the algorithm. The same degradation of convergence is not observed when, e. g., the edge of a chart is in the region of importance for the ground state. Additionally, the random walk must not allow the angles to roam outside of their intended range. $\theta_n \in (0, \pi)$, $\phi_n \in (0, 2\pi)$. This is typically handled by performing the following reflection steps:

- If $d = \theta_n - \pi > 0$, then $\theta_n \leftarrow \pi - d$
- If $\theta_n < 0$, then $\theta_n \leftarrow -\theta_n$
- If $d = \phi_n - 2\pi > 0$, then $\phi_n \leftarrow 2\pi - d$
- If $\phi_n < 0$, then $\phi_n \leftarrow -\phi_n$.

11.5 Exercises

1. Derive the right-hand side of equation (11.2) from (11.1) by thinking of the scalar quantity $g^{\mu\nu}\partial_\mu\partial_\nu$ as a dot product of the gradient in a row vector with the gradient column vector transformed by a covariant metric tensor

$$g^{\mu\nu}\partial_\mu\partial_\nu = \begin{pmatrix} \frac{\partial}{\partial x} & \frac{\partial}{\partial y} \end{pmatrix} \begin{pmatrix} a & b \\ b & c \end{pmatrix} \begin{pmatrix} \frac{\partial}{\partial x} \\ \frac{\partial}{\partial y} \end{pmatrix}$$

keeping in mind that a, b and c are constants when operating with the partial derivatives.

2. Prove parts (a) and (c) of Theorem 34.
3. Consider the following multidimensional Gaussian integral:

$$I(A) = \int d^n x \, \exp\left(-\frac{1}{2}\sum_{i,j=1}^{n} A_{ij} x_i x_j\right).$$

The argument of the exponential function is a quadratic form,

$$\sum_{i,j=1}^{n} A_{ij} x_i x_j = x^T A x,$$

where x is a column vector with components $x_1, x_2, \dots, x_n$. We assume that the matrix A is nonsingular and symmetric, $A^T = A$. Then A^{-1} exists and we can find a matrix P such that

$$P^T A P = \Lambda$$

where Λ is diagonal with all eigenvalues, λ_i, greater than zero.

(a) Make the following change of variable $y = Px$ insert into $x^T A x$ and show that

$$I(A) = \int d^n y \, \exp\left(-\frac{1}{2}\sum_{i=1}^{n} \lambda_i y_i^2\right).$$

To do this, you need to find the Jacobian matrix

$$J = \frac{\partial(x_1, x_2, \dots, x_n)}{\partial(y_1, y_2, \dots, y_n)}$$

because the volume element in n dimensions transforms as follows:

$$d^n x = (J J^T)^{1/2} d^n y.$$

But this is simple because $x = P^T y$ and P is a member of a special group (review the diagonalization procedure in Chapter 1).

(b) Use the group properties of the exponential function to write the integral as follows:

$$I(A) = \prod_{i=1}^{n} \int dy_i \, \exp\left(-\frac{1}{2}\lambda_i y_i^2\right)$$

where $\prod$ is the product symbol,

$$\prod_{i=1}^{n} \int dy_i \exp\left(-\frac{1}{2}\lambda_i y_i^2\right) = \left[\int dy_1 \exp\left(-\frac{1}{2}\lambda_1 y_1^2\right)\right]$$
$$\times \left[\int dy_2 \exp\left(-\frac{1}{2}\lambda_2 y_2^2\right)\right] \times \cdots$$
$$\times \left[\int dy_n \exp\left(-\frac{1}{2}\lambda_n y_n^2\right)\right].$$

(c) Each term in the product is a standard Gaussian integral of the form

$$\int_{-\infty}^{\infty} e^{-ax^2}\, dx = \pi^{1/2} a^{-1/2}.$$

Use this result and a property of determinants to show that

$$I(A) = (2\pi)^{n/2}[\det(A)]^{-1/2}.$$

4. Consider the generalized Gaussian integral of the form

$$I(A,b) = \int d^n x \exp\left(-\frac{1}{2}\sum_{i,j=1}^{n} A_{ij}x_i x_j + \sum_{i=1}^{n} b_i x_i\right)$$

where A is nonsingular and symmetric. Prove that

$$I(A,b) = (2\pi)^{n/2}[\det(A)]^{-1/2} \exp\left(\frac{1}{2}\sum_{i,j=1}^{n} A_{ij}^{-1} b_i b_j\right)$$

by following the steps below. In one dimension, we just complete the square to eliminate the linear term and use standard Gaussian integral theorems. In $\mathbb{R}^n$, we can eliminate the linear terms with a change of variables $x \to y$ that subtracts the minimum of the quadratic form from the y vector.

(a) Begin by showing that if $A^T = A$, then $(A^{-1})^T = A^{-1}$. Perhaps the best starting point is to consider the transpose of $1 = A^{-1}A$ where 1 is, of course, the $n \times n$ unit matrix.

(b) Show that

$$\frac{\partial}{\partial x_k}\left(-\frac{1}{2}\sum_{i,j=1}^{n} A_{ij}x_i x_j + \sum_{i=1}^{n} b_i x_i\right) = 0$$

yields a linear system of equations,

$$Ax = b.$$

(c) Define the vector $y = x - A^{-1}b$ so that we shift the minimum of the quadratic form to the origin of the y coordinates. Eliminate x from the generalized quadratic form and show that

$$-\frac{1}{2}x^T Ax + bx = -\frac{1}{2}y^T Ay + \frac{1}{2}bA^{-1}b.$$

You will need to use the result in part (a).

(d) Use a previous result to complete the proof.

5. The following are useful theorems for sampling covariant multivariate Gaussian random numbers. They represent an extension of the theorem in exercise 1.18 from the natural numbers to the integers and rational numbers as exponents.
   (a) Prove that for two nonsingular $n \times n$ matrices A and B, $(AB)^{-1} = B^{-1}A^{-1}$.
   (b) Let $\Lambda = P^T AP$ be the diagonal matrix containing the eigenvalues of a symmetric positive definite $n \times n$ matrix A. Use the result in part (a) to prove $P\Lambda^{-1}P^T = A^{-1}$.
   (c) Let $A = B^2$, and let $\Lambda = P^T AP$ be as in part (b). Prove that $B = P\Lambda^{1/2}P^T$.
6. The following is an exercise regarding a particle of mass m on the surface of a sphere of radius r.
   (a) Show that the Laplace–Beltrami operator on the 2-sphere mapped by polar and azimuthal angles is the well-known Legendrian,

$$\Delta_{\mathrm{LB}} = I^{-1}\left\{\frac{\partial^2}{\partial\theta^2} + \frac{1}{\sin^2\theta}\frac{\partial^2}{\partial\phi^2} + \frac{\cos\theta}{\sin\theta}\frac{\partial}{\partial\theta}\right\},$$

   where $I = mr^2$ is the moment of inertia.
   (b) Show that the Itô calculus updates for the two angles θ and ϕ are

$$\theta_{n+1} = \theta_n + \frac{1}{I^{1/2}}\int_{t_n}^{t_{n+1}} dW_s + \frac{1}{2I^{1/2}}\int_{t_n}^{t_{n+1}} \frac{\cos\theta}{\sin\theta}\,ds,$$

$$\phi_{n+1} = \phi_n + \frac{1}{I^{1/2}}\int_{t_n}^{t_{n+1}} \frac{1}{\sin\theta}\,dW_s.$$

   (c) Evaluate the two integrals for the implementation of the Euler–Maruyama method.
   (d) Show that to implement the Milshtejn method one just needs to add

$$-\frac{\cos\theta}{2I\sin^2\theta}(W_n^1 W_n^2 - \Omega_n^{12})$$

   to the ϕ update.
   (e) Recall that the range for the polar angle θ is $\theta \in (0, \pi)$. Mathematically, the open interval insures the chart excludes problematic points. What do you anticipate

will happen in a simulation if the angles become too close to either end of the interval?

7. Repeat all the parts of the previous exercise, but use the stereographic projections from the north map.
8. Consider a particle of mass m in a ring of radius r.
   (a) The ring is an embedding in an Euclidean plane $x^2 + y^2 = r^2$. Use the strategies in Chapter 9 to derive the metric tensor (a one-by- one matrix in this case) when the space is mapped by the polar angle θ.
   (b) Use the result in the previous exercise to show that in terms of the angle $\theta \in (0, 2\pi)$ the kinetic energy operator becomes

$$-\frac{\hbar^2}{2mr^2}\frac{\partial^2}{\partial\theta^2}.$$

   (c) Show that $\psi_j = e^{ij\theta}$ is an eigenfunction of the operator in part (b). Here, j is a integer, including zero, −1, +1, etc. and $i = \sqrt{-1}$.
   (d) Normalize $\psi_j = e^{ij\theta}$, then show that

$$\int_0^{2\pi} \psi_k^* \psi_j d\theta = \delta_{k,j}.$$

   (e) Consider the particle in a ring subjected to a charge-dipole interaction. The potential energy is $U = U_0 \cos\theta$. Derive the following matrix elements:

$$\langle k|U|j\rangle = \frac{U_0}{2\pi}\int_0^{2\pi} e^{-ik\theta}\cos\theta e^{ij\theta} d\theta = \frac{U_0}{2}(\delta_{k,j+1}, \delta_{k,j-1}).$$

   (f) Show that the Hamiltonian matrix for this particular potential U becomes

$$\langle k|H|j\rangle = \frac{\hbar^2}{2mr^2} j^2 \delta_{k,j} + \langle k|U|j\rangle.$$

   (g) Use the Python code suggested here to compute the Hamiltonian matrix $\langle k|H|j\rangle$ for $k, j \in \{-n, -n+1, \ldots, 0, \ldots, n\}$ for a given value of n, then add your own code to diagonalize it. Find the ground state energy for $U_0, r = 1$ and $m = 919.0$, by incrementing the number of states until the answer has converged to 5 decimal places.

```
import numpy as np
n = 4
size = 2*n+1
mass = 919.0
h = np.zeros((size,size))
```

```
for j in range(size):
    k = j-n
    if (j < size-1):
        h[j,j+1] = 0.5
        h[j+1,j] = 0.5
    h[j,j] = 0.5*k*k/mass
```

(h) Change the mass to 1.0 atomic units and the initial value of n to 2, then increment n until the answer has converged to 5 decimal places. How do you explain the difference in convergence behavior when changing the value of the mass?

9. Modify the basic diffusion Monte Carlo method introduced in Chapter 5 to find the ground state of a particle of mass m in a ring of radius r subject to a $U_0 \cos\theta$ potential energy. This will require that both north and south charts are used in the course of the simulation. Follow the steps below. For simplicity, select U_0 and r to be unity.
   (a) Derive the update on the stereographic projection ξ. You may begin with equation (9.19).
   (b) Find an expression for U in terms of the stereographic projection ξ. Then find the value of ξ at which U is at minimum.
   (c) Write a Python code for three separate functions, that return U, the square root of $g^{\mu\nu}$ and $\mathcal{G}^{\mu}$, respectively.
   (d) Run the code using $\Delta t = 1$, a mass of 919.0 atomic units and compare the ground state energy estimate with the converged value from the previous exercise.
   (e) Change the mass to 1.0 atomic units, $\Delta t = 0.01$ and run the code. How many steps does it take to throw an exception? Can you diagnose the problem?
   (f) Insert the following two lines:

```
if (abs(x[i]) > 2.0*radius):
    x[i] = 4.0/x[i]    # switch chart
```

   properly indented so that it is immediately before the end of the `for i in range(size):` population loop nested inside the step loop and run the code again.

12 Hamilton's principle

Even today, apart from some changes and reinforcements, the foundation laid down by [Newton] provides us with the most natural and didactically simplest approach to mechanics.

Arnold Sommerfeld

Perhaps one of the most far reaching principle in physics, Hamilton's principle occupies a central role in gauge theory. We find it interesting that so many modern theories, even those among which there seems to be conflicting evidence as, e. g., in the case of general relativity and quantum field theory, can be cast in terms of some action optimization. In this chapter, we introduce the concept of functionals broadly defined for applications in both classical and nonrelativistic quantum mechanics. The functional is called the Lagrangian, a function of two sets of variables. The first is a set of functions and all their first derivatives taken with respect to the independent variables. In Newtonian mechanics, there is only one independent variable: t. In quantum mechanics, the set of independent variables are the coordinates of space time. In practice, one defines an action integral over the set of independent variables of the system, and seek to find equations satisfied by the dependent variables, e. g., position and velocity in classical mechanics, the wave function, its first derivatives, and the corresponding complex conjugate in quantum mechanics. These are called the Euler–Lagrange equations. After demonstrating how these are derived using numerous examples, we introduce a strategy to derive a class of molecular dynamics algorithms to integrate the classical laws of motion numerically.

12.1 Functionals

Definition 50 (Functional). A functional S is a map from a set of n smooth functions to the real numbers,

$$S : \{f^\mu(x,y,\ldots,t), \partial_x f^\mu(x,y,\ldots,t),\ldots,\partial_t f^\mu(x,y,\ldots,t)\}_{\mu=1}^{n} \to \mathbb{R}$$

defined as an integral over a finite domain $\mathcal{D}$,

$$S = \int_{\mathcal{D}} \mathcal{L}[f^\mu(x,y,\ldots,t), \partial_x f^\mu(x,y,\ldots,t),\ldots,\partial_t f^\mu(x,y,\ldots,t)]\, dx, dy, \ldots, dt.$$

The function $\mathcal{L}[f^\mu,\ldots,t]$ is called the Lagrangian density, or simply a Lagrangian if the only independent variable is t. Its arguments are a set of dependent variables $f^\mu, \partial_x f^\mu, \ldots, \partial_t f^\mu$ over a space time domain $\mathcal{D}$ of independent variables $x, y, \ldots, t$, or just a time domain in classical mechanics. The Lagrangian is a function of functions, but can also explicitly depend on the independent variables $x, y, \ldots, t$. A key feature for functionals to work as intended is in regards to the set of dependent variable $f^\mu, \partial_x f^\mu, \ldots, \partial_t f^\mu$;

https://doi.org/10.1515/9783111610207-012

These must be mutually independent. Of course, the same must be also true for the set of independent variables $x, y, \ldots, t$. Functionals are extremely useful tools, for it seems that the laws of classical and quantum physics can be derived by formulating an appropriate Lagrangian, and finding extrema for the functional S it defines. The following examples from classical and quantum mechanics explain how the process works in general.

Example: A classical free particle in $\mathbb{R}^3$. Consider the following Lagrangian:

$$\mathscr{L} = \frac{1}{2}m(\dot{x}(t)^2 + \dot{y}(t)^2 + \dot{z}(t)^2) \tag{12.1}$$

where

$$\dot{x} = \frac{dx}{dt},$$

etc. Here, a particle of mass m is moving with velocity $v = (\dot{x}(t), \dot{y}(t), \dot{z}(t))$. The Lagrangian only depends explicitly on three functions $\dot{x}(t)$, $\dot{y}(t)$, $\dot{z}(t)$ and implicitly on t. The functional S is called the classical action over a time interval $t \in [a, b]$,

$$S = \int_a^b \mathscr{L}\, dt = \frac{1}{2}m \int_a^b (\dot{x}(t)^2 + \dot{y}(t)^2 + \dot{z}(t)^2)\, dt.$$

We may remember from introductory physics that a free particle "feels no force," and consequently, should not be accelerated. Therefore, the velocity components should be constants, however, that is not required to define S. If the velocity vector is a constant, it turns out that the functional S is at minimum. It is possible to use intuitive arguments to understand this result. First, let us note that the free particle action does not have an upper bound, but for a finite time interval $[a, b]$ it does have a lower bound. Suppose at $t = a$ the particle is at x_0, y_0, z_0, and at $t = b$ it is at x_1, y_1, z_1. The instantaneous velocity components change with time, but have to do so about the average speed $\langle v \rangle$ dictated by the two points in space and time $\langle v \rangle = d/(b-a)$ where $d = [(x_1 - x_0)^2 + (y_1 - y_0)^2 + (z_1 - z_0)^2]^{1/2}$ is the distance traveled. If the functions $\dot{x}(t)$, $\dot{y}(t)$, $\dot{z}(t)$ are constant, the integral is trivial and we can get a simple closed form for the action,

$$S_{\min} = \frac{1}{2}m\langle v \rangle^2(b-a). \tag{12.2}$$

Because $\mathscr{L}$ depends on the square of the instantaneous velocity, any variations from the average velocity at any point along the straight line increases S from its minimum value because it is always true that $\langle v^2 \rangle \geq \langle v \rangle^2$, the average of the squares is always equal or greater than the square of the average. Recall that for applications in classical mechanics, the dependent variables are the coordinates of points in state space as defined in Chapter 10 and the only independent variable is time. The next three examples are all classical Lagrangians. Example: A damped classical particle in $\mathbb{R}^3$. The Lagrangian in this case is

$$\mathscr{L} = \frac{1}{2}m(\dot{x}(t)^2 + \dot{y}(t)^2 + \dot{z}(t)^2)e^{\gamma t} \tag{12.3}$$

and it is an explicit function of the dependent variables, i. e., the velocity components and the independent variable t. In this case, the velocity is no longer a constant even when S is at its minimum value. The components of v drop exponentially when the path is optimized to minimize S.

Example: Consider the case of a harmonic oscillator in $\mathbb{R}$:

$$\mathscr{L} = \frac{1}{2}m\dot{x}^2 - \frac{1}{2}kx^2.$$

The Lagrangian depends on the functions x and $\dot{x}$. Here, we are subtracting the potential energy from the kinetic energy. It is natural to ask why not integrating the total energy instead of subtracting one type from the other. At this point, we simply make some intuitive arguments and provide a more formal explanation later on. If the Lagrangian is not explicitly dependent on time, the total energy $E = U + K$ of the system is conserved for those paths in state space that are at the action minimum. This includes the present example, where the particle reaches one of two classical turning points when the spring is extended or contracted at the maximum permitted by the total energy. The particle comes to a stop and reverses direction in the following instant. Therefore, at the turning points the total energy is all potential, the kinetic energy is zero. By contrast, when the particle is located at the position of minimum potential energy, which we can arbitrarily set to zero (but we justify this carefully later), then the total energy is all kinetic in nature. At all other points E is the sum of both kinetic and potential. The free particle case tells us we should look for a minimum of the action. It turns out that the best way to minimize the action is achieved when the two types of contributions are subtracted rather than added.

Example: A classical particle subject to a potential U in a manifold $\mathcal{M}$. The relevant Lagrangian is

$$\mathscr{L}[x^\mu(t)] = \frac{1}{2}g_{\mu\nu}\dot{x}^\mu\dot{x}^\nu - U \tag{12.4}$$

where the set of dependent variables $\{x^\mu\}$ map points in $\mathcal{M}$ onto $\mathbb{R}^n$, $g_{\mu\nu}$ is the Hessian metric tensor, and U may depend on both the $\{x^\mu\}$ set and the set of time derivatives $\{\dot{x}^\mu\}$.

Example: The Schrödinger–Lagrangian density for a single particle in $\mathbb{R}$ subject to a potential U is

$$\begin{aligned}&\mathscr{L}(\psi, \psi^*, \partial_x\psi, \partial_x\psi^*, \partial_t\psi, \partial_t\psi^*)\\&\quad = -\frac{\hbar^2}{2m}(\partial_x\psi)\partial_x\psi^* + i\frac{\hbar}{2}(\psi^*\partial_t\psi - \psi\partial_t\psi^*) - U\psi\psi^*\end{aligned} \tag{12.5}$$

where ψ^* is the complex conjugate of ψ and $\mathscr{L}$ does not explicitly depend on the independent variables x and t, but is a function of the field variable ψ, its complex conjugate and their respective partial derivatives. The functional in this case is a double integral,

$$S = \int_{\mathbb{R}} dx \int_a^b dt \mathscr{L}(\psi, \psi^*, \partial_x\psi, \partial_x\psi^*, \partial_t\psi, \partial_t\psi^*).$$

It is assumed that ψ^*, ψ, and their respective derivatives are all independent of each other.

12.1.1 The Euler–Lagrange equations

We now move our attention to the problem of finding the extreme values for functionals. These are defined as a specific set of n smooth functions such that $S : \{f^\mu(x,y,\dots,t), \partial_x f^\mu(x,y,\dots,t),\dots,\partial_t f^\mu(x,y,\dots,t)\}_{\mu=1}^n \to \mathbb{R}$ yields a minimum. For example, considering the classical free particle in $\mathbb{R}^3$, one can choose any arbitrary set of functions of $t, x, y, z, \dot{x}, \dot{y}, \dot{z}$, and calculate a value for the classical action S from them. The action is perfectly defined, and in fact unlike the present task, in the next chapter we will be interested in a certain class of trajectories that do not necessarily yield a minimum action. Borrowing from ordinary calculus, the idea that at a function minimum the gradient vanishes, we may seek a minimum for S by setting all its partials with respect to the functions $f\{f^\mu(x,y,\dots,t), \partial_x f^\mu(x,y,\dots,t),\dots,\partial_t f^\mu(x,y,\dots,t)\}_{\mu=1}^n$ to zero. However, the derivative of a functional with respect to the functions in its domain is a more advanced concept not typically covered in a regular calculus course. To illustrate the principle, it is best to first consider the case of a classical particle in $\mathbb{R}$ subjected to a potential U:

$$\mathscr{L} = \frac{1}{2}m\dot{x}^2 - U$$

and let us seek functions $x(t)$, $\dot{x}(t)$ such that the variation of S is zero,

$$\delta S = \delta \int_a^b \left(\frac{1}{2}m\dot{x}^2 - U\right) dt = 0$$

by perturbing the trajectory with a "small" $x'(t) = x(t) + \varepsilon(t)$ correction, where ε must satisfy the conditions $\varepsilon(a) = \varepsilon(b) = 0$. Then $\dot{x}'(t) = \dot{x}(t) + \dot{\varepsilon}(t)$. Inserting these into the equation for δS, we get

$$\delta S = \int_a^b \left[\frac{1}{2}m(\dot{x}(t) + \dot{\varepsilon}(t))^2 - U(x(t) + \varepsilon(t))\right] dt$$
$$- \int_a^b \left(\left[\frac{1}{2}m\dot{x}^2 - U(x')\right]\right) dt.$$

The first term on the right is the action along the perturbed trajectory, and we are looking for the difference between the resulting perturbed action relative to the unperturbed one. To get an explicit expression, we expand both the kinetic energy, and potential energy into a power series about $x(t)$ keeping only the linear terms in $\epsilon(t)$ and its time derivative. We drop terms that are quadratic or higher in both ε and $\dot{\varepsilon}$ since these can be made arbitrarily small. The result is

$$\delta S = \int_a^b \left[\frac{1}{2}m(\dot{x}^2 + 2\dot{x}\dot{\varepsilon}(t)) - U(x(t)) + \varepsilon(t)\partial_x U(x(t)) + \cdots\right] dt$$
$$- \int_a^b \left(\left[\frac{1}{2}m\dot{x}^2 - U(x)\right]\right) dt.$$

After canceling the common terms, we end up with

$$\delta S = m \int_a^b (\dot{x}\dot{\varepsilon}(t))dt - \int_a^b \varepsilon(t)\partial_x U(x(t))\, dt.$$

The first integral can be simplified using integration by parts. Letting $\int udv = uv - \int vdu$, by $dv = \dot{\varepsilon}$, we get

$$\int_a^b (\dot{x}\dot{\varepsilon})dt = \dot{x}\,\varepsilon|_a^b - \int_a^b (\ddot{x}\varepsilon(t))dt.$$

The conditions $\varepsilon(a) = \varepsilon(b) = 0$ cause the boundary terms on the right to vanish. When we recombine the surviving terms and we set δS to zero, we end up with

$$\delta S = -\int_a^b \varepsilon(t)[m(\ddot{x}) + \partial_x U(x)]\, dt = 0.$$

It is straightforward to prove that for an integral of the type $\int A(t)\varepsilon(t)\, dt$ to be zero for any arbitrary nonzero function ε in the interval $t \in [a, b]$ it is necessary that $A(t) = 0\ \forall t \in [a, b]$. Using this fact, we are presented with Newton's second law for the system

$$m(\ddot{x}') + \partial_x U(x') = 0,$$

$$m(\ddot{x}') = F_x$$

where $F_x = -\partial_x U$ is the force, and $\ddot{x}$ the acceleration.

The generalized version of the above theorem is as follows. The solution of

$$\delta S = \delta \int_{\mathcal{D}} \mathcal{L}[f^\mu(x^\nu), \partial_x f^\mu(x^\nu), \ldots, \partial_t f^\mu(x^\nu)]\, dx^\nu\, dt = 0$$

are the Euler–Lagrange equations,

$$\partial_\nu \left[\frac{\partial \mathcal{L}}{\partial(\partial_\nu f^\mu)} \right] = \frac{\partial \mathcal{L}}{\partial f^\mu} \tag{12.6}$$

where ν is the $n+1$ independent variables index that runs from 0 up to n, where x^0 is typically assigned to t and μ runs from 1 to l, the number of dependent variables. Therefore, there are l such equations, one for each of the f^μ functions.

The proof of the result in equation (12.6) is the same used to derive the Euler–Lagrange equations for the simpler cases and adds no new insights. Instead, it is more instructive and useful to learn how to use this result by deriving the laws of motion from given Lagrangians. We do this for the examples we introduced earlier.

Example: A classical free particle in $\mathbb{R}^3$. Here, $\mathcal{L}$ is a function of three functions only $\partial_0 f^1, \partial_0 f^2, \partial_0 f^3 = \dot{x}(t), \dot{y}(t), \dot{z}(t)$. Equation (12.6) gives us three separate equations, one for x, one for y, and one for z,

$$\frac{d}{dt}\left[\frac{\partial \mathcal{L}}{\partial \dot{x}}\right] = \frac{\partial \mathcal{L}}{\partial x}.$$

But note that $\mathcal{L}$ in equation (12.1) does no depend on x, y, or z. Therefore, the relevant partial derivatives are

$$\frac{\partial \mathcal{L}}{\partial \dot{x}} = m\dot{x}, \quad \frac{\partial \mathcal{L}}{\partial x} = 0,$$

with four other corresponding equations for the derivatives with respect to $y, z, \dot{y}$, and $\dot{z}$. The right-hand side of each of the three Euler–Lagrange equations is zero. This result is consistent with the introductory physics concept that a free particle "feels no force,"

$$\frac{d}{dt}\left[\frac{\partial \mathcal{L}}{\partial \dot{x}}\right] = m\ddot{x} = 0,$$

and consequently, it is not accelerated. Rather, it follows a straight line along an initial velocity vector $v_0 = \dot{x}(a)\mathbf{i} + \dot{y}(a)\mathbf{j} + \dot{z}(a)\mathbf{k}$.

Example: A damped classical particle in $\mathbb{R}^3$. From equation (12.3), it follows that

$$\frac{\partial \mathcal{L}}{\partial \dot{x}} = m\dot{x}e^{\gamma t}, \quad \frac{\partial \mathcal{L}}{\partial x} = 0$$

along with four other corresponding equations for the remaining components. It follows that

$$\frac{d}{dt}\left[\frac{\partial \mathscr{L}}{\partial \dot{x}}\right] = m(\ddot{x}e^{\gamma t} + \gamma \dot{x}e^{\gamma t}) = 0.$$

In this case, the velocity is not a constant even though there are "no forces" on the particle. We derive the differential equation satisfied by x. Those for y and z are solved exactly the same way. The equation is

$$\ddot{x} + \gamma \dot{x} = 0$$

after we divide both sides by $me^{\gamma t}$. Neither of these two quantities are zero at any time. Using known initial values for x and $\dot{x}$, say x_0 and $\dot{x}_0$, and letting $\dot{x} = v_x$ we can rewrite this as a first-order ordinary differential equation and obtain

$$\dot{v}_x = -\gamma v_x$$

a separable differential equation that is solved by straightforward integration,

$$v_x = \dot{x}_0 e^{-\gamma t}.$$

Integrating once more and applying the initial conditions, we get

$$x = x_0 + \frac{\dot{x}_0}{\gamma}(1 - e^{-\gamma t}).$$

Now note that as $t \to \infty$ the velocity drops to zero and the value of the position is finite.

Example: The Lagrangian for the classical particle in a n-dimensional manifold is in equation (12.4). The n Euler–Lagrange equations read

$$\partial_t\left[\frac{\partial \mathscr{L}}{\partial \dot{x}^\mu}\right] = \frac{\partial \mathscr{L}}{\partial x^\mu}$$

for $\mu = 1$ to n. From equation (12.4), we get

$$\frac{\partial \mathscr{L}}{\partial \dot{x}^\mu} = g_{\mu\nu}\dot{x}^\nu \quad \frac{\partial \mathscr{L}}{\partial x^\mu} = \frac{1}{2}(\partial_\mu g_{\beta\nu})\dot{x}^\beta \dot{x}^\nu - \partial_\mu U.$$

This gives

$$\partial_t\left[\frac{\partial \mathscr{L}}{\partial \dot{x}^\mu}\right] = g_{\mu\nu}\ddot{x}^\nu + \partial_\beta g_{\mu\nu}\dot{x}^\beta \dot{x}^\nu,$$

So, the equations of motion become

$$g_{\mu\nu}\ddot{x}^\nu + \partial_\beta g_{\mu\nu}\dot{x}^\beta \dot{x}^\nu = \frac{1}{2}(\partial_\mu g_{\beta\nu})\dot{x}^\beta \dot{x}^\nu - \partial_\mu U$$

This is already a useful result, but it is common practice to rewrite space derivatives of the metric tensor in terms of the Christoffel connection coefficients. To do so, one rewrites the last equation by swapping β and ν in the derivative terms on the left-hand side to obtain the following expression:

$$g_{\mu\nu}\ddot{x}^{\nu} + \partial_{\nu}g_{\mu\beta}\dot{x}^{\beta}\dot{x}^{\nu} = \frac{1}{2}(\partial_{\mu}g_{\beta\nu})\dot{x}^{\beta}\dot{x}^{\nu} - \partial_{\mu}U.$$

Since these are summation indices we are not changing the final result, adding the last two equations and dividing by two yield

$$g_{\mu\nu}\ddot{x}^{\nu} + \frac{1}{2}\partial_{\beta}g_{\mu\nu}\dot{x}^{\beta}\dot{x}^{\nu} + \frac{1}{2}\partial_{\nu}g_{\mu\beta}\dot{x}^{\beta}\dot{x}^{\nu} = \frac{1}{2}\partial_{\mu}g_{\beta\nu}\dot{x}^{\beta}\dot{x}^{\nu} - \partial_{\mu}U$$

now we combine all the derivatives of the metric tensor to get

$$g_{\mu\nu}\ddot{x}^{\nu} + \frac{1}{2}[\partial_{\beta}g_{\mu\nu} + \partial_{\nu}g_{\mu\beta} - \partial_{\mu}g_{\beta\nu}]\dot{x}^{\beta}\dot{x}^{\nu} = -\partial_{\mu}U,$$

and finally we multiply both sides by $g^{\alpha\nu}$ and sum over μ to arrive at Newton's second law in Riemannian manifolds,

$$\ddot{x}^{\alpha} + \Gamma^{\alpha}_{\beta\nu}\dot{x}^{\beta}\dot{x}^{\nu} = -g^{\alpha\nu}\partial_{\nu}U$$

where $\Gamma^{\alpha}_{\beta\nu} = \frac{1}{2}g^{\alpha\nu}[\partial_{\beta}g_{\mu\nu} + \partial_{\nu}g_{\mu\beta} - \partial_{\mu}g_{\beta\nu}]$ are the Christoffel connection coefficients. The $g^{\alpha\nu}\partial_{\nu}U$ terms sum up to the α component of the contravariant force. For free particles in manifolds, the expressions

$$\ddot{x}^{\alpha} + \Gamma^{\alpha}_{\beta\nu}\dot{x}^{\beta}\dot{x}^{\nu} = 0$$

are known as the geodesic equations. Geodesics are shortest distance curves between two points in a manifolds. In $\mathbb{R}^n$, the geodesic are straight lines since the connection coefficients $\Gamma^{\alpha}_{\beta\nu}$ vanish, and consequently, the particle is not accelerated.

Example: The Schrödinger–Lagrangian for a single particle in $\mathbb{R}$ subject to a potential U is as in equation (12.5) Because we have two dependent variables ψ and ψ^* and their corresponding first derivatives, we end up with two equations. Taking derivatives with respect to ψ^* and $\partial_x\psi^*$ yield the Schrödinger equation, whereas taking derivative with respect to ψ and $\partial_x\psi$ yields its complex conjugate. For the former case, we get

$$\frac{\partial\mathcal{L}}{\partial(\partial_x\psi^*)} = -\frac{\hbar^2}{2m}\partial_x\psi \quad \frac{\partial\mathcal{L}}{\partial\psi^*} = i\frac{\hbar}{2}\partial_t\psi - U\psi,$$
$$\frac{\partial\mathcal{L}}{\partial(\partial_t\psi^*)} = -i\frac{\hbar}{2}\psi.$$

The Euler–Lagrange equation for this case reads

$$\partial_x\left[\frac{\partial\mathscr{L}}{\partial(\partial_x\psi^*)}\right]+\partial_t\left[\frac{\partial\mathscr{L}}{\partial(\partial_t\psi^*)}\right]=\frac{\partial\mathscr{L}}{\partial\psi^*}$$

and inserting the expressions for the two partial derivatives in this gives

$$-\frac{\hbar^2}{2m}\frac{\partial^2\psi}{\partial x^2}-i\frac{\hbar}{2}\partial_t\psi=i\frac{\hbar}{2}\partial_t\psi-U\psi,$$

which can be rearranged trivially into the regular form of the time dependent Schrödinger equation. All of these examples demonstrate how to use equation (12.6) and show the broad applicability of the least action principle.

12.2 Energy conservation

Definition 51 (Classical canonical momentum). The derivative of the classical Lagrangian with respect to the time derivative $\dot{x}^\mu$ is called the canonical momentum conjugate to x^μ,

$$p_\mu=\frac{\partial\mathscr{L}}{\partial\dot{x}^\mu}.$$

With this definition, the Euler–Lagrange equations of classical mechanics read

$$\dot{p}_\mu=\frac{\partial\mathscr{L}}{\partial x^\mu}. \tag{12.7}$$

Example: Given the free particle Lagrangian in equation (12.1), we compute

$$p_x=\frac{\partial\mathscr{L}}{\partial(\dot{x})}=m\dot{x}$$

with equivalent expressions for the y and z components. The last result should be familiar. It is called the mechanical momentum, but it is not sufficiently general for our purpose.

Example: (Angular momentum) Consider the following change of coordinates in $\mathbb{R}^3$ from Cartesian to spherical polar:

$$x=r\cos\phi\sin\theta,\quad y=r\sin\phi\sin\theta,\quad z=r\cos\theta,$$

which is useful whenever the potential is isotropic, meaning it only depends on r. Then, by transforming $m(\dot{x}^2+\dot{y}^2+\dot{z}^2)/2-U$, we get

$$\mathscr{L}=\frac{1}{2}m(\dot{r}^2+r^2\dot{\theta}^2+r^2\sin^2\theta\,\dot{\phi}^2)-U(r). \tag{12.8}$$

There are three canonical momenta. The radial momentum looks similar to the mechanical one,

$$p_r = \frac{\partial \mathscr{L}}{\partial(\dot{r})} = m\dot{r}$$

but the other two components have some extra coefficients,

$$p_\theta = \frac{\partial \mathscr{L}}{\partial(\dot{\theta})} = mr^2\dot{\theta}, \quad p_\phi = \frac{\partial \mathscr{L}}{\partial(\dot{\phi})} = mr^2 \sin^2\theta\,\dot{\phi}.$$

Equation (12.7) leads to a conservation of momentum principle that can be stated as follows. If the Lagrangian is unchanged after a space translation, the conjugate canonical momentum is conserved. While this statement is certainly true, there are examples where canonical momentum vectors are conserved even when the derivative of the Lagrangian does not vanish. The best example is the angular momentum of a particle in $\mathbb{R}^3$ in an isotropic (also called central) potential. The Euler–Lagrange equations derived from (12.8) are

$$\begin{aligned}
\dot{p}_r &= -\frac{\partial U}{\partial r} \\
\dot{p}_\phi &= 0 \\
\dot{p}_\theta &= mr^2 \sin\theta\cos\theta\,\dot{\phi}^2.
\end{aligned}$$

It seems that one of the two components of the angular momentum is not conserved, but we should be more careful here. When we express the angular momentum vector L along the **i**, **j**, and **k** directions, we can demonstrate that $\dot{L} = 0$. Here, we are representing L along two unit vectors that point along the angular degrees of freedom e_θ and e_ϕ,

$$L = (mr^2\dot{\theta})e_\theta + (mr^2\sin^2\theta\,\dot{\phi})e_\phi,$$

but unlike the unit vector along the Cartesian axis, e_θ and e_ϕ, change direction over time.

Definition 52 (The Hamiltonian). The Legendre transform of the Lagrangian function is called the Hamiltonian function

$$\mathscr{H} = p_\mu \dot{x}^\mu - \mathscr{L}$$

where the independent velocity components in $\mathscr{H}$ are exchanged for their conjugate canonical momenta.

Therefore, in classical mechanics, the Hamiltonian is a function of canonical momenta and position for a system as functions of time. The union of these two sets of independent variables is called the system's *phase space*. It is possible to derive the Legendre transform of a Lagrangian density into a Hamiltonian tensor and find Hamilton's equations for fields from it. This idea has been explored to compute quantum scattering matrices.

Example: Consider the dampened particle Lagrangian in equation (12.3), then the canonical momentum conjugate to x is

$$p_x = m\,e^{\gamma t}\,\dot{x},$$

with equivalent expressions for the remaining components. It follows that

$$p_\mu \dot{x}^\mu = p_x x + p_y y + p_z z = m\,e^{\gamma t}\,(\dot{x}^2 + \dot{y}^2 + \dot{z}^2),$$

however, we want to exchange $\dot{x}$ in favor of p_x, and we want to do so in both the $p_\mu \dot{x}^\mu$ and the $\mathscr{L}$ terms of the Hamiltonian,

$$\begin{aligned}
\mathscr{H} &= m\,e^{\gamma t}\left(\frac{p_x^2}{m^2 e^{-2\gamma t}} + \frac{p_y^2}{m^2 e^{-2\gamma t}} + \frac{p_z^2}{m^2 e^{-2\gamma t}}\right)\\
&\quad - \frac{1}{2} m e^{\gamma t}\left(\frac{p_x^2}{m^2 e^{-2\gamma t}} + \frac{p_y^2}{m^2 e^{-2\gamma t}} + \frac{p_z^2}{m^2 e^{-2\gamma t}}\right)\\
&= -\frac{1}{2me^{\gamma t}}(p_x^2 + p_y^2 + p_z^2),
\end{aligned}$$

where we have used expressions like this one $\dot{x} = p_x e^{-\gamma t} m^{-1}$ to eliminate the velocity components in favor of the conjugate momenta components, and performed some simplifications. Note that the Hamiltonian in a mechanical system is $K + U$ the total energy.

Example: Consider the Lagrangian in equation (12.4) for a particle in a manifold. Note that

$$p_\mu = \frac{\partial \mathscr{L}}{\partial \dot{x}^\mu} = g_{\mu\nu}\dot{x}^\nu$$

and, therefore,

$$\begin{aligned}
\dot{x}^\nu &= g^{\mu\nu} p_\mu,\\
p_\mu \dot{x}^\mu &= g^{\mu\nu} p_\mu p_\nu
\end{aligned}$$

whereas $\mathscr{L}$ transforms to

$$\frac{1}{2} g^{\mu\nu} p_\mu p_\nu - U,$$

since μ and ν are summed over and we used $g_{\mu\nu} g^{\mu\nu} = 1$. It follows that

$$\mathscr{H} = \frac{1}{2} g^{\mu\nu} p_\mu p_\nu + U.$$

Therefore, the Hamiltonian of classical mechanics is the total energy in manifolds.

We are now ready to explore what makes a mechanical system conserve energy.

Theorem 38. *The time derivative of $\mathscr{H}$ satisfies the following expression*

$$\dot{\mathscr{H}} = -\frac{\partial \mathscr{L}}{\partial t}.$$

Proof. Consider the total time derivative of $\mathscr{L}(\dot{x}^\mu, x^\mu, t)$,

$$\frac{d\mathscr{L}}{dt} = \frac{\partial \mathscr{L}}{\partial \dot{x}^\mu}\ddot{x}^\mu + \frac{\partial \mathscr{L}}{\partial x^\nu}\dot{x}^\mu + \frac{\partial \mathscr{L}}{\partial t}.$$

Now we use the definition of canonical momentum to replace the partial derivative of $\mathscr{L}$ in the first term on the right-hand side, and $\dot{p}_\mu = g_{\mu\nu}\ddot{x}^\nu + \partial_\beta g_{\mu\nu}\dot{x}^\beta \dot{x}^\nu$ to replace the second partial derivative

$$\frac{d\mathscr{L}}{dt} = p_\mu \ddot{x}^\mu + \dot{p}\dot{x}^\mu + \frac{\partial \mathscr{L}}{\partial t}.$$

Note that this can be rearranged to read

$$\frac{\partial \mathscr{L}}{\partial t} = -\frac{d}{dt}[p\dot{x}^\mu - \mathscr{L}] = -\dot{\mathscr{H}}.$$ □

This theorem states that if $\mathscr{L}$ is explicitly time dependent, the system it describes does not conserve energy. Conversely, if $\partial\mathscr{L}/\partial t = 0$, then the sum of kinetic and potential energy is a constant. We can rephrase the energy invariance result by stating that, if $\mathscr{L}$ is invariant under an infinitesimal time translation $t' = t + dt$, then the Hamiltonian is a conserved quantity. There is much more to invariance than what we discuss here and the subject is central to gauge theory, so this topic will be revisited in later chapters.

12.3 Equivalent Lagrangians

Thus far we have used the theory of functionals to find the equations of "motion" given a Lagrangian $\mathscr{L}$. We can also ask the reverse question. Given some equations of motion, can we find the $\mathscr{L}$ that generated them? It turns out we can always do so, however, the solution is not unique. The field of equivalent Lagrangians is quite interesting and the main objective here is to finally explain in a mathematical way why it is always okay to add a constant energy term to $\mathscr{L}$ without affecting the resulting dynamics. Recall in Chapter 5 we used a Dirac delta vector space to represent the population of replicas of a quantum system, but we ended up with a constant, but infinite energy that we disregard. The statement: "*The laws of physics look exactly the same in a constant (uniform) energy field as they do in its absence,*" saved us from what might have looked like an embarrassing divergence. With the least action principle in our growing toolbox, it is quite easy to prove this result.

Definition 53 (Equivalent Lagrangians). Two Lagrangian functions $\mathscr{L}$ and $\mathscr{L}' = \mathscr{L}$ are said to be identical if they produce identical Euler–Lagrange equations.

Theorem 39. *Let f be a function of x^μ and t, but not $\dot{x}$ and let $\dot{f}$ be the total time derivative of f. The two Lagrangians $\mathscr{L}$ and $\mathscr{L}' = \mathscr{L} + \dot{f}$ are identical.*

Proof. Since f depends on x^μ and t, but not $\dot{x}$, its total time derivative is

$$\frac{df}{dt} = \frac{\partial f}{\partial x^\mu}\dot{x}^\mu + \frac{\partial f}{\partial t}.$$

The Euler–Lagrange equations for $\mathscr{L}'$ are constructed in the usual way, starting with the velocity derivatives

$$\frac{\partial \mathscr{L}'}{\partial \dot{x}^\mu} = \frac{\partial \mathscr{L}}{\partial \dot{x}^\mu} + \frac{\partial \dot{f}}{\partial \dot{x}^\mu}$$

but the last partial derivative is $\partial f/\partial x^\mu$ as can be seen by evaluating the partial derivative of df/dt with respect to $\dot{x}^\mu$. It follows that

$$\begin{aligned}\frac{d}{dt}\left(\frac{\partial \mathscr{L}'}{\partial \dot{x}^\mu}\right) &= \frac{d}{dt}\left(\frac{\partial \mathscr{L}}{\partial \dot{x}^\mu}\right) + \frac{d}{dt}\left(\frac{\partial \dot{f}}{\partial \dot{x}^\mu}\right)\\ &= \frac{d}{dt}\left(\frac{\partial \mathscr{L}}{\partial \dot{x}^\mu}\right) + \frac{d}{dt}\left(\frac{\partial f}{\partial x^\mu}\right)\\ &= \frac{d}{dt}\left(\frac{\partial \mathscr{L}}{\partial \dot{x}^\mu}\right) + \frac{\partial^2 f}{\partial t \partial x^\mu} + \frac{\partial^2 f}{\partial x^\nu \partial x^\mu}\dot{x}^\nu\end{aligned}$$

whereas, the second term of the Euler–Lagrange equations for $\mathscr{L}'$ develops as follows:

$$\begin{aligned}\frac{\partial \mathscr{L}'}{\partial x^\mu} &= \frac{\partial \mathscr{L}}{\partial x^\mu} + \frac{\partial \dot{f}}{\partial x^\mu}\\ &= \frac{\partial \mathscr{L}}{\partial x^\mu} + \frac{\partial^2 f}{\partial x^\mu \partial x^\nu}\dot{x}^\nu + \frac{\partial^2 f}{\partial x^\mu \partial t}.\end{aligned}$$

When we combine the two pieces, we end up with

$$\begin{aligned}\frac{d}{dt}\left(\frac{\partial \mathscr{L}}{\partial \dot{x}^\mu}\right) + \frac{\partial^2 f}{\partial t \partial x^\mu} + \frac{\partial^2 f}{\partial x^\nu \partial x^\mu}\dot{x}^\nu &= \frac{\partial \mathscr{L}}{\partial x^\mu} + \frac{\partial^2 f}{\partial x^\mu \partial x^\nu}\dot{x}^\nu + \frac{\partial^2 f}{\partial x^\mu \partial t}\\ \frac{d}{dt}\left(\frac{\partial \mathscr{L}}{\partial \dot{x}^\mu}\right) &= \frac{\partial \mathscr{L}}{\partial x^\mu}.\end{aligned}$$

These are exactly the Euler–Lagrange equations one derives from $\mathscr{L}$. Therefore, $\mathscr{L}' \equiv \mathscr{L}$ are identical as we wanted to show. □

For example, a constant $\dot{f}$ would trivially yield the same equations of motion, which finally answers the nagging question surrounding the arbitrary choice of origin for the energy scale. For a less trivial example, consider

$$\mathscr{L} = \frac{1}{2}m\dot{x}^2 - U$$

and

$$\mathscr{L}' = \mathscr{L} + 3tx^2\dot{x} + x^3. \tag{12.9}$$

The Euler–Lagrange equation derived from $\mathscr{L}$ is

$$m\ddot{x} = -\frac{\partial U}{\partial x}. \tag{12.10}$$

For $\mathscr{L}'$, we get

$$\frac{\partial \mathscr{L}'}{\partial \dot{x}} = m\dot{x} + 3tx^2,$$
$$\frac{d}{dt}\left(\frac{\partial \mathscr{L}'}{\partial \dot{x}}\right) = m\ddot{x} + 6tx\dot{x} + 3x^2$$

for the first part of the equation and

$$\frac{\partial \mathscr{L}'}{\partial x} = -\frac{\partial U}{\partial x} + 6tx\dot{x} + 3x^2$$

by evaluating the derivative of (12.9) with respect to x. The last two equations must equal each other by equation (12.6), but the extra terms cancel leaving us with the equations of motion in (12.10). The fact that more than one Lagrangian or Lagrangian density can yield the same equations of motion seems to suggest that energy functionals are not elements of reality, but merely mathematical tools used to formulate models with which we compute physical observables. There are many unanswered questions regarding equivalent Lagrangians, and the subject is much deeper than what we have briefly covered here. The nonuniqueness of Lagrangians for a given set of Euler–Lagrange equations is connected with invariance but should not be confused with it. The latter is useful when defined much more liberally. At this point, we introduce a powerful strategy used to derive integration algorithms for the Euler–Lagrange equations of classical mechanics in generic spaces.

12.4 The leap-frog algorithm in manifolds

The benefits of employing holonomic constraints in classical mechanics are not as vast as they are in quantum simulations. The main motivation for using adiabatic approximations to separate high frequency degrees of freedom comes from the demanding convergence to the quantum limit. However, applications using alternative coordinate systems as, e. g., dihedral angles that map conformation space for the dynamics of carbon chains abound, and these are examples of Riemannian manifolds. Polymer dynamics and the protein folding problems are prime examples of this frontier in computational and theoretical chemistry. Our method does not make use of Lagrange multiplier to introduce constraints [62]. Consequently, it can be used to derive algorithms for much more general spaces that are not necessarily holonomic embeddings, but could be, e. g., parameter spaces of a Lie group of transformations, or some of the alternative coordinate systems we have studied in Chapter 10. The approach begins by partitioning the time interval (t_a, t_b) into a finite set of k equally spaced time intervals $(t_a, t_1, t_2, \dots, t_b)$, where $t_0 = t_a$ and $t_k = t_b$. Then the action functional becomes a sum,

$$S = \sum_{j=0}^{k-1} \mathcal{L}(x_j, x_{j+1}).$$

In the discrete version of the Lagrangian $\mathcal{L}$, the $2n$ degrees of freedom are x_j and x_{j+1}; there are no time derivatives. Therefore, the Euler–Lagrange equations are obtained by setting the derivatives of the intermediate positions x_j to zero with $x_0 = a$ and $x_k = b$ fixed. Here, a and b are two points in the n-dimensional manifold $\mathcal{M}$, and x_j is a shorthand for the n variables that maps the point j in $\mathcal{M}$ onto $\mathbb{R}^n$, where $x_j = \{x^{(1)}(t_j), x^{(2)}(t_j), \dots, x^{(n)}(t_j)\}$. Consider, for instance, a two-time interval partition,

$$S = \mathcal{L}(x_0, x_1) + \mathcal{L}(x_1, x_2).$$

Since $k = 2$, x_0 and x_2 are fixed and we only have one degree of freedom, i. e., x_1. The Euler–Lagrange equation $\delta S = 0$ for that case is

$$\frac{\partial}{\partial x_1^\mu} \mathcal{L}(x_0, x_1) + \frac{\partial}{\partial x_1^\mu} \mathcal{L}(x_1, x_2) = 0. \tag{12.11}$$

The general k partition is just a simple extension of this result,

$$S = \mathcal{L}(x_0, x_1) + \mathcal{L}(x_1, x_2) + \cdots + \mathcal{L}(x_{j-1}, x_j) + \mathcal{L}(x_{j1}, x_{j+1}) + \cdots + \mathcal{L}(x_{k-1}, x_k), \tag{12.12}$$

and $\delta S = 0$ becomes a gradient with $n(k-1)$ degrees of freedom,

$$\sum_{j=1}^{k-1} \left\{ \frac{\partial}{\partial x_j^\mu} \mathcal{L}(x_{j-1}, x_j) + \frac{\partial}{\partial x_j^\mu} \mathcal{L}(x_j, x_{j+1}) \right\} = 0.$$

We are now ready to prove this important result.

Theorem 40. *The general solution of $\delta S = 0$ over a k time interval partition is*

$$\nabla_1 \mathcal{L}(x_j, x_{j+1}) + \nabla_2 \mathcal{L}(x_j, x_{j+1}) = 0 \tag{12.13}$$

where ∇_1 is the partial derivative operator with respect to the first argument of $\mathcal{L}$ and, ∇_2 with respect to the second.

Proof. The $k = 2$ case has been shown in equation (12.11). For a three-time interval partition equation (12.12) reads

$$S = \mathcal{L}(a, x_1) + \mathcal{L}(x_1, x_2) + \mathcal{L}(x_2, b),$$

where $a = x_0$ and $b = x_3$. There are two degrees of freedom: x_1 and x_2. Moreover, x_1 is only an argument in the first two terms of S and x_2 is only an argument in the last two terms of S. In a k-time partition, any given x_j can only be an argument in two consecutive terms of S: $\mathcal{L}(x_{j-1}, x_j)$ and $\mathcal{L}(x_j, x_{j+1})$. □

A two-stage propagator based on the last theorem can be constructed by defining the discrete version of the canonical momentum as follows:

$$p_j = -\Delta t \nabla_1 \mathcal{L}(x_j, x_{j+1}), \tag{12.14}$$

and

$$p_{j+1} = \Delta t \nabla_2 \mathcal{L}(x_j, x_{j+1}), \tag{12.15}$$

where the Δt is multiplied into the gradient to obtain units of momentum. The expression for p_j can be used to advance x_{j+1} and the second allows for the computation of p_{j+1} once x_j and x_{j+1} are known. When combined, these represent a map between two points in phase space $x_j, p_j \to x_{j+1}, p_{j+1}$ that preserves its symplectic symmetry. To clarify the derivation, we begin with Cartesian coordinates in $\mathbb{R}$. The Lagrangian in its discrete form becomes

$$\mathcal{L}(x_j, x_{j+1}) = \frac{1}{2} m \left(\frac{x_{j+1} - x_j}{\Delta t} \right)^2 - \frac{U(x_j) + U(x_{j+1})}{2}. \tag{12.16}$$

It is key to average the potential energy over the two consecutive points to maintain a symmetric dependence of $\mathcal{L}$ on the two variables x_k and x_{k+1}. If this step is not taken, the algorithms derived from $\mathcal{L}$ will not be energy conserving. Applying equations (12.14) and (12.15) to (12.16) yield

$$p_j = m \left(\frac{x_{j+1} - x_j}{\Delta t} \right) + \frac{\Delta t}{2} \frac{\partial}{\partial x_j} U(x_j), \tag{12.17}$$

$$p_{j+1} = m \left(\frac{x_{j+1} - x_j}{\Delta t} \right) - \frac{\Delta t}{2} \frac{\partial}{\partial x_{j+1}} U(x_{j+1}). \tag{12.18}$$

For the monodimensional case, we can solve analytically for x_{j+1} from (12.17),

$$x_{j+1} = x_j + \frac{p_j \Delta t}{m} - \frac{(\Delta t)^2}{2m} \frac{\partial}{\partial x_j} U(x_j), \tag{12.19}$$

and insert the result back into (12.18),

$$p_{j+1} = p_j - \frac{\Delta t}{2} \left(\frac{\partial}{\partial x_{j+1}} U + \frac{\partial}{\partial x_j} U \right). \tag{12.20}$$

It is straightforward to generalize this method. First, in $\mathbb{R}^n$ equations (12.17) and (12.18) simply become

$$p_{\mu,j} = m\delta_{\mu\nu}\left(\frac{x^\nu_{j+1} - x^\nu_j}{\Delta t}\right) + \frac{\Delta t}{2}\frac{\partial}{\partial x^\mu_j}U \tag{12.21}$$

$$p_{\mu,j+1} = m\delta_{\mu\nu}\left(\frac{x^\nu_{j+1} - x^\nu_j}{\Delta t}\right) - \frac{\Delta t}{2}\frac{\partial}{\partial x^\mu_{j+1}}U \tag{12.22}$$

where $\delta_{\mu\nu}$ is the Euclidean metric tensor. Our equations have the same mass for all dimensions. However, it is straightforward to define a Hessian metric tensor as a diagonal matrix, the same way we have in Chapter 10. Alternatively, a system with unequal masses can be understood as a special case of the next example, when $g_{\mu\nu}$ is uniform but not necessarily diagonal as, e. g., when relative coordinates for the n particle system are considered:

$$p_{\mu,j} = g_{\mu\nu}\left(\frac{x^\nu_{j+1} - x^\nu_j}{\Delta t}\right) + \frac{\Delta t}{2}\frac{\partial}{\partial x^\mu_j}U,$$

$$p_{\mu,j+1} = g_{\mu\nu}\left(\frac{x^\nu_{j+1} - x^\nu_j}{\Delta t}\right) - \frac{\Delta t}{2}\frac{\partial}{\partial x^\mu_{j+1}}U.$$

The equations for x^ν_{j+1} in this case can be cast into a matrix-vector form and can be solved with standard linear algebra methods by finding the inverse of the metric tensor once.

Things get more complicated when the metric tensor is not uniform. Then the symmetric definitions of $\mathcal{L}$ has to contain the endpoint average of $g_{\mu\nu}$ for the same reasons articulated above in regards to U:

$$\mathcal{L}(x_j, x_{j+1}) = \frac{1}{4}(g_{\mu\nu,j} + g_{\mu\nu,j+1})\left(\frac{x_{j+1} - x_j}{\Delta t}\right)^2 - \frac{U(x_j) + U(x_{j+1})}{2}. \tag{12.23}$$

Its two derivatives p_k and p_{k+1} become

$$\begin{aligned} p_{\mu,j} = {} & \frac{1}{2}(g_{\mu\nu,j} + g_{\mu\nu,j+1})\left(\frac{x^\nu_{j+1} - x^\nu_j}{\Delta t}\right) \\ & - \frac{\Delta t}{4}\left(\frac{\partial}{\partial x^\mu_j} g_{\kappa\nu,j}\right)\left(\frac{x^\nu_{j+1} - x^\nu_j}{\Delta t}\right)\left(\frac{x^\kappa_{j+1} - x^\kappa_j}{\Delta t}\right) + \frac{\Delta t}{2}\frac{\partial}{\partial x^\mu_j}U, \end{aligned} \tag{12.24}$$

$$\begin{aligned} p_{\mu,j+1} = {} & \frac{1}{2}(g_{\mu\nu,j} + g_{\mu\nu,j+1})\left(\frac{x^\nu_{j+1} - x^\nu_j}{\Delta t}\right) \\ & + \frac{\Delta t}{4}\left(\frac{\partial}{\partial x^\mu_{j+1}} g_{\kappa\nu,j+1}\right)\left(\frac{x^\nu_{j+1} - x^\nu_j}{\Delta t}\right)\left(\frac{x^\kappa_{j+1} - x^\kappa_j}{\Delta t}\right) - \frac{\Delta t}{2}\frac{\partial}{\partial x^\mu_{j+1}}U. \end{aligned} \tag{12.25}$$

Equation (12.24) is nonlinear in the position difference vectors; therefore, a numerical implementation of this general case requires an initial estimate of x_{j+1}^{ν} (typically obtained by ignoring the quadratic terms), followed by iteration until self consistency is achieved. As we shall see in the numerical example, this family of leap-frog algorithms are energy preserving and the energy fluctuations scale with the square of the time interval Δt.

12.4.1 A classical molecular dynamics example

As a simple example, consider the following code:

```
import numpy as np
import random as rnd
rnd.seed()    # seed with system time
mass,deltat = 919.0,0.05
xj  = 2.1   # initial position
pj  = 0.0   # initial momentum
uj,duj = pot(xj)
k = 0
steps = 1 + 10*int(85/deltat)  # Compute the trajectory steps
while k < steps:
    xj +=  deltat*(pj - deltat*duj/2.0)/(mass) # update the position
    ujp1,dujp1 = pot(xj)  #
    pj +=  - deltat*(dujp1 + duj)/2.0
    uj = ujp1
    duj = dujp1
    kej = pj*pj/(2.0*mass) # kinetic energy
    energy = kej + uj     # potential + kinetic
    k += 1
```

By setting the initial value of the momentum to zero, the code integrates Newton's equations for a particle of 919.0 atomic units of mass starting from a classical turning point. It computes the value of U and its derivative, and then inside the while loop it computes x_{j+1} using (12.19) in the first line. Then $U(x_{j+1})$ and its derivative are computed and returned from the potential energy function and these are used to compute p_{j+1} according to (12.20).

The potential energy function is a monodimensional Lennard–Jones model:

$$U(r) = 4\epsilon\left[\left(\frac{\sigma}{r}\right)^{12} - \left(\frac{\sigma}{r}\right)^{6}\right]. \qquad (12.26)$$

The function is designed to return both the value of U and its derivative,

$$\frac{dU}{dr} = -\frac{24\epsilon}{r}\left[\left(\frac{\sigma}{r}\right)^{12} - \left(\frac{\sigma}{r}\right)^{6}\right]. \tag{12.27}$$

The Python code for pot() is below:

```
def pot(x):
    e = 1.0
    s = 2.0/x
    s2 = s*s
    s4 = s2*s2
    s6 = s2*s4
    s12 = s6*s6
    u = 4.0*e*(s12 - s6)
    du = -24.0*e*(2.0*s12 - s6)/x
    return u,du
```

The main function finishes the while the loop section of the code by updating terms for the iteration to the next time interval and computes the total energy. Since the Lagrangian is not explicitly dependent on time, the energy should be conserved. Therefore, plotting E versus time for several values of Δt is the first test for any molecular dynamics code. In Figure 12.1, we display such graph for two values of Δt: 1 and 0.1 atomic units. The energy does oscillate as the dashed line shows, however, upon reducing Δt by a factor of 10, the energy envelope decreases by approximately a factor of 100 indicating a quadratic convergence for the energy. More importantly, the energy envelope is horizontal in the graph, indicating that the energy is indeed conserved. By adding a few

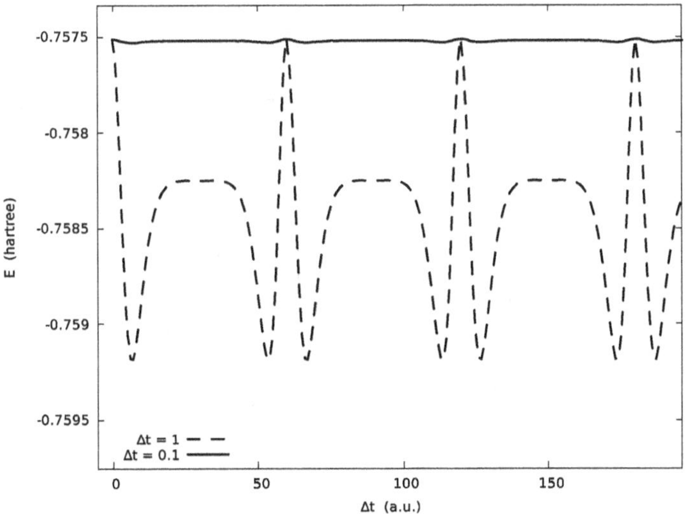

Figure 12.1: The energy of a particle in a Lennard–Jones potential as a function of simulation time.

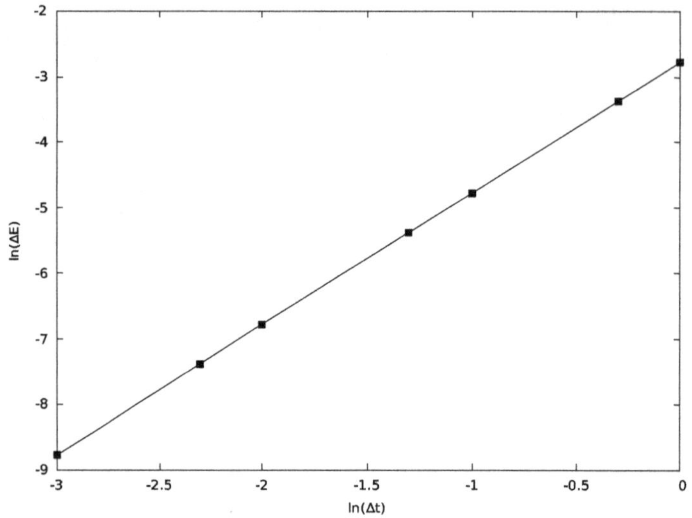

Figure 12.2: Natural logarithm of the energy fluctuation plotted versus the natural logarithm of Δt.

lines of code, we compute the maximum E_{max} and minimum energy E_{min} computed for the trajectory. At the end of the run, we compute $\Delta E = E_{\text{max}} - E_{\text{min}}$, and its natural log. A graph of $\ln(\Delta E)$ versus $\ln(\Delta t)$ can be found in Figure 12.2. The data points are fit with a straight line, the slope of which is very close to 2.0.

12.5 Exercises

1. Consider the particle in a viscous fluid characterized by a drag coefficient γ. The pertinent Lagrangian is in equation (12.3).
   (a) Derive the three Euler–Lagrange equations satisfied by the trajectory $x(t), y(t), z(t)$ subject to the initial conditions

$$x(t=0) = x_0, \quad y(t=0) = y_0, \quad z(t=0) = z_0$$
$$\dot{x}(t=0) = \dot{x}_0, \quad \dot{y}(t=0) = \dot{y}_0, \quad \dot{z}(t=0) = z_0.$$

   (b) Show that

$$\frac{d}{dt}\left[\frac{\partial \mathscr{L}}{\partial \dot{x}}\right] = \frac{\partial \mathscr{L}}{\partial x}$$

   yields

$$m(\ddot{x}e^{\gamma t} + \gamma \dot{x}e^{\gamma t}) = 0.$$

(c) Show that

$$v_x = \dot{x}_0 e^{-\gamma t}$$

by evaluating two indefinite integrals,

$$\int \frac{dv_x}{v_x} = -\gamma \int dt + C,$$

exponentiating both sides and applying the boundary condition to find an expression for the constant e^C.

(d) Use the same strategy and the solution of part (c) to show that

$$x = x_0 + \frac{v_0}{\gamma}(1 - e^{-\gamma t}).$$

(e) Take the limit as $t \to 0$ of $v_t(t)$ and $x(t)$ and show that the initial conditions are satisfied. Then take the limit of $v_t(t)$ and $x(t)$ as $t \to \infty$.

(f) Is the energy of the particle a conserved quantity? Answer this question by comparing the energy at $t = 0$ and its limit as $t \to \infty$. Comment on what you observe: If the energy is not conserved, where did the energy come from (or go)?

(g) Insert the equation for x in part (d) and the corresponding ones for y and z back into $\mathscr{L}$ and find an expression for the action S as time proceed from t_0 to t_1. Show that even if t_1 approaches infinity the action remains finite.

2. Consider a classical particle of mass m subjected to a uniform gravitational force $U = gz$ in $\mathbb{R}^3$.
   (a) Write the Lagrangian $\mathscr{L}$ for the system.
   (b) Find the Euler–Lagrange equations satisfied by x, y and z.
   (c) Solve the Euler–Lagrange equations using $x_0, y_0, z_0, \dot{x}_0, \dot{y}_0$, and $\dot{z}_0$ as the initial conditions.
   (d) Find equations for the canonical momenta as functions of time from the solutions in part (c).
   (e) Substitute the equations for x, y, and z into the expression for the total energy,

$$\mathscr{H} = \frac{1}{2m}(p_x^2 + p_y^2 + p_z^2) + U.$$

   (f) Find out if the energy is conserved by direct substitution of the solutions in parts (c) and (d) in the Hamiltonian to compute $E(t)$.

3. Consider a free particle of mass m in a ring of radius r as a simple example of a manifold $\mathcal{M}$. Recall that the ring in an embedding on the xy plane, $x^2 + y^2 = r^2$.
   (a) Use polar coordinates $x = r\cos\theta, y = r\sin\theta$ and derive the Lagrangian $\mathscr{L}[\dot{\theta}, \theta]$ for the system.
   (b) What is the canonical momentum for this system? Write an expression for it.
   (c) Is the canonical momentum conserved?

4. Why does the factor of 1/2 disappear when evaluating the partial derivative of the kinetic energy

$$K = \frac{1}{2} g_{\mu\nu} \dot{x}^{\mu} \dot{x}^{\nu}$$

with respect to $\dot{x}^{\mu}$? To answer the question, consider a generic $n = 2$ case,

$$K = \frac{1}{2} g_{\mu\nu} \dot{x}^{\mu} \dot{x}^{\nu} = \frac{1}{2} \begin{pmatrix} \dot{x}^{(1)} & \dot{x}^{(2)} \end{pmatrix} \begin{pmatrix} g_{11} & g_{12} \\ g_{12} & g_{22} \end{pmatrix} \begin{pmatrix} \dot{x}^{(1)} \\ \dot{x}^{(2)} \end{pmatrix}$$

expand the quadratic form, then evaluate the partial derivatives with respect to $\dot{x}^{(1)}$ and $\dot{x}^{(2)}$. Confirm that the factor of 1/2 should not show in the final result,

$$\frac{\partial K}{\partial \dot{x}^{\mu}} = g_{\mu\nu} \dot{x}^{\nu}.$$

5. Transform

$$\frac{1}{2} m(\dot{x}^2 + \dot{y}^2 + \dot{z}^2)$$

into the spherical polar equivalent and show that the result is the kinetic energy term of $\mathscr{L}$ in equation (12.8).

6. Find the expressions for the Lagrangian, the canonical momenta, and derive the Hamiltonian for $\mathbb{S}^2$ when mapped with the stereographic projections from the north chart.
7. Show that $\dot{p}_{\mu} = g_{\mu\nu} \ddot{x}^{\nu} + \partial_{\beta} g_{\mu\nu} \dot{x}^{\beta} \dot{x}^{\nu}$ by using the Euler–Lagrange equations derived from $\mathscr{L}$ in equation (12.4).
8. This exercise leads you to prove that the angular momentum is conserved for a particle of mass m in $\mathbb{R}^3$ subjected to a central potential.
   (a) Use Cartesian coordinates to find expressions for the canonical momentum vector p and the angular momentum vector $L = r \times p$. Your expression should be in terms of m, x, y, z and the velocity components $\dot{x}, \dot{y}, \dot{z}$.
   (b) Derive the following result from your expression in (a):

$$\dot{L} = m[(y\ddot{z} - z\ddot{y})\mathbf{i} + (z\ddot{x} - x\ddot{z})\mathbf{j} + (x\ddot{y} - y\ddot{x})k].$$

   (c) Prove that for a central potential $U(r)$,

$$m\ddot{x} = -\frac{1}{r}\left(\frac{\partial U}{\partial r}\right)x,$$

   with the corresponding equations for the y and z components. Recall that $r = (x^2 + y^2 + z^2)^{1/2}$.

(d) Insert the result from part (c) into the expression for $\dot{L}$ in part (b) and prove that $\dot{L} = 0$, i. e., the angular momentum is conserved.

9. In this exercise, we explore equivalent Lagrangian densities.
   (a) Derive the Euler–Lagrange equations from the following Lagrangian density:

$$\mathscr{L} = -\frac{\hbar^2}{2m}(\partial_x \psi)(\partial_x \psi^*) - \psi^*(U - E)\psi$$

   where ψ^* is the complex conjugate of ψ and $\partial_x \psi = \partial\psi/\partial x$. The configuration space is $\mathbb{R}$.

   (b) Show that the following Lagrangian density:

$$\mathscr{L}' = -\frac{\hbar^2}{2m}(\partial_x \psi)(\partial_x \psi^*) + \psi\partial_x\psi + \psi^*\partial_x\psi^* - \psi^*(U - E)\psi$$

   derives the same Euler–Lagrange equation as $\mathscr{L}$ in part (a).

   (c) The two added terms in $\mathscr{L}'$, namely $\psi\partial_x\psi$ and $\psi^*\partial_x\psi^*$ have a very important physical meaning, and the exercise in part (b) is a less rigorous way of proving that these properties are conserved, without using Noether's theorem [99]. What are these physical properties?

   (d) Prove that adding any functional f to $\mathscr{L}$ in part (a) that satisfies the following condition:

$$\frac{\partial^2 f}{\partial x \partial(\partial_x \psi)} = \frac{\partial f}{\partial \psi}$$

   along with its complex conjugate

$$\frac{\partial^2 f}{\partial x \partial(\partial_x \psi^*)} = \frac{\partial f}{\partial \psi^*}$$

   produces an equivalent Lagrangian.

10. This exercise suggests additional ways to explore the molecular dynamics code in the chapter.
    (a) Add a print function such as

```
print("%12.6f %12.6f %12.6f" %(k*deltat,xj,pj))
```

    to the molecular dynamics code inside the while loop, then graph p_j versus x_j. Compare the resulting graph with the one in Figure 12.3.

    (b) Add the following two lines to the code immediately before the while loop:

```
emin = 0.0
emax = -1.0
```

    and the following lines inside the while loop after the energy is computed:

```
if (energy < emin):
    emin = energy
if (energy > emax):
    emax = energy
```

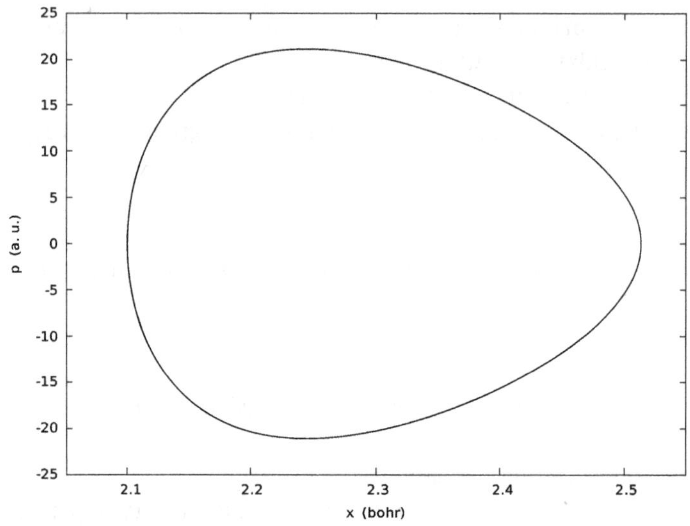

Figure 12.3: Trajectory in phase space for a particle of 919.0 atomic units of mass in a Lennard–Jones potential.

then add code to compute ΔE at the end of the trajectory. Compute trajectories with various values of Δt in the range $(0.001, 1)$ and reproduce the plot in Figure 12.2.

(c) Extend the trajectory length so it is ten times as long, use a $\Delta t = 0.5$, and verify that the energy envelope remain horizontal by graphing E versus t.

11. Consider the following Lagrangian for a particle in a potential U in $\mathbb{R}$:

$$\mathscr{L} = \left(\frac{1}{2}m\dot{x}^2 - U\right)e^{\gamma t}.$$

(a) Derive the Euler–Lagrange equation. This models a particle in a potential U experiencing drag as, e. g., a pendulum in air, or in more viscous fluids.

(b) The traditional approach to integrate the Euler–Lagrange equation in part (a) is to use the Runge–Kutta algorithm. To derive it, one assumes the acceleration,

$$a_j = -\frac{1}{m}\left(\frac{\partial U}{\partial x_j} - \gamma p_j\right)$$

is approximately constant for a sufficiently small Δt.

$$p_{j+1} = m \int_{t_j}^{t_j+1} \ddot{x}\, dt \approx m a_j \Delta t,$$

then assumes the momentum p_{j+1} is also constant over the same interval,

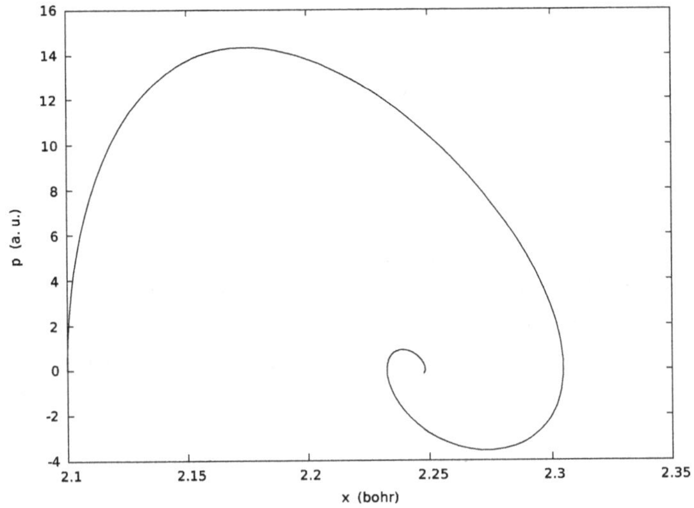

Figure 12.4: Phase space trajectory for a particle in a Lennard–Jones potential inside a dampening field.

$$x_{j+1} = x_j + \frac{1}{m} \int_{t_j}^{t_j+1} p_{j+1}\, dt \approx x_j + \frac{p_{j+1}}{m} \Delta t.$$

Replace the while loop code with the snippet below and run it with a $\gamma = 0.1$, $\Delta t = 0.1$, and set `steps = 1 + int(85/deltat)`. Compare the phase space trajectory with the one in Figure 12.4:

```
while k < steps:
  a = -duj/mass - gamma*pj/mass
  pj +=  a*deltat*mass
  xj +=  pj*deltat/mass
  uj,duj = pot(xj)
  energy = pj*pj/(2.0*mass) + uj
  k += 1
  time = deltat*k
```

(c) Add code to print the energy and inspect graphs of p_j, x_j, and E versus time. Explain in your own words what is taking place.

13 The Feynman path integral

Students should learn to study at an early stage the great works of the great masters instead of making their minds sterile through the everlasting exercises of college, which are of no use whatever, except to produce a new Arcadia where indolence is veiled under the form of useless activity...

Eugenio Beltrami

In this chapter, we discuss some applications of an alternate representation of quantum mechanics. We have seen that both the laws of classical and quantum mechanics can be derived by optimizing the action of a Lagrangian. R. P. Feynman uses the idea of the classical action to derive the path integral representation of wave functions, by turning the least action principle of classical mechanics on its head. His proposal is to consider not only the classical path, but all possible paths from two given points, $a = (x_a, t_a)$ and $b = (x_b, t_b)$ to represent the wave function of a particle. He then postulates that each one of the infinitely many possible paths contributes to the representation depending on the values of its corresponding classical action. We begin by formalizing the idea using the so-called *time-slice* representation of paths, and we derive the Schrödinger equation following in more detail the outline in Section 4.1 of Feynmann and Hibbs [101].

It is possible to find analytical solutions to the infinite-dimensional integral for a handful of simple cases. However, computing wave functions in general using the real time path integral is a formidable challenge. On the other hand, the imaginary time path integral representation is amenable to stochastic simulations and finds an important interpretation in statistical mechanics. It has been shown, e. g., that the imaginary time path integral can be used to compute thermodynamic properties in the N, V, T ensemble [107–110] such as the internal energy

$$\mathcal{U} = \frac{\sum_{i=0}^{\infty} E_i \exp(-\beta E_i)}{\sum_{i=0}^{\infty} \exp(-\beta E_i)}, \tag{13.1}$$

the corresponding heat capacity,

$$\mathcal{C}_V = \left(\frac{\partial \mathcal{U}}{\partial T}\right)_T$$

etc., without a priori knowledge of the eigenenergies or having to compute them as we demonstrate in this chapter. Here, $\beta = 1/(k_B T)$, T is the temperature in Kelvin, k_B, Boltzmann's constant, E_i are the system energy levels, and

$$\mathcal{Q} = \sum_{i=0}^{\infty} \exp(-\beta E_i) \tag{13.2}$$

is the partition function. The Metropolis algorithm is the main tool used to carry out imaginary time path integral simulations; therefore, we introduce it briefly as well. Finally, in the $T \to 0$ limit, the Path Integral Ground State method [111–117] is an important

https://doi.org/10.1515/9783111610207-013

alternative strategy to compute ground state properties, though we do not discuss it in this textbook.

13.1 Introduction

The path integral is an operator-free representation of the time evolution of quantum states based on the following two principles formulated by R. P. Feynman in his 1948 paper [118].

- *If a set of ideal position measurements is made to determine if a particle follows a certain path in a region R of space, as time moves forward, then the probability p(R) that the outcome of the position measurements is affirmative is the results of the sum of the contribution from all the possible paths in the region.*
- *Each path contributes the same amount in amplitude to the probability but the phase contribution of each path is the action*

$$S[\dot{x}(t), x(t)] = \int dt\, \mathcal{L}(\dot{x}, x) \tag{13.3}$$

associated with the path as the time integral of the Lagrangian $\mathcal{L}(\dot{x}, x)$.

Both principles are consistent with the postulates of quantum mechanics and give us a way of representing the probability $p(R)$. The path integral formulation of quantum mechanics is the most transparent way to see classical mechanics emerge from the laws of quantum physics. In the second principle, the action taken by the system is not the least action that defines the classical trajectory we explored in the previous chapter. Instead, all paths can contribute to $p(R)$. However, as we construct the probability $p(R)$, the paths that lie closest to the classical path will interfere constructively for the most part with one another, whereas paths that are far away from the classical path will interfere destructively for the most part with the other paths nearby.

To construct the probability $p(R)$, Feynman partitions the time into n finite segments equally spaced by Δt. Then the action can be broken up as a sum,

$$S[\dot{x}(t), x(t)] = \sum_{i=0}^{n-1} S(x_{i+1}, x_i) \approx \Delta t \sum_{i=0}^{n-1} \mathcal{L}\left(\frac{x_{i+1} - x_i}{\Delta t}, \frac{x_{i+1} + x_i}{2}\right), \tag{13.4}$$

where x_0, t_0 is the starting point, n is the number of *time slices*, and x_{n+1}, t_{n+1} is the endpoint of the path. In Figure 13.1, we sketch the process for a free particle. The classical path is the solid line, while the other two are examples of possible paths obtained by partitioning time as we did in the previous chapter. For the dotted line, the set of points

$$\{(x_0, t_0), (x_1, t_1), (x_2, t_2), (x_3, t_3)\}$$

uses three *time slices* to represent a possible path. The paths are not smooth for any finite n, but become smooth and differentiable as n tends toward infinity. Moreover,

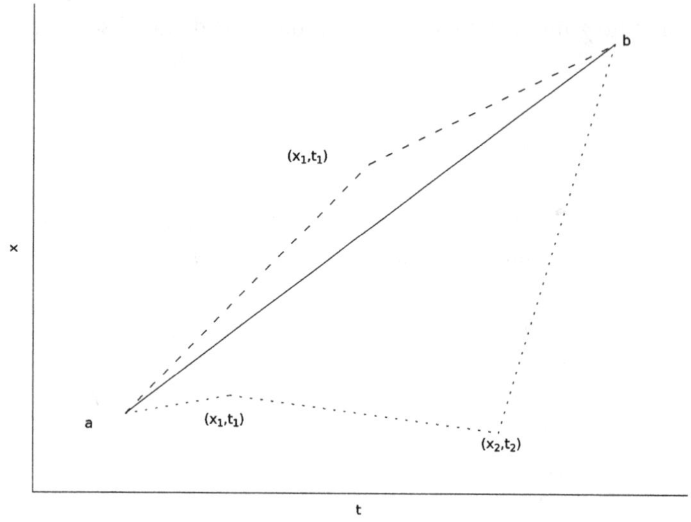

Figure 13.1: The time-slice representation of paths between point a and b using $n = 0, 1$, and 2 slices.

Δt becomes small in the same limit, we can approximate the integral for the action by assuming that the Lagrangian is constant, and we expand it about a point in the linear path $x_i, t_i \to x_{i+1}, t_{i+1}$. The probability $p(R)$ is obtained in the limit $\Delta t \to 0$ of the following multidimensional integral,

$$p(R) = \lim_{n\to\infty} A \int_R dx_1 \cdots \int_R dx_n \exp\left[\frac{i}{\hbar}\sum_{k=0}^{n-1} S(x_{k+1}, x_k)\right], \tag{13.5}$$

where A is a normalization constant obtained by setting the integral of $p(R)$ over all the space equal to one. The relationship between the probability $p(R)$ and the time-dependent wave function is

$$p(R) = \int_R dx\, \Psi^\dagger(x,t)\, \Psi(x,t). \tag{13.6}$$

Equation (13.5) provides a way to formulate the path integral expressions for the wave function and its complex conjugate

$$\Psi(x_n, t) = \lim_{\Delta t\to 0} A \int_R dx_{n-1} \int_R dx_{n-2} \ldots \exp\left[\frac{i}{\hbar}\sum_{k=-\infty}^{n-1} S(x_{k+1}, x_k)\right], \tag{13.7}$$

$$\Psi^\dagger(x_n, t) = \lim_{\Delta t\to 0} A \int_R dx_{n+1} \int_R dx_{n+2} \cdots \exp\left[\frac{i}{\hbar}\sum_{k=n+1}^{\infty} S(x_{k+1}, x_k)\right]. \tag{13.8}$$

These remarkable equations tell us that the probability that a position measurement at time t yields $x \in R$ is produced by the sum of all past histories (which produce Ψ), and all future histories (which produce $\Psi^\dagger$). The restrictions on the regions of integration R for Ψ and $\Psi^\dagger$ depend on all past (and future!) position measurements. The reference to future measurement impacting on the present measurement should be understood as nothing more than a manifestation of Heisenberg's uncertainty principle as it applies to the time-energy pair, since these are incompatible properties.

The derivation of the path integral [101, 102] representation of quantum mechanics is highly heuristic, and guided for the most part by Feynman's great intuitive abilities. Nevertheless, Feynman's path integral representation of quantum mechanics becomes accepted immediately since he demonstrates that equation (13.7) satisfies the time dependent Schrödinger equation. It is possible to use equation (13.7) and find a path integral representation of the matrix elements of the time evolution operator in the position representation

$$\begin{aligned} K(x,t=0;x',t) &= \langle x'|\exp\left(-\frac{i}{\hbar}Ht\right)|x\rangle \\ &= \lim_{\Delta t\to 0} A\int_{-\infty}^{\infty} dx_1\cdots\int_{-\infty}^{\infty} dx_n \exp\left[\frac{i}{\hbar}\sum_{k=0}^{n-1} S(x_{k+1},x_k)\right], \end{aligned} \tag{13.9}$$

where $x_0 = x$, $x_n = x'$, H is the Hamiltonian operator and the factor outside the integral is

$$A = \left(\frac{i2\pi\hbar}{m}\right)^{-n/2}. \tag{13.10}$$

for a particle of mass m in $\mathbb{R}^1$.

Feynman's path integral formulation can be used on a number of interesting problems, including the fully relativistic quantum field theory and the scattering of elementary particles. Keeping in mind the goal of overcoming the exponential growth of algorithms based on vector spaces, we continue to look for opportunities to develop stochastic processes that we can simulate. To implement such representation of quantum mechanics into a stochastic approach, it would be necessary to use random numbers distributed according to the exponential in equation (13.9). Unfortunately, stochastic methods based on the real time path integral are plagued with the infamous *sign problem*. Generating distributions of random numbers with

$$\exp\left[i\sum_{k=1}^{n} S(x_{k+1},x_k)/\hbar\right] \tag{13.11}$$

as the importance sampling function turns out to be challenging for any finite propagation time. The same oscillations that constructively and destructively interfere path contributions make any attempt to generate the correct distribution of random numbers futile. Distributions of random numbers must be positive definite. Without importance sampling, the statistical fluctuation of Monte Carlo integration for all but the smallest number of dimensions becomes overwhelming. Nevertheless, since the mid-1980s, chemists have found a large number of applications for Feynman's path integrals. The majority of these are in statistical mechanics, and we will move our attention to the imaginary time propagation by path integral, since in that case the sign problem is eliminated, simply because the phase in equation (13.11) becomes a positive definite function. Before we discuss these methods, we need to demonstrate how Feynman places the path integral representation of quantum mechanics in more firm mathematical grounds by demonstrating that it is compatible with Schrödinger's theory.

13.2 Schrödinger's equation from the path integral

From the path integral definition of the time evolution operator $K(x_b, t_b; x_a, t_a)$ in equation (13.9), the wave function at x, t, can be obtained by propagating it from time t_a as follows:

$$\psi(x,t) = \int_{-\infty}^{\infty} K(x,t;x_a,t_a)\,\psi(x_a,t_a)\,dx_a. \tag{13.12}$$

The first goal is to derive an equation for $K(x, t; x_a, t_a)$ represented by an arbitrary number of time slices, for Lagrangians of the type,

$$\mathscr{L} = \frac{m}{2}\dot{x}^2 - U(x).$$

By allowing the number of slices, or the partition size to approach infinity, the path integral converges to the exact expression of the time evolution matrix element, which is the kernel of equation (13.12). By using the expression of $K(x_b, t_b; x_a, t_a)$ from a single slice, the second goal is to derive the time dependent Schrödinger equation. More precisely, the result we want to prove is the following.

Theorem 41. *The time evolution operator satisfies the following identity,*

$$K(x_b,t_b;x_a,t_a) = \lim_{\epsilon\to 0} A^n \int_{-\infty}^{\infty} dx_1 \int_{-\infty}^{\infty} dx_n \cdots$$
$$\times \int_{-\infty}^{\infty} dx_{n-1} \exp\left\{\frac{im}{2\hbar\epsilon}\sum_{k=1}^{n}(x_k - x_{k-1})^2\right\} \exp\left\{-\frac{i}{\hbar}\epsilon\sum_{k=1}^{n} U\left(\frac{x_k + x_{k-1}}{2}\right)\right\}.$$

Proof. We begin with the $n = 2$ case first, and recall that the path represented by the points $x = x_a, t_a, x_1, \epsilon, x_2, 2\epsilon, \ldots, x_b, n\epsilon$ is connected point-to- point by a straight line. The propagator is a single integral,

$$K_2(x_b, t_b; x_a, t_a) = \int_{-\infty}^{\infty} dx_1\, K_1(x_b, t_b; x_1, \epsilon)\, K_1(x_1, \epsilon; x_a, t_a),$$

where

$$K_1(x_1, \epsilon; x_a, t_a) = A \exp\left[\frac{im}{2\hbar} \int_{t_a}^{t_a+\epsilon} \dot{x}^2 dt\right] \exp\left[\frac{i}{\hbar} \int_{t_a}^{t_a+\epsilon} U[x(t)]dt\right],$$

$$K_1(x_b, t_b; x_1, \epsilon) = A \exp\left[\frac{im}{2\hbar} \int_{t_a+\epsilon}^{t_b} \dot{x}^2 dt\right] \exp\left[\frac{i}{\hbar} \int_{t_a+\epsilon}^{t_b} U[x(t)]dt\right].$$

When we approximate the path $x(t)$ between t_a and $t_a + \epsilon$, we obtain

$$\dot{x} = \frac{x_1 - x_a}{\epsilon},$$
$$x = x_a + \frac{x_1 - x_a}{\epsilon}(t - t_a),$$

and we approximate the potential with its value at the end-point of the first slice,

$$U[x(t)] \approx U[x(t_a + \epsilon)] = U\left[x_a + \frac{x_1 - x_a}{\epsilon}(\epsilon)\right] = U[x_1].$$

Alternatively, we could use the beginning point representation $U[x(t)] \approx U[x_{i-1}]$ on the ith slice or as Feynman and Hibbs choose to do, the mid-point representation $U[x(t)] \approx U[(x_i + x_{i-1})/2]$. All three representations of the path integral are equivalent (in $\mathbb{R}^n$). In any case, the time integrals become trivial and one has

$$K_1(x_1, \epsilon; x_a, t_a) \approx A \exp\left[\frac{im}{2\hbar} \frac{(x_1 - x_a)^2}{\epsilon}\right] \exp\left[\frac{i}{\hbar}\epsilon U\left(\frac{x_1 + x_a}{2}\right)\right].$$

Applying the same approximations to the second time-slice, we get

$$K_1(x_b, t_b; x_1, \epsilon) = A \exp\left[\frac{im}{2\hbar} \frac{(x_b - x_1)^2}{\epsilon}\right] \exp\left[\frac{i}{\hbar}\epsilon U\left(\frac{x_b + x_1}{2}\right)\right],$$

and the $n = 2$ approximation to $K(x_b, t_b; x_a, t_a)$ becomes

$$K_2(x_b,t_b;x_a,t_a) = A^2 \int_{-\infty}^{\infty} dx_1 \exp\left[\frac{im}{2\hbar}\frac{(x_1-x_a)^2}{\epsilon}\right]\exp\left[\frac{i}{\hbar}\epsilon U\left(\frac{x_1+x_a}{2}\right)\right]$$
$$\times \exp\left[\frac{im}{2\hbar}\frac{(x_b-x_1)^2}{\epsilon}\right]\exp\left[\frac{i}{\hbar}\epsilon U\left(\frac{x_b+x_1}{2}\right)\right],$$
$$K_2(x_b,t_b;x_a,t_a) = A^2 \int_{-\infty}^{\infty} dx_1 \exp\left\{\frac{im}{2\hbar\epsilon}\left[(x_1-x_a)^2+(x_b-x_1)^2\right]\right\}$$
$$\times \exp\left[\frac{i}{\hbar}\epsilon\left\{U\left(\frac{x_1+x_a}{2}\right)+U\left(\frac{x_b+x_1}{2}\right)\right\}\right].$$

Even with the crude approximation made to the time integral of U, in general this single integral will have no known analytical solution unless U is some quadratic function of x. The general n case can be trivially constructed following exactly the same process as we did for $n = 2$. □

Next, we demonstrate that propagating by an infinitesimal amount of time ϵ, it is possible to derive the Schrödinger equation from this last expression as well as the proper integration factor A. Following Section 4-1 of Feynman and Hibbs [101], we begin by using only one slice for the propagator: $t = t_a$, $t_b = t + \epsilon$, $\epsilon \to 0$ to rewrite the integral in equation (13.12) for $\psi(x, t + \epsilon)$. There remains one integral to evaluate, since the value of $x - x_a$ can still be arbitrarily large even as $t_b - t_a$ becomes infinitely small. It is convenient to rewrite the last two equations following closely the definitions of t and t_b. The single slice approximation to $K[x, (t_a + \epsilon); x_a, t_a]$ becomes

$$K(x,t;x_a t_a) \approx A\exp\left\{\frac{im}{2\hbar\epsilon}(x-x_a)^2\right\}\exp\left\{-\frac{i}{\hbar}\epsilon U\left(\frac{x-x_a}{2}\right)\right\},$$

and inserting this into the expression for $\psi(x, t + \epsilon)$ we derive

$$\psi(x,t+\epsilon) = A\int_{-\infty}^{\infty} \exp\left\{\frac{im}{2\hbar\epsilon}(x-x_a)^2\right\}\exp\left\{-\frac{i}{\hbar}\epsilon U\left(\frac{x+x_a}{2}\right)\right\}\psi(x_a,t_a)\,dx_a.$$

We next change variables in the integral letting

$$\eta = x_a - x.$$

Basically, we are making a u substitution, with $dx_a = d\eta$. The integral becomes

$$\psi(x,t+\epsilon) = A\int_{-\infty}^{\infty} d\eta\,\exp\left\{\frac{im\eta^2}{2\hbar\epsilon}\right\}\times\exp\left\{-\frac{i}{\hbar}\epsilon U\left(x+\frac{\eta}{2}\right)\right\}\psi(x+\eta,t). \tag{13.13}$$

We now perform the following operations with equation (13.13):

1. Expand $\psi(x, t+\epsilon)$ into a power series up to first order in ϵ.
2. Expand

$$\exp\left\{-\frac{i}{\hbar}\epsilon U\left(x+\frac{\eta}{2}\right)\right\}$$

in a power series up to first order in ϵ.

3. If η is too large, the

$$\exp\left\{\frac{im\eta^2}{2\hbar\epsilon}\right\}$$

term oscillates violently averaging the integral to zero. Therefore, we can expand the wave function $\psi(x+\eta, t)$ and $\epsilon U(x+\eta/2)$ in powers of η, keeping terms up to η^2 in the expansions.

4. Collect all the terms of the same order and derive the desired expressions. Alas! We are using a slightly differently definition of A compared to Feynman and Hibbs.

One does not expand the function $\exp\{im\eta^2/(2\hbar\epsilon)\}$ in equation (13.13) because the variable ϵ sets the scale for η: $\eta^2 \propto \epsilon$. The variable η can take any value in $\mathbb{R}$, however, the probability that a value of η is outside the range, say, $(-5\epsilon, 5\epsilon)$ is vanishingly small. Moreover, as $\eta \to 0$, the function approaches a Dirac delta distribution, and this feature is critical for the propagator to reproduce $\psi(x, t)$ as the time interval vanishes. In more technical terms, one does not expand the Gaussian function because it is part of the "path integral measure."

The expansions of $\psi(x, t+\epsilon)$ and the potential energy term into a power series up to first order in ϵ are

$$\psi(x, t+\epsilon) \approx \psi(x,t) + \epsilon\frac{\partial\psi(x,t)}{\partial t} + \cdots,$$

$$\exp\left\{-\frac{i}{\hbar}\epsilon U\left(x+\frac{\eta}{2}\right)\right\} \approx 1 - \frac{i}{\hbar}\epsilon U\left(x+\frac{\eta}{2}\right) + \cdots.$$

Expanding $\psi(x+\eta, t)$ and $\epsilon U(x+\frac{\eta}{2})$ in powers of η up to the second order in η yields

$$\psi(x+\eta, t) \approx \psi(x,t) + \eta\frac{\partial\psi(x,t)}{\partial\eta} + \frac{1}{2}\eta^2\frac{\partial^2\psi(x,t)}{\partial\eta^2} + \cdots,$$

$$\epsilon U\left(x+\frac{\eta}{2}\right) \approx \epsilon U(x) + \epsilon\eta\frac{\partial U(x)}{\partial\eta} + \frac{1}{2}\epsilon\eta^2\frac{\partial^2 U(x)}{\partial\eta^2} + \cdots.$$

Collecting all these terms and plugging them back into equation (13.13) yield the following formidable expression:

$$
\begin{aligned}
&\psi(x,t)+\epsilon\frac{\partial\psi(x,t)}{\partial t}\\
&\simeq A\int_{-\infty}^{\infty}d\eta\,\exp\left\{\frac{im\eta^2}{2\hbar\epsilon}\right\}\left[\psi(x,t)+\eta\frac{\partial\psi(x,t)}{\partial\eta}+\eta^2\frac{1}{2}\frac{\partial^2\psi(x,t)}{\partial\eta^2}\right]\\
&-\epsilon\frac{i}{\hbar}A\int_{-\infty}^{\infty}d\eta\,\exp\left\{\frac{im\eta^2}{2\hbar\epsilon}\right\}\left[U(x)+\eta\frac{\partial U(x)}{\partial\eta}+\eta^2\frac{1}{2}\frac{\partial^2U(x)}{\partial\eta^2}\right]\\
&\times\left[\psi(x,t)+\eta\frac{\partial\psi(x,t)}{\partial\eta}+\eta^2\frac{1}{2}\frac{\partial^2\psi(x,t)}{\partial\eta^2}\right].
\end{aligned}
\tag{13.14}
$$

To handle this expression, we begin by inspecting only those terms that do not depend on any power of η and ϵ on both sides of the equal sign. The resulting zero order terms are

$$
\psi(x,t)\approx A\int_{-\infty}^{\infty}d\eta\,\exp\left\{\frac{im\eta^2}{2\hbar\epsilon}\right\}\psi(x,t),
$$

which can be solved analytically,

$$
\begin{aligned}
\psi(x,t)&\approx A\,\psi(x,t)\int_{-\infty}^{\infty}d\eta\,\exp\left\{\frac{im\eta^2}{2\hbar\epsilon}\right\},\\
1&\approx A\int_{-\infty}^{\infty}d\eta\,\exp\left\{\frac{im\eta^2}{2\hbar\epsilon}\right\},
\end{aligned}
$$

for the normalization constant A,

$$
A=\left(\frac{m}{2\pi i\hbar\epsilon}\right)^{1/2}.
\tag{13.15}
$$

Next, we cancel out the zero-order terms in equation (13.14) and we are left with

$$
\begin{aligned}
&\epsilon\frac{\partial\psi(x,t)}{\partial t}\\
&\simeq A\int_{-\infty}^{\infty}d\eta\,\exp\left\{\frac{im\eta^2}{2\hbar\epsilon}\right\}\left[\eta\frac{\partial\psi(x,t)}{\partial\eta}+\eta^2\frac{1}{2}\frac{\partial^2\psi(x,t)}{\partial\eta^2}\right]\\
&-\frac{i}{\hbar}\epsilon A\int_{-\infty}^{\infty}d\eta\,\exp\left\{\frac{im\eta^2}{2\hbar\epsilon}\right\}\left[U(x)+\eta\frac{\partial U(x)}{\partial\eta}+\eta^2\frac{1}{2}\frac{\partial^2U(x)}{\partial\eta^2}\right]\\
&\times\left[\psi(x,t)+\eta\frac{\partial\psi(x,t)}{\partial\eta}+\eta^2\frac{1}{2}\frac{\partial^2\psi(x,t)}{\partial\eta^2}\right].
\end{aligned}
\tag{13.16}
$$

Remembering that $\epsilon\eta$ scales like $\epsilon^{3/2}$ and $\epsilon\eta^2$ like ϵ^2 allows us to drop two terms inside the second integral of equation (13.16). The resulting first-order terms are

$$\epsilon\frac{\partial\psi(x,t)}{\partial t} \approx A\int_{-\infty}^{\infty} d\eta\,\exp\left\{\frac{im\eta^2}{2\hbar\epsilon}\right\}\left[\eta\frac{\partial\psi(x,t)}{\partial\eta}+\eta^2\frac{1}{2}\frac{\partial^2\psi(x,t)}{\partial\eta^2}\right]$$
$$-\frac{i}{\hbar}\epsilon A\,U(x)\psi(x,t)\int_{-\infty}^{\infty} d\eta\,\exp\left\{\frac{im\eta^2}{2\hbar\epsilon}\right\}.$$

The term proportional to η in the first integral vanishes by symmetry:

$$\epsilon\frac{\partial\psi(x,t)}{\partial t} \approx A\frac{1}{2}\left[\frac{\partial^2\psi(x,t)}{\partial\eta^2}\right]\int_{-\infty}^{\infty} d\eta\,\eta^2\,\exp\left\{\frac{im\eta^2}{2\hbar\epsilon}\right\}$$
$$-\frac{i}{\hbar}\epsilon A\,U(x)\psi(x,t)\int_{-\infty}^{\infty} d\eta\,\exp\left\{\frac{im\eta^2}{2\hbar\epsilon}\right\}.$$

Now the last integral is simply A^{-1}, whereas

$$A\int_{-\infty}^{\infty} d\eta\,\eta^2\,\exp\left\{-\frac{m\eta^2}{2i\hbar\epsilon}\right\} = \frac{i\hbar\epsilon}{m}.$$

Therefore,

$$\epsilon\frac{\partial\psi(x,t)}{\partial t} \approx i\hbar\epsilon\,\frac{1}{2m}\left[\frac{\partial^2\psi(x,t)}{\partial\eta^2}\right] - \frac{i}{\hbar}\epsilon\,U(x)\psi(x,t),$$

which can be trivially rearranged into

$$i\hbar\frac{\partial\psi(x,t)}{\partial t} \approx -\frac{\hbar^2}{2m}\left[\frac{\partial^2\psi(x,t)}{\partial\eta^2}\right] + U(x)\psi(x,t),$$

the time-dependent Schrödinger equation.

While the derivation of the Schrödinger equation from the path integral representation of the time evolution operator is mathematically correct, there remains controversy around the real-time path integral representing a type of stochastic time evolution operator. The main issue concerns the "path integral measure," the argument being that such an object cannot be properly defined. Integral measures are positive definite functions. Perhaps the first hint of trouble is the normalization constant A in equation (13.15). This is a multivalued function since $i^{1/2}$ has infinitely many branches. However, the analytic continuation of the path integral into imaginary time defines a random walk called the Wiener process, a propagator similar to the Itô integrals encountered in Chapter 11. In the next section, we see that the imaginary time finds a very useful physical interpretation consistent with the diffusion Monte Carlo method of Chapter 5. Imaginary time can be understood as some inverse temperature. Then, as τ approaches infinity, T approaches zero, and the system is predominately in the ground state.

13.3 The imaginary time path integral

In $\mathbb{R}^d$, the derivation of the imaginary time ($t \to -i\beta\hbar$) path integral is straightforward. We are going to need two mathematical theorems from the calculus on vector spaces. The first is the so-called *resolution of the identity*, which for the $d = 1$ case looks like

$$1 = \int_{-\infty}^{\infty} dx |\, x\rangle\langle x|. \tag{13.17}$$

This result is relatively simple to prove, starting with

$$1 = \int_{-\infty}^{\infty} dx\, \psi_n^* \psi_n, \tag{13.18}$$

and using the bracket representation for ψ and ψ^* to get

$$1 = \int_{-\infty}^{\infty} dx \langle n|x\rangle\langle x|n\rangle. \tag{13.19}$$

The theorem now follows by bringing the integral sign inside the brackets, (both $|\, n\rangle$ and $\langle n|$ are independent of x),

$$1 = \langle n| \left(\int_{-\infty}^{\infty} dx |\, x\rangle\langle x| \right) |\, n\rangle. \tag{13.20}$$

Since $\langle n \mid n\rangle = 1$, the quantity inside the bracket must also be one. The second theorem is the trace invariance of a matrix under a similarity transformation (see Theorem 10). For Hilbert spaces, n is infinite and the theorem is amended with the following caveat; all the sums involved must be convergent.

With these results, we can begin by expressing the partition function of eq. (13.2) in the following way:

$$\begin{aligned} \mathcal{Q} &= \sum_{n=0}^{\infty} \exp(-\beta E_n) = \sum_{n=0}^{\infty} \langle n| \exp(-\beta \hat{H}) |\, n\rangle \\ &= \int_{-\infty}^{\infty} dx \langle x| \exp(-\beta \hat{H}) |\, x\rangle, \end{aligned} \tag{13.21}$$

where the trace invariance theorem is used in the first line and the resolution of identity in the second. The symbol $\langle x|\exp(-\beta\hat{H})|x\rangle$ is known as the element of the *density matrix*. The density matrix element can be shown to take a simple form for a free particle,

$$\rho_f(x,x',\beta) = \langle x'|\exp(-\beta\hat{K})|x\rangle = \left[\frac{m}{2\pi\hbar^2\beta}\right]^{1/2} \exp\left[-\frac{m}{2\beta\hbar^2}(x-x')^2\right], \tag{13.22}$$

where $\hat{K}$ is the kinetic energy operator. The reader can verify that this expression satisfies Bloch's equation

$$-\frac{\hbar^2}{2m}\frac{\partial^2}{\partial x^2}\rho_f(x,x',\beta) = -\frac{\partial\rho_f(x,x',\beta)}{\partial\beta}. \tag{13.23}$$

The imaginary time interval 0, $\beta\hbar$ is subdivided into n equal subintervals by introducing n imaginary time slices,

$$\exp(-\beta\hat{H}) = \lim_{n\to\infty}\prod_{i=1}^{n}\exp\left(-\frac{\beta}{n}\hat{H}\right), \tag{13.24}$$

where the limit is needed to create smooth paths and sample all possible paths, just as it is done for the real time version of the representation. Upon inserting the resolution of the identity $n-1$ times, one arrives at

$$\begin{aligned}\mathcal{Q} = \lim_{n\to\infty} &\int_{-\infty}^{\infty} dx \int_{-\infty}^{\infty} dx_1 \int_{-\infty}^{\infty} dx_2 \cdots \int_{-\infty}^{\infty} dx_{n-1} \\ &\times \langle x|\exp\left(-\frac{\beta}{n}\hat{H}\right)|x_1\rangle\langle x_1|\exp\left(-\frac{\beta}{n}\hat{H}\right)|x_2\rangle\cdots\langle x_{n-1}|\exp\left(-\frac{\beta}{n}\hat{H}\right)|x\rangle.\end{aligned} \tag{13.25}$$

In subdividing the imaginary time, we increase the temperature associated with each of the matrix elements. This allows one to approximate the matrix element by a classical expression. Introducing the notation

$$v = \lim_{n\to\infty} n\frac{x_i - x_{i-1}}{\beta\hbar}, \tag{13.26}$$

into the classical Hamiltonian

$$\mathcal{H} = \frac{1}{2m}p^2 + V(x), \tag{13.27}$$

we are able to write the high temperature density matrix element as

$$\begin{aligned}&\langle x_{i-1}|\exp(-\beta\hat{H}/n)|x_i\rangle \\ &\simeq \left[\frac{m}{2\pi\hbar^2\beta}\right]^{1/2}\exp\left[-\frac{nm}{2\beta\hbar^2}(x_i - x_{i-1})^2 - \frac{\beta}{n}U(x_i)\right].\end{aligned} \tag{13.28}$$

To arrive at this result, we split $\exp(-\beta\hat{H}/n)$ into its kinetic and potential operators, $\hat{H} = \hat{K} + \hat{U}$, break the exponential function in two pieces,

$$\exp(-\beta\hat{H}/n) = \exp(-\beta\hat{K}/n)\exp(-\beta\hat{U}/n),$$

use equation (13.23), and recombine into a single exponential. The kinetic and potential operators do not commute, of course. Therefore, the expression we derive neglects quadratic and higher order terms in β/n and yields an approach that converges linearly with the number of time slices n.

Inserting the last expression into equation (13.25) yields the primitive path integral representation of the partition function,

$$\begin{aligned} \mathcal{Q} &= \lim_{n\to\infty}\left[\frac{m}{2\pi\hbar^2\beta}\right]^{n/2} \\ &\times \int_{-\infty}^{\infty} dx \int_{-\infty}^{\infty} dx_1 \cdots \int_{-\infty}^{\infty} dx_{n-1} \exp\left[-\frac{nm}{2\beta\hbar^2}\sum_{i=1}^{n}(x_i - x_{i-1})^2 - \frac{\beta}{n}\sum_{i=1}^{n}U(x_i)\right], \end{aligned} \tag{13.29}$$

where it is understood that $x_0 = x_n = x$. Upon setting $n = 1$, one obtains the classical partition function. The set of points in space $\{x_0, x_1, \ldots, x_n\}$ is called the *Brownian bridge*. If the paths are closed, meaning $x_0 = x_n$, the set is also referred in the chemical physics literature as a configuration of a ring polymer for obvious reasons.

To simplify the expressions for the thermodynamic properties, the following notation is introduced:

$$W(x, x_1, \ldots, x_{n-1}, \beta) = \left(\frac{m}{2\pi\beta\hbar^2}\right)^{n/2} \exp[-\mathcal{S}(x, x_1, \ldots, x_{n-1}, \beta)], \tag{13.30}$$

$$\mathcal{S}(x, x_1, \ldots, x_{n-1}, \beta) = \frac{nm}{2\beta\hbar^2}\sum_{i=1}^{n}(x_i - x_{i-1})^2 + \frac{\beta}{n}\sum_{i=1}^{n}U(x_i). \tag{13.31}$$

The average for a physical property estimated with the expression A is

$$\langle A(x, x_1, \ldots, x_{n-1}, \beta)\rangle = \frac{\int_{-\infty}^{\infty} dx \int_{-\infty}^{\infty} dx_1 \cdots \int_{-\infty}^{\infty} dx_{n-1} AW}{\int_{-\infty}^{\infty} dx \int_{-\infty}^{\infty} dx_1 \cdots \int_{-\infty}^{\infty} dx_{n-1} W}, \tag{13.32}$$

where we have suppressed the dependencies of W and A in the second line for clarity. Using $\langle E\rangle = -\partial \ln Q/\partial\beta$, one arrives at the *T estimator* for the energy,

$$\langle E\rangle = \frac{n}{2\beta} + \left\langle\frac{\partial\mathcal{S}}{\partial\beta}\right\rangle, \tag{13.33}$$

$$\langle E\rangle = \frac{n}{2\beta} + \left\langle -\frac{nm}{2\beta^2\hbar^2}\sum_{i=1}^{n}(x_i - x_{i-1})^2 + \frac{1}{n}\sum_{i=1}^{n}U(x_i)\right\rangle. \tag{13.34}$$

However, it is known that the variance of the T estimator grows with n and in $\mathbb{R}^d$. If the gradient of U is available, a much better energy estimator is obtained by using the virial theorem,

$$\langle E\rangle = \left\langle \frac{1}{2n}\sum_{i=1}^{n} x_i \frac{dU(x_i)}{dx} + \frac{1}{n}\sum_{i=1}^{n} U(x_i) \right\rangle. \tag{13.35}$$

To calculate the heat capacity, one simply uses

$$\frac{C_V}{k_B} = -\frac{1}{(k_B T)^2}\frac{\partial \langle E\rangle}{\partial \beta}. \tag{13.36}$$

The derivative of the virial energy expression in (13.35) yields a superior estimator for the heat capacity (see exercise 8):

$$\begin{aligned} k_B T^2 C_V = & \left\langle \left[\frac{1}{2n}\sum_{i=1}^{n} x_i \frac{dU(x_i)}{dx} + \frac{1}{n}\sum_{i=1}^{n} U(x_i) \right] \right. \\ & \times \left. \left[-\frac{nm}{2\beta^2\hbar^2}\sum_{i=1}^{n}(x_i - x_{i-1})^2 + \frac{1}{n}\sum_{i=1}^{n} U(x_i) \right] \right\rangle \\ & - \left\langle \frac{1}{2n}\sum_{i=1}^{n} x_i \frac{dU(x_i)}{dx} + \frac{1}{n}\sum_{i=1}^{n} U(x_i) \right\rangle \\ & \times \left\langle -\frac{nm}{2\beta^2\hbar^2}\sum_{i=1}^{n}(x_i - x_{i-1})^2 + \frac{1}{n}\sum_{i=1}^{n} U(x_i) \right\rangle. \end{aligned} \tag{13.37}$$

All the averages are in a form suitable for computation with the Metropolis algorithm. This is the topic of the next section.

13.4 The Metropolis algorithm

When the cumulative probability of a random distribution is not available in analytical form, transforming uniform random numbers to sample the probability density function is not impossible. The method of choice in those cases is based on rejection methods. The basic idea is very simple. It is possible, e. g., to draw a pair of random numbers uniformly in the domain and codomain of the probability density function one wishes to sample, and then accept the move if the point is in the area under the curve and rejected the move otherwise as it is shown in Figure 13.2.

In this section, we introduce a powerful and widely used tool to sample generic probability density functions like W in equation (13.30) based on the rejection method called the Metropolis algorithm. The Metropolis algorithm is very simple and can be used to sample essentially any positive definite distribution in any number of dimensions d, with the computational effort scaling with d or with U. Just like the diffusion

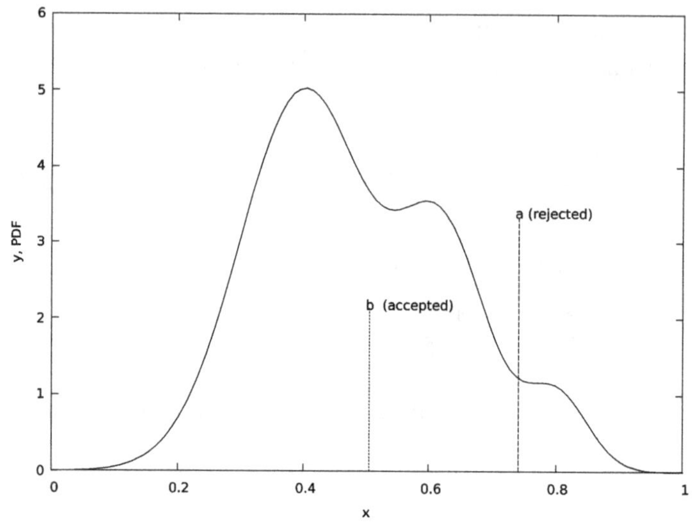

Figure 13.2: Example of a probability density function in $\mathbb{R}$ illustrating the rejection method.

Monte Carlo method, we seek to update the position and generate a sequence of random numbers $x_0, x_1, x_2, \ldots$ that are distributed according to the probability density function $f(x)$ after a sufficiently long period. Unlike the diffusion Monte Carlo, we do not need a population of replicas. A move is an update from x to x' using any distribution $T(x'|x)$, where T is a conditional probability to move the system at x' given that it is at x. This step is normally (but not exclusively) achieved by drawing a uniform random number in the interval $(x - \Delta/2, x + \Delta/2)$ where Δ is some parameter optimized to maximize efficiency. To derive an expression for $P_a(x'|x)$, the probability that a move $x \to x'$ is accepted, the *detailed balance* condition is introduced

$$P_a(x'|x)T(x'|x)f(x') = f(x)T(x|x')P_a(x|x'). \tag{13.38}$$

This is a sufficient, but not necessary condition for the random walk to reach equilibrium. The acceptance probability is then constructed using the following ratio:

$$q(x'|x) = \frac{T(x|x')f(x')}{T(x'|x)f(x)}.$$

One frequently used possibility is to set

$$P_a(x'|x) = \min\{1, q(x'|x)\}$$

but other combinations involving $q(x'|x)$ have been suggested. Furthermore, if $T(x'|x)$ is uniform,

$$T(x'|x) = \frac{1}{\Delta},$$

then

$$P_a(x'|x) = \min\left\{1, \frac{f(x')}{f(x)}\right\}. \tag{13.39}$$

The following Python code demonstrates how a distribution in one dimension can be sampled using the Metropolis algorithm:

```
import numpy as np
import random as rnd
rnd.seed()    # seed with system time
i = 0
x = rnd.random()    # start with 0 < x < 1
fx = f(x)
delta = 0.8    # adjusted
rejected = 0
accepted = 0
steps = 2000000
while i < steps:
    xt = x + delta*(rnd.random() - 0.5)  # attempt
    y = rnd.random()  #
    ratio = f(xt)/fx
    if (ratio > y):
        x = xt              # accept the move
        fx = ratio*fx
        accepted += 1
    else:
        rejected += 1    # reject the move
    i += 1
    if (i > steps/2):
        print("%12.8f" %x)
print("# accepted ", accepted*100/steps, "%")
print("# rejected ", rejected*100/steps, "%")
```

The code is missing the definition of $f(x)$. We tested it using a three- term Gaussian function

$$f(x) = \sum_{i=1}^{3} a_i e^{-(x-b_i)/s_i} \tag{13.40}$$

with the following parameters:

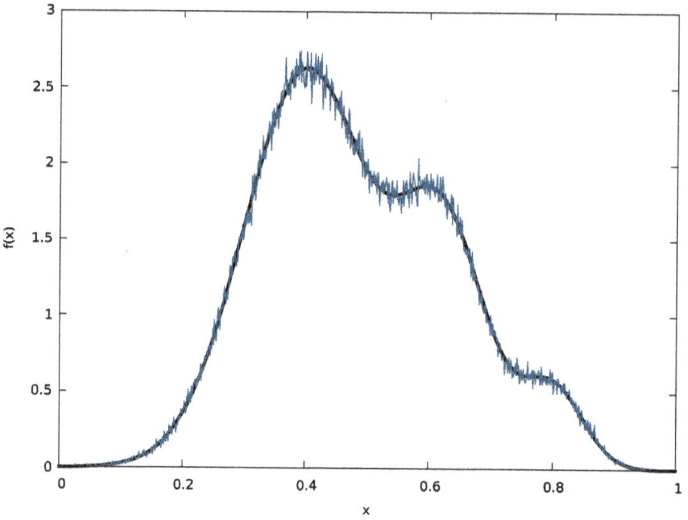

Figure 13.3: The probability density function $f(x)$ (thick black line) juxtaposed to its stochastic estimate (steel blue lines connecting the histogram points). The probability density function $f(x)$ is normalized so that the area under it equals 1.0.

$$(a_1, a_2, a_3) = 5.0, 3.0, 1.0$$
$$(b_1, b_2, b_3) = 0.4, 0.621, 0.8 \quad (s_1, s_2, s_3) = 0.02, 0.01, 0.005.$$

The graph of $f(x)$ with these parameters is in Figure 13.2. The coding of $f(x)$ is an exercise left to the reader. The walk is initiated by a random values of x in $(0, 1)$. Inside the while loop, the first line attempts a move $x' = x + \Delta(\xi - 0.5)$ where ξ is a uniform random number in $(0, 1)$. After drawing a second uniform random number $y \in (0, 1)$, the ratio $q = f(x')/f(x)$ in equation (13.38) is computed. The branching statements that follow are to accept the move if $q > y$ and reject otherwise. The parameter Δ is adjusted by running the code several times until the percent accepted moves is between 40 % and 60 %.

The simplicity of the approach is evident, but there are some additional considerations when implementing the Metropolis algorithm. The random walk samples the desired probability density functions only in the asymptotic limit. This usually translates into burning a large number of random moves, typically in the millions, before the desired properties can be estimated. This process is informally referred to as the “warming cycle.” This is why values of x are printed out after the first million moves. These values are collected into a file and processed into a 1000 bin histogram and shown in Figure 13.3. The other disadvantage of the Metropolis algorithm is that the walk it generates is highly correlated. This usually means that a number of independent and sufficiently long simulations have to be carried out to obtain the correct statistical fluctuations of estimated physical properties.

It is straightforward to extend the approach to sample a multidimensional probability densities including those of systems of particles. One simply performs the move acceptance/rejection cycle for every degree of freedom. A complete cycle through all degrees of freedom is called a pass. Then the variable `step` keeps track of how many passes are performed. Typically, one million passes are needed to reach equilibrium and another million passes are necessary to break correlations and to obtain statistically significant estimates. When we implement Metropolis using high throughput computing, we run a warming cycle of one million passes ones. Then, the ending configurations are replicated k time so that k independent walks are started. This is possible once a walk is equilibrated since, in that case results are independent of the starting configuration.

Additionally, for systems of particles, the random walk may not converge to the proper distribution at temperatures where the system is near a phase transition or a phase change. Special techniques, such as parallel tempering, or infinite swapping, have to be implemented to improve the algorithm performance. Finally, path integral applications require that the number of time slices be systematically incremented to verify that estimates converge to the quantum limit. In the exercise section, we train the user to implement the primitive time sliced approach and explain one method that can be used to accelerate the convergence relative to the number of time slices. However, there are more powerful methods to accelerate convergence, and many of them implement vector spaces to represent the Brownian bridge.

13.4.1 Exercises

1. The probability density function $f(x)$ in the code is a sum of three Gaussian curves,

$$f(x) = \sum_{i=1}^{3} a_i e^{-(x-b_i)^2/s_i} \qquad x \in \mathbb{R}.$$

The parameters a_i, b_i, s_i are real as well.

(a) Normalize $f(x)$.

(b) Show that

$$\langle x \rangle = \frac{\sum_{i=1}^{3} a_i b_i \sqrt{s_i}}{\sum_{i=1}^{3} a_i \sqrt{s_i}}.$$

(c) Derive this result

$$\langle x^2 \rangle = \frac{\sum_{i=1}^{3} a_i (s_i + b_i^2) \sqrt{s_i}}{\sum_{i=1}^{3} a_i \sqrt{s_i}}.$$

(d) Write the Python code to compute $f(x)$ using the parameters provided in the chapter.

(e) Modify the code in the chapter so that $\langle x^2\rangle$, $\langle x\rangle$ and $\sigma_x = \sqrt{\langle x^2\rangle - \langle x\rangle^2}$ are computed analytically from your results in parts (a) and (b), and then estimated stochastically after the warm-up period.

2. Consider the following classical Hamiltonian for a system in $\mathbb{R}$:

$$\mathcal{H} = \frac{1}{2m}p^2 + U$$

where p is the canonical conjugate momentum and U is the system potential. The classical partition function is

$$Q_c = \int_{-\infty}^{\infty} dp \int_{-\infty}^{\infty} dx\, e^{-\beta\mathcal{H}}.$$

This corresponds to an ensemble of systems each containing a single particle of mass m.

(a) Derive the following expression for the classical internal energy:

$$\mathcal{U}_c = \frac{\int_{-\infty}^{\infty} dp \int_{-\infty}^{\infty} dx\, \mathcal{H}\, e^{-\beta\mathcal{H}}}{\int_{-\infty}^{\infty} dp \int_{-\infty}^{\infty} dx\, e^{-\beta\mathcal{H}}}$$

from the ansatz

$$\mathcal{U}_c = -\frac{1}{Q_c}\frac{\partial Q_c}{\partial\beta}.$$

(b) Derive the *configuration integral*

$$Q_c = \left(\frac{2\pi m}{\beta}\right)^{1/2} \int_{-\infty}^{\infty} dx e^{-\beta U}.$$

(c) Use the harmonic oscillator potential $U = kx^2/2$ to find an expression for Q_c and $\mathcal{U}_c$ in terms of m and β. Does your answer agree with the equipartition theorem of classical statistical mechanics?

(d) Derive an expression for C_V from the result in part (c). Then take the limit as $T \to 0$ of C_V and $\mathcal{U}_c$. Recall that it is in this limit (low T) that classical statistical mechanics fails. The internal energy should approach the zero point energy of the system and the heat capacity should approach zero.

3. Derive equation (13.37).

4. Classical mechanics (or thermodynamics) emerges as a limit for large values of the mass and at high temperatures. To see this, consider an ensemble of 1-d harmonic oscillators at temperature T. The quantum canonical partition function is a sum over quantum state energies

$$Q = \sum_{n=0}^{\infty} \exp(-\beta E_n). \tag{13.41}$$

If the explicit form of the spectrum of the Hamiltonian operator is used, the canonical partition function can be reexpressed as a geometric series,

$$Q = \sum_{n=0}^{\infty} \exp\left[-\beta\hbar\omega\left(n + \frac{1}{2}\right)\right]. \tag{13.42}$$

(a) Show that the partition function becomes

$$Q = \frac{\exp(-\frac{1}{2}\beta\hbar\omega)}{1 - \exp(-\beta\hbar\omega)}. \tag{13.43}$$

(b) The internal energy $\mathcal{U}$ and heat capacity can be calculated analytically. Derive its expression

$$\mathcal{U} = \hbar\omega\left[\frac{1}{2} + \frac{\exp(-\beta\hbar\omega)}{1 - \exp(-\beta\hbar\omega)}\right].$$

(c) Take the limit as $\beta \to \infty$ (the same as $T \to 0$). Comment on the result.

(d) Derive the expression for the heat capacity for the same system

$$C_V = -\frac{1}{k_B T^2}\frac{\partial \mathcal{U}}{\partial \beta} = \frac{(\hbar\omega)^2}{k_B T^2}\left\{\frac{\exp(-\beta\hbar\omega)}{[1 - \exp(-\beta\hbar\omega)]^2}\right\}$$

then show that as T becomes large ($\beta \to 0$) C_V approaches the correct equipartition value. It is simpler to use $x = \beta\hbar\omega$ to rewrite the large T limit as follows:

$$\lim_{x\to 0} k_B \frac{x^2 e^{-x}}{(1 - e^{-x})^2}$$

and then use l' Hôpital's rule twice.

5. Recast the expression for U and C_V into a more useful form for a general spectrum by deriving equation (13.1) and

$$C_V = \frac{1}{k_B T^2}\left[\frac{\sum_{n=0}^{\infty} E_n^2 \exp(-\beta E_n)}{\sum_{n=0}^{\infty} \exp(-\beta E_n)} - \left(\frac{\sum_{n=0}^{\infty} E_n \exp(-\beta E_n)}{\sum_{n=0}^{\infty} \exp(-\beta E_n)}\right)^2\right]. \tag{13.44}$$

6. Verify by substitution that the free particle density matrix satisfies equation (13.23).
7. The imaginary time interval $0, \beta\hbar$ is subdivided into n equal sub-intervals. As n tends to infinity, one can write

$$\exp(-\beta\hat{H}) = \lim_{n\to\infty} \prod_{i=1}^{n} \exp\left(-\frac{\beta}{n}\hat{H}\right). \tag{13.45}$$

It is very important to realize that the resulting path integral is different from the one we derived earlier. First, the argument of the exponential is no longer imaginary, it is real, and second, the Lagrangian has now been replaced by the Hamiltonian. How did this come to pass? Starting with this real time integral,

$$K(x_a, t_b; x_a, t_a)$$
$$= \int_{-\infty}^{\infty} dx_1 \int_{-\infty}^{\infty} dx_n \cdots \int_{-\infty}^{\infty} dx_{n-1} \exp\left\{\frac{im}{2\hbar\epsilon}\sum_{i=1}^{n}(x_i - x_{i-1})^2\right\}$$
$$\times \exp\left\{-\frac{i}{\hbar}\epsilon\sum_{i=1}^{n} U\left(\frac{x_i + x_{i-1}}{2}\right)\right\}.$$

show that by letting

$$\epsilon \to -i\frac{\beta}{n}$$

$$\int_{-\infty}^{\infty} K(x_a, t_b; x_a, t_a)\, dx_a \to Q.$$

To get the correct value for the normalization constant, go back to the definition of the density matrix. This result demonstrates the connection between the real time propagator and its analytical continuation in imaginary time.

8. Derive equation (13.37) from equation (13.35).

14 Ring polymer molecular dynamics

"Obvious" is the most dangerous word in mathematics.

Eric Temple Bell

The star of this chapter is an object called the position autocorrelation function. It is a dot product of the position vector for each particle at time t with the same vector measured at $t = 0$. When this property is computed in a quantum simulation and then transformed into the frequency domain, features that correspond to quantum energy level differences can be identified. The Fourier transform of simulated position autocorrelation functions are approximations of absorption spectra. This fact is made obvious when explicit expression of the quantum position-position autocorrelated function for a quantum system in the eigenbasis of the unperturbed system is derived. The simulations can be classical in nature such as, e. g., those we considered in Chapter 12, or as we discuss in this chapter, can be excellent approximations of the true quantum system. There are two classes of dynamic simulations, which are capable of capturing quantum effects: ring polymer molecular dynamics [119–129] and path integral centroid dynamics [130–133]. As the last name suggests these approaches are derived from the imaginary time path integral representation. In the present chapter, we focus on the implementation of the ring polymer molecular dynamics approach, and how it can be used to compute excellent approximations to the position autocorrelation function.

14.1 Signal processing

We have already seen some applications of the Fourier transform in Chapter 5, where the tool is used to simplify a class of partial differential equations. Another use of the Fourier transform is to convert a time signal into its dynamic spectrum. In particular, the position autocorrelation function obtained from a classical molecular dynamics simulation can be transformed into the frequency domain, where distinct features identify the "natural frequencies" of the system. We begin by showing that for a particle of mass m in a harmonic oscillator with force constant k in $\mathbb{R}$, the Fourier transform of the classical position autocorrelation function has a peak precisely at $\omega = \sqrt{k/m}$. Recall that for a given initial position x_0 and velocity v_0 the trajectory $x(t)$ is

$$x(t) = x_0 \cos \omega t + v_0 \sin \omega t.$$

The classical position autocorrelation function C_{xx}, is a scalar function defined as the dot product of the position vector at time t with the starting position vector. In $\mathbb{R}$, it becomes a trivial product

$$C_{xx}(t) = x_0 x(t) = x_0^2 \cos \omega t + x_0 v_0 \sin \omega t. \tag{14.1}$$

https://doi.org/10.1515/9783111610207-014

Now let us consider its Fourier transform,

$$\tilde{C}_{xx}(\omega') = \int_{-\infty}^{\infty} C_{xx}(t)e^{-i\omega' t}\,dt. \tag{14.2}$$

This integral is clearly divergent. There is no physical argument suggesting that $C_{xx}(t)$ should drop to zero at either the positive or negative end of the time domain. One generally imposes a sampling window of some type to handle the problem. We consider a couple of them to see how each choice impacts the results.

We can simply truncate the signal such that $C_{xx}(t) = 0$ for $t < 0$ and for $t > t_1$ and (14.1) for $t \in [0, t_1]$. By using the following identities,

$$\cos\omega t = \frac{e^{i\omega t} + ie^{-i\omega t}}{2}, \qquad \sin\omega t = \frac{e^{i\omega t} - ie^{-i\omega t}}{2i},$$

combining the powers, evaluating the resulting integrals, and extracting the real and imaginary parts of $\tilde{C}_{xx}(\omega')$, we arrive at the following two expressions:

$$\begin{aligned}
&\mathbb{R}[\tilde{C}_{xx}(\omega')] \\
&\quad = \frac{1}{2}x_0^2\left[\frac{\sin(\omega-\omega')t_1}{\omega-\omega'} - \frac{\cos(\omega+\omega')t_1 - 1}{\omega+\omega'}\right] \\
&\quad + \frac{1}{2}x_0v_0\left[\frac{\sin(\omega-\omega')t_1}{\omega-\omega'} + \frac{\cos(\omega+\omega')t_1 - 1}{\omega+\omega'}\right],
\end{aligned} \tag{14.3}$$

$$\begin{aligned}
&\mathbb{J}[\tilde{C}_{xx}(\omega')] \\
&\quad = \frac{1}{2}x_0^2\left[-\frac{\cos(\omega-\omega')t_1 - 1}{\omega-\omega'} + \frac{\sin(\omega+\omega')t_1}{\omega+\omega'}\right] \\
&\quad + \frac{1}{2}x_0v_0\left[-\frac{\cos(\omega-\omega')t_1 - 1}{\omega-\omega'} - \frac{\sin(\omega+\omega')t_1}{\omega+\omega'}\right].
\end{aligned} \tag{14.4}$$

It is immediately obvious that two denominators approach zero as ω' approaches ω, and the remaining two approach zero as ω' approaches $-\omega$. However, the function has a finite value in those limits. Since we are only interested in frequencies greater than zero we just consider the following limit:

$$\begin{aligned}
&\lim_{\omega'\to\omega} \mathbb{R}[\tilde{C}_{xx}(\omega')] \\
&\quad = \frac{1}{2}x_0^2\left(t_1 - \frac{\cos 2\omega t_1 - 1}{2\omega}\right) + \frac{1}{2}x_0v_0\left(t_1 + \frac{\cos 2\omega t_1 - 1}{2\omega}\right),
\end{aligned} \tag{14.5}$$

$$\begin{aligned}
&\lim_{\omega'\to\omega} \mathbb{J}[\tilde{C}_{xx}(\omega')] \\
&\quad = \left(\frac{1}{2}x_0^2 - \frac{1}{2}x_0v_0\right)\left[\frac{\sin 2\omega t_1}{2\omega}\right].
\end{aligned} \tag{14.6}$$

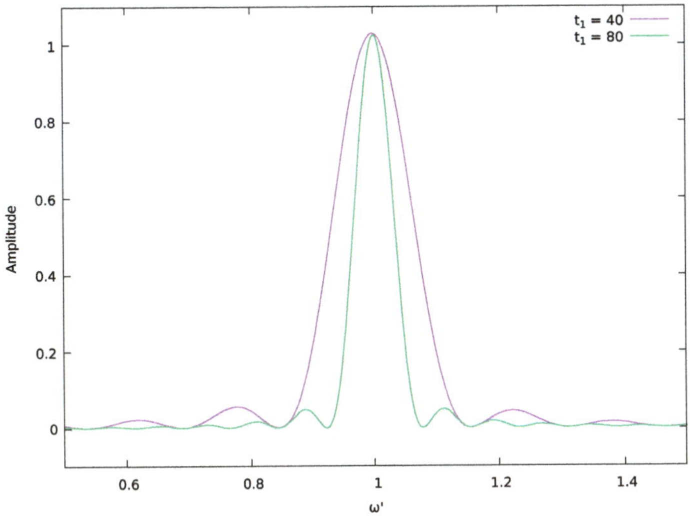

Figure 14.1: The power spectrum of $C_{xx}(t)/t_1$ for a special case of the particle in a harmonic oscillator truncated at two different times: 0, t_1.

The modulus $|\tilde{C}_{xx}(\omega')|^2/t_1^2$ is plotted in Figure 14.1 for the special case $x_0^2/2 = 1$, $v_0 = 0$, and $\omega = 1$ for two values of t_1. It is common practice to divide $|\tilde{C}_{xx}(\omega')|^2$ by $t_1 C_{xx}(t = 0)$ to rescale the amplitude of the main peaks. This process normalizes the Fourier transform so that features in the frequency domain do not exceed one. At this point, we can make two observations from this example:

1. The main feature in the graph is a peak at $\omega' = \omega$, with a width that depends on t_1.
2. There are satellite peaks in the graphs with amplitudes that are independent of t_1.

The satellite peaks appear because by truncating the signal, we have convoluted a square pulse together with the signal that we are trying to analyze. It can be shown that the Fourier transform of a square pulse has a multitude of peaks. Clearly, spurious peaks are not desirable. There are a number of window-like functions that handle the divergence of the integral and suppress spurious peaks in the spectrum of the signal, e. g., the Gaussian window, defined as

$$\tilde{C}_{xx}(\omega') = \int_{-\infty}^{\infty} e^{-t^2/t_1^2} C_{xx}(t) e^{-i\omega' t}\, dt \tag{14.7}$$

for various values of t_1. The Fourier transform of the Gaussian window is a Gaussian in the frequency domain,

$$\tilde{g}(\omega') = \int_{-\infty}^{\infty} e^{-t^2/t_1^2} e^{-i\omega' t}\, dt,$$

and can be solved by combining the exponents,

$$\tilde{g}(\omega') = \int_{-\infty}^{\infty} e^{-t^2/t_1^2 - i\omega' t}\, dt,$$

and completing the square,

$$-t^2/t_1^2 - i\omega' t = -\frac{1}{t_1^2}\left(t + \frac{i\omega' t_1^2}{2}\right)^2 - \frac{(\omega')^2 t_1^2}{4}.$$

The result is

$$\tilde{g}(\omega') = e^{-(\omega')^2 t_1^2/4} \int_{-\infty}^{\infty} \exp\left[-\frac{1}{t_1^2}\left(t + \frac{i\omega' t_1^2}{2}\right)^2\right] dt$$

$$\tilde{g}(\omega') = t_1 \sqrt{\pi} e^{-(\omega')^2 t_1^2/4}. \tag{14.8}$$

Therefore, there should be no spurious peaks in the spectrum of $C_{xx}(t)$ for the harmonic oscillator. We confirm that statement by working through the Fourier transform of $C_{xx}(t)$ for the harmonic oscillator with a Gaussian window. Using the expressions for the sine and cosine in terms of complex exponential,

$$\tilde{C}_{xx}(\omega') = x_0^2 \int_{-\infty}^{\infty} e^{-t^2/t_1^2} \frac{e^{i\omega t} + ie^{-i\omega t}}{2} e^{-i\omega' t}\, dt$$

$$+ x_0 v_0 \int_{-\infty}^{\infty} e^{-t^2/t_1^2} \frac{e^{i\omega t} - ie^{-i\omega t}}{2} e^{-i\omega' t}\, dt,$$

$$\tilde{C}_{xx}(\omega') = \frac{1}{2} x_0^2 \left[\int_{-\infty}^{\infty} e^{-t^2/t_1^2 + i(\omega - \omega')t}\, dt\right.$$

$$\left. + i \int_{-\infty}^{\infty} e^{-t^2/t_1^2 - i(\omega + \omega')t}\, dt\right]$$

$$+ \frac{1}{2} x_0 v_0 \left[\int_{-\infty}^{\infty} e^{-t^2/t_1^2 + i(\omega - \omega')t}\, dt,\right.$$

$$\left. - i \int_{-\infty}^{\infty} e^{-t^2/t_1^2 - i(\omega + \omega')t}\, dt\right].$$

There are two squares to complete,

$$-t^2/t_1^2 + i(\omega - \omega')t = -\frac{1}{t_1^2}(t + a_1)^2 + b_1$$

$$a_1 = -i\frac{t_1^2(\omega - \omega')}{2}$$

and

$$b_1 = \frac{1}{t_1^2}a_1^2 = -\frac{t_1^2(\omega - \omega')^2}{4}.$$

The second square is

$$-t^2/t_1^2 - i(\omega + \omega')t = -\frac{1}{t_1^2}(t + a_2)^2 + b_2$$

where

$$a_2 = i\frac{t_1^2(\omega + \omega')}{2},$$
$$b_2 = \frac{1}{t_1^2}a_2^2 = -\frac{t_1^2(\omega + \omega')^2}{4}.$$

These allow us to write

$$\begin{aligned}\tilde{C}_{xx}(\omega') = \frac{1}{2}x_0^2\Bigg[&\int_{-\infty}^{\infty} e^{-(t+a_1)^2/t_1^2 - t_1^2(\omega-\omega')^2/4}\,dt \\ &+ i\int_{-\infty}^{\infty} e^{-(t+a_2)^2/t_1^2 - t_1^2(\omega+\omega')^2/4}\,dt\Bigg] \\ + \frac{1}{2}x_0v_0\Bigg[&\int_{-\infty}^{\infty} e^{-(t+a_1)^2/t_1^2 - t_1^2(\omega-\omega')^2/4}\,dt, \\ &- i\int_{-\infty}^{\infty} e^{-(t+a_2)^2/t_1^2 - t_1^2(\omega+\omega')^2/4}\,dt\Bigg].\end{aligned}$$

After some simplifications, the last expression becomes

$$\begin{aligned}\tilde{C}_{xx}(\omega') = &\frac{1}{2}x_0^2 t_1\sqrt{\pi}\left[e^{-t_1^2(\omega-\omega')^2/4} + ie^{-t_1^2(\omega+\omega')^2/4}\right] \\ &+ \frac{1}{2}x_0v_0t_1\sqrt{\pi}\left[e^{-t_1^2(\omega-\omega')^2/4} - ie^{-t_1^2(\omega+\omega')^2/4}\right].\end{aligned} \tag{14.9}$$

The domains of the Gaussian integrals are straight paths in the complex plane. However, this does not bring any additional complications because the imaginary portion of the domain is constant. Here, too, there is a main peak in the negative frequency range $\omega' = -\omega$. The function $e^{-t_1^2(\omega-\omega')^2/2}$ is plotted in Figure 14.2 for two values of t_1. Of course, there are no satellite peaks as expected. Moreover, upon closer inspection, the main

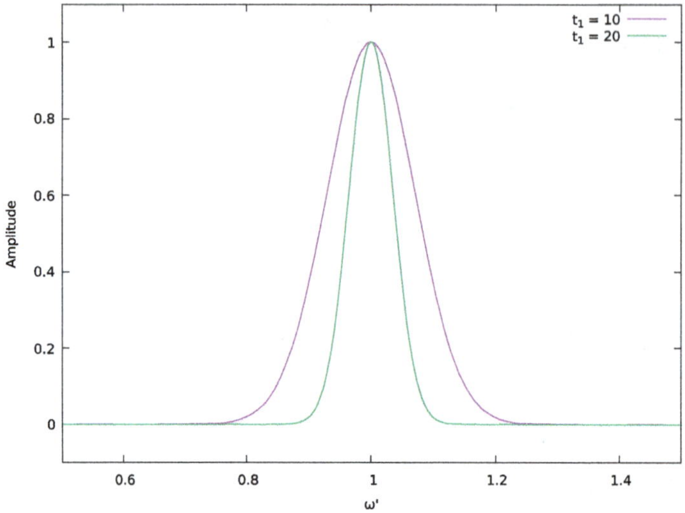

Figure 14.2: The power spectrum of $C_{xx}(t)/(t_1\sqrt{\pi})$ for a special case of the particle in a harmonic oscillator obtained with two different Gaussian windows.

feature obtained with a Gaussian window is as sharp as that of a truncated sample, but t_1 is a factor of four smaller. In practice, Gaussian windows do not eliminate side peaks when the signal is collected at discrete values of time, and they do not represent the best window in that case, but have physical meaning since they represent better models for pulsed monochromatic radiation.

Signal processing is ubiquitous in engineering and scientific applications. The Python numpy class has a built-in `numpy.fft.fft()` function that transforms an input signal stored in an array using the fast Fourier transform algorithm. During the production run, generating the ensemble average of autocorrelation function, generally, is far more expensive than its spectral analysis, and because our applications do not require large volumes of transformations the gain of using fast Fourier transform is minimal, and the code based on the trapezoid rule given later in the chapter is sufficient. Nonetheless, we discuss how the fast Fourier transform (FFT) can be implemented for completeness. The FFT becomes useful whenever high resolution is required, however, `numpy.fft.fft()` function pads the input array with zeros whenever the size of the array is not a power of 2, which amounts to truncating the signal. Additionally, caution has to be used when consulting other references on signal processing. Most applied mathematics and engineering textbooks define the Fourier transform the same way we do here. However, some authors use $e^{i\omega' t}$ as the kernel of the transform, and many textbooks use the frequency $e^{i2\pi\nu' t}$ in units of inverse time instead of radians over time units as we do here.

The output of `numpy.fft.fft()` is in Hertz if the time units are seconds, e. g., but the units of the harmonic oscillator natural frequency $\sqrt{k/m}$ are rad/s. Additionally, there are subtle differences between the numerical evaluation of the continuous version of the Fourier transform and its discrete version that manifest themselves typically in the appearance of spurious peaks. The biggest problem arises when the function being transformed is not band limited, i. e., when the time interval Δt is such that one can find frequencies above the corresponding Nyquist frequency $\omega_N = \pi/\Delta t$. When a sine wave with frequency $\omega > \omega_N$ is sampled, the frequency spectrum in $(0, \omega_N)$ contains a spurious peak at $\omega_N - \omega$, a phenomenon known as *aliasing*.

14.2 The ring polymer Hamiltonian

Here, we concentrate on how the autocorrelation function $C_{xx}(t)$ in equation (14.1) is generated in the first place. One can use molecular dynamics with a numerical integrator like the one in Chapter 12 and simply compute the dot product $x(0) \cdot x(t)$ for a sufficiently long time. The Fourier transform of $C_{xx}(t)$ yields the natural frequencies of the system, and these can be used to estimate the ground state by using the harmonic oscillator approximation. The same could be accomplished by diagonalizing the Hessian of U. However, in some cases the Hessian matrix is not available or its too expensive to compute and transforming $C_{xx}(t)$ is a cheaper, or perhaps the only alternative. On the other hand, for floppy or large complex systems, signal processing from ensembles of classical trajectories with starting conditions sampled from the Boltzmann distribution at some temperature T provide a wealth of useful information beyond the harmonic analysis. The power spectrum from a molecular dynamics simulation represents a classical model for the ground to first excited state transition. As always, we are interested in learning how quantum effects can impact such information. It turns out that one can use the imaginary time version of the path integral representation of quantum mechanics and introduce the necessary corrections. To see how this has been developed, let us recall that, given a classical Hamiltonian of the type,

$$\mathcal{H} = \frac{1}{2m} \sum_{i=1}^{n} p_i^2 + U \tag{14.10}$$

for n-point-like particles of mass m in $\mathbb{R}$, one can rewrite the canonical partition function in terms of a configuration integral

$$\begin{aligned} Q_n &= \int_{-\infty}^{\infty} dp_1 \cdots \int_{-\infty}^{\infty} dp_n \int_{-\infty}^{\infty} dx_1 \cdots \int_{-\infty}^{\infty} dx_n \, e^{-\beta\mathcal{H}} \\ &= \left(\frac{2m\pi}{\beta}\right)^{n/2} \int_{-\infty}^{\infty} dx_1 \cdots \int_{-\infty}^{\infty} dx_n \, e^{-\beta U}. \end{aligned}$$

Craig and Manolopoulos compare the last equation with equation (13.28) written as

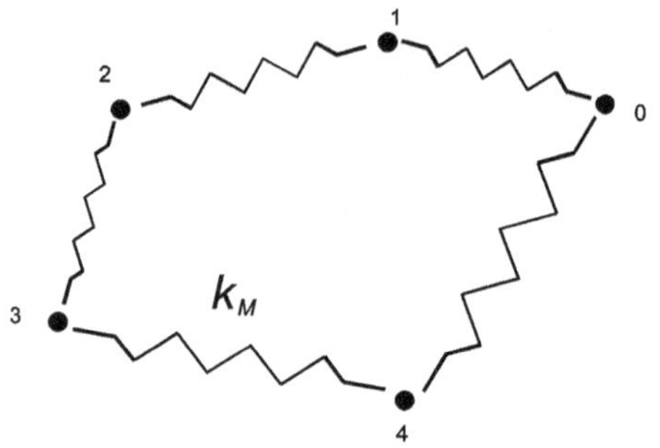

Figure 14.3: A sketch of a configuration of the ring polymer with $n = 5$ beads.

$$Q_n = \left(\frac{2m\pi}{\beta_n}\right)^{n/2} \int_{-\infty}^{\infty} dx_0 \int_{-\infty}^{\infty} dx_1 \cdots \int_{-\infty}^{\infty} dx_{n-1}\, e^{-\beta_n U_{\text{eff}}}$$

where $\beta_n = \beta/n$ and the effective potential

$$U_{\text{eff}} = \sum_{i=1}^{n} \frac{m}{2\beta_n^2\hbar^2}(x_i - x_{i-1})^2 + U(x_i) \tag{14.11}$$

is that of a ring-like system (see Figure 14.3) connected by harmonic springs all with the same force constant

$$k_M = \frac{m}{\beta_n^2\hbar^2},$$

together with the potential U. Their idea is simple, the ring-polymer "classical" Hamiltonian $\mathcal{H}_n$ must have the same form as (14.10), with $U \to U_{\text{eff}}$. Then all there is to do is sample starting positions and momenta from $e^{-\beta_n \mathcal{H}_{\text{eff}}}$, run the molecular dynamics simulations all of the same length, compute the ensemble average $\langle C_{xx}(t)\rangle$ and its Fourier transform to obtain a simulated spectrum. Craig and Manolopoulos show that the ensemble average $\widetilde{\langle C_{xx}\rangle}(\omega')$ obtained with a sufficient number of "beads" n is an excellent approximation of the Kubo transform of the position autocorrelation function of the quantum system, i. e., a particle on mass m in the field of U in $\mathbb{R}$ in the present example. The position autocorrelation function corresponds to an ensemble average of the dipole operator that is proportional to the main interaction between light and matter that absorbs or emits radiation as it transitions from state to state. The correlation function can be derived from first-order perturbation theory.

To see how the Fourier transform of $\langle C_{xx}(t)\rangle$ is related to the absorption spectrum of the system, we derive an expression for the Kubo transform of the position-position autocorrelation in the energy representation. In $\mathbb{R}$, the quantum thermal expectation value of a position dependent operator A can be estimated with the following path integral expression $\langle A\rangle = \lim_{n\to\infty}\langle A\rangle_n$,

$$\langle A\rangle_n = \frac{1}{(2\pi\hbar)^n Q_n}\int dp_1 \cdots \int dx_n\, e^{-\beta_n H_n(p,x)} A_n(x), \tag{14.12}$$

where $A_n(x) = \sum_{j=1}^{n} A(x_j)/n$, and n is the number of slices (or beads), the $2n$-fold integral is over the $p_1, \ldots, p_n$ and $x_1, \ldots, x_n$ domain of the ring polymer all from $-\infty$ to $+\infty$. The quantum thermal expectation value of the product of two position-dependent operators $\hat{A}$ and $\hat{B}$ is the $t \to 0$ limit of the Kubo-transformed real-time correlation function of the two operators A and B, defined as

$$C_{AB}(t) = \frac{1}{\beta Q}\int_0^{\beta} d\lambda\, \mathrm{Tr}[e^{-(\beta-\lambda)\hat{H}}\hat{A}\, e^{-\lambda\hat{H}}\, e^{i\hat{H}t/\hbar}\hat{B}\, e^{-i\hat{H}t/\hbar}] \tag{14.13}$$

where Tr is the trace of the matrix inside the brackets, and $\hat{H}$ is the Hamiltonian operator for the system. The Kubo-transformed real-time position autocorrelation function is obtained by setting $\hat{A}$ to $d\hat{x}/dt$ and $\hat{B}$ to $\hat{x}$, the position operator.

In the Heisenberg picture, the time evolution of the position operator $\hat{x}$ is governed by the following differential equation:

$$\frac{d\hat{x}}{dt} = \frac{1}{i\hbar}[\hat{x}, \hat{H}], \tag{14.14}$$

that has the following general solution:

$$\hat{x}(t) = e^{i\hat{H}t/\hbar}\hat{x}(0)\, e^{-i\hat{H}t/\hbar}. \tag{14.15}$$

The autocorrelation function is constructed over an ensemble at β of initial ($t = 0$) positions. The imaginary version of (14.15) obtained by setting $t = -i\hbar\lambda$ is used to evolve $\hat{x}(\beta)$ by integrating from $\lambda = 0$ up to $\lambda = \beta$. This operation generates the needed ensemble of starting conditions. Equation (14.13) results when the dot product of the evaluation of $\hat{x}$ in imaginary time and the evaluation of $\hat{x}$ in real time is written explicitly in the eigenbasis,

$$C_{xx}(t) = \frac{1}{\beta Q}\sum_{k,k'}\{(e^{-\beta E_k} - e^{-\beta E_{k'}})\,\langle k'|\hat{x}|k\rangle^2 e^{-i(E_k - E_{k'})t/\hbar}\} \tag{14.16}$$

where

$$Q = \sum_{j=0}^{\infty} e^{-\beta E_j}$$

is the canonical partition function, and $\langle k|\hat{x}|k'\rangle$ is the matrix element of the position operator in the energy representation. It is obvious that the Fourier transform of $C_{xx}(t)$ contains features peaking at $\omega = (E_k - E_{k'})/\hbar$ with amplitudes proportional to the position matrix element between state k and k' squared, and two Boltzmann factors. In the

$\beta \to 0$ limit, the Kubo transform position-position autocorrelation function becomes the classical equivalent we explored in the previous section.

14.3 A test case

To demonstrate how ring polymer molecular dynamics can be implemented in chemical physics research and compare it with the classical molecular dynamics approach, we have considered a particle of mass $m = 919.0$ atomic units in a Lennard–Jones potential. See equations (12.26) and (12.27). The parameters ϵ and σ are 0.123 hartree and 1.2597 bohr, respectively. Together with the mass m, the system is a reasonable model for the vibrational states of the H_2 molecule. For completeness, we mention here that H_2 does not have a permanent dipole moment, and as such is does not have a vibrational absorption spectrum. However, the analysis we carry out here remains a valuable tool to approximate the energies of the first few vibrationally excited states of the system.

To begin, we compute the first few energy levels by using the discrete variable representation of Colbert and Miller. (See Chapter 2.) The dependent variable x ranges from σ to $\sigma + 5$ bohr. The size of the basis set is increased systematically from 500 to 2000 to verify that the most important energy levels have converged. These can be found in the second column of Table 14.1 along with their relative weights $p(E_i) = e^{-\beta E_i}/Q$ in column 4 for $\beta = 100$ hartree^{-1}, corresponding to a temperature of 3158 K.

Table 14.1: The most important energy levels at $\beta = 100$ hartree^{-1} for the test system. The first entry in the $E_j - E_{j-1}$ column is the zero point energy, $E_0 - E_{\min}$, where $E_{\min} = -\epsilon$ is the lowest energy value of U.

j	E_j	e^{-100E_j}	$p(E_j)$	$E_j - E_{j-1}$
0	−0.0880	6625	0.9904	0.0350
1	−0.0407	58.60	0.008760	0.0473
2	−0.0143	4.199	0.0006277	0.0264
3	−0.00284	1.329	0.0001987	0.0115

The code below uses the eigenvalues from the converged discrete variable representation to estimate the canonical ensemble partition function, the internal energy, and the heat capacity. We graph these properties in Figure 14.4. These preliminary steps are used to select the best temperature at which we run the classical and ring polymer molecular dynamics simulations. To best compare the two methods, we seek a sufficiently low temperature so that quantum effects are important, but not so low that the convergence of the path integral requires substantial computational resources:

```
temp,kb = 300.0, 3.166811e-6
k = 0
```

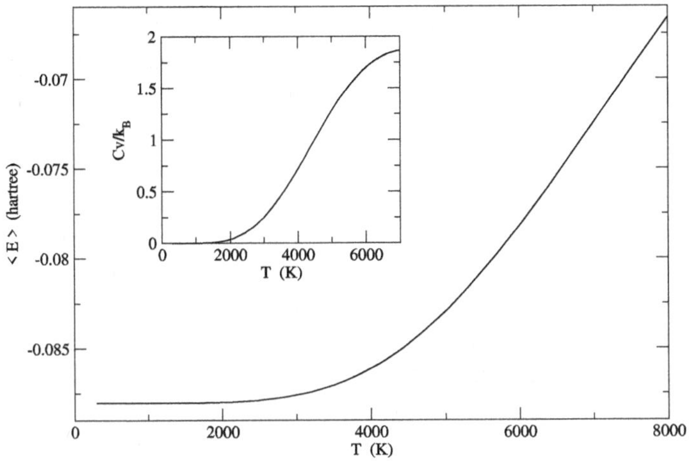

Figure 14.4: The canonical ensemble average energy and C_V/k_B (inset) for the particle in a Lennard–Jones potential.

```
while k < 200:        # loop over values of T
    beta = 1.0/(kb*temp)
    sume,sume2, sumq = 0.0,0.0,0.0
    i=0
    while i < 4:   # loop over the relevant energy levels
        qt = np.exp(-beta*ei)
        sumq += qt
        sume += qt*ei
        sume2 += qt*ei*ei
        i += 1
    sume = sume/sumq
    cv = (sume2/sumq  - sume*sume)*beta*beta
    print ("%12.7f  %12.7f  %12.7f %12.7f " %(temp,beta,sume,cv))
    k += 1
```

At this point, it is imperative to comment on a few subtleties regarding the energy levels computed with the discrete variable representation. Unlike the harmonic oscillator problem or the quartic potential, we explore in Chapters 6 and 7, the Lennard–Jones potential, and many other more realistic models have a finite value at the $x \to \infty$ asymptote. Constructing the grid of positions from σ to $\sigma + 5$ bohr yields only four bound states, those found in Table 14.1. The real physical system, of course, has an infinite number of bound states below the zero energy. However, to capture those one has to enlarge the range of x. For example, extending the grid to $\sigma + 8$ bohr allows us to better approximate another bound state in the 10^{-5} hartree range. There is an adverse effect when the grid in extended: The convergence with respect to the basis set size is impacted, and even for this simple monodimensional case, converging the eigenvalues above those we have tabulated becomes eventually untractable. But such effort is not really necessary.

When systems characterized by a dissociative U are simulated either classically, or via the path integral at sufficiently elevated temperatures, dissociation events begin to creep into existence. These events are elements of reality; the physical systems is sufficiently hot to undergo dissociation with a "measurable" probability. As a matter of fact, there exists a small (though incommensurable) probability that the system dissociate even at $T = 0$ K. When a configuration is sufficiently close to a dissociated state, the simulation overflows and comes to a grinding halt. To eliminate such states, which do not in fact represent the system under study, it is common practice to enclose the system in a finite volume of space. This can be done analytically by including a term in U that rises rapidly at a sufficiently large value of x, say x_B, or a hard wall at the same carefully chosen location. While this remedy may seem to compromise the outcome of the simulations, careful studies have shown that, for systems of point particles, only the liquid to gas region is affected by changing x_B, and that there is a simple physical interpretation of the additional term in U. It has been shown, e. g., that the fictitious barrier added to U impacts the computed thermodynamic properties the same way as surrounding the system of point particles by a buffer gas.

The value of $\epsilon = 0.123$ hartree for the present test case corresponds to an enormous temperature of $T_\epsilon \approx 3 \times 10^6$ K. Dissociation events should become measurable around $T_d = T_\epsilon/10 \approx 3 \times 10^5$ K in simulations with millions of moves. What we mean by measurable here, is that any given simulation that starts with a bound configuration will very likely lead to a dissociation before conclusion if U does not include a barrier. For the test case, we have chosen a value of T almost 100 times smaller than T_d, which in practice means that the states above zero do not bare any weight within the accuracy we wish to achieve, as reflected in the data in Table 14.1. That is the main reason why the innermost loop over the energy levels stops at 3.

As a first estimate of the energy levels, we use classical molecular dynamics simulation. We modify slightly the leap-frog code in Chapter 12, simply run a single trajectory, and compute its Fourier transform using the Bartlett window, introduced in the exercises. We choose to start the position at $x_{\min} = 1.414$ bohr, and we use the `numpy.gauss(0,s)` with zero mean and $s = \sqrt{m/\beta}$ ($\beta = 100$ hartree^{-1}) to generate an initial value of the momentum p. The value of Δt is set at 0.5 hartree^{-1} and the classical trajectory is propagated for 1024 steps. The trajectory $x(t)$ goes through about five cycles, which are sufficient to estimate the main frequency peak after the autocorrelation function is transformed using the code below:

```
nyquist = np.pi/deltat   # Nyquist frequency  (rad hartree)
nfp = 2000        # number of frequency points
frequency = 0.0
t1 = steps*deltat
deltaf = 0.03*nyquist/nfp
```

```
i = 0
while i < nfp:
    sum1 = 0.0
    sum2 = 0.0
    k = 0
    while k < steps:    # integration loop
        wb = 1.0 - np.abs(((k+1) - steps/2.0)/(steps/2.0))
                                                      # Bartlett window
        time = deltat*(k+1)
        ft1 = ax[k]*np.cos(frequency*time)
        ft2 = ax[k]*np.sin(frequency*time)
        sum1 += deltat*wb*ft1
        sum2 += deltat*wb*ft2
        k += 1
    power = (sum1*sum1 + sum2*sum2)/(t1*t1)
    i += 1
    print("%12.7f %12.7f " %(frequency,power))
    frequency += deltaf
```

The array `ax[k]` contains the position-position autocorrelation function $C_{xx}(t) = x(0)x(t)$. The code uses the trapezoid rule to approximate the integral in equation (14.2). With a single trajectory, we determine a main peak at a frequency of $\omega = 0.06654$ rad hartree, which agrees qualitatively with its harmonic approximations 0.06940 rad hartree. To check for a quantitative agreement, we would have to compute the second derivative of U at the minimum, divide its value by the mass, and take the square root of the result. This is left as an exercise.

We can now estimate a $0 \to 1$ transition to be 0.06940 hartree, which would place harmonic estimate of E_0 at $\hbar\omega/2 - \epsilon \approx -0.0897$ hartree. This compares favorably with the E_0 in Table 14.1. The agreement between the harmonic estimate for $E_1 = E_0 + \hbar\omega$ at -0.0232 hartree, however, is not great. The best way to appreciate the power of the ring polymer molecular dynamics approach is to experience it firsthand in the exercises.

14.4 Exercises

1. Provide the missing details for the derivation of equations (14.3) through (14.6).
2. Provide the missing details for the derivation of equation (14.9).
3. Find the analytical expression of the Fourier transform for a square pulse

$$p(t) = \begin{cases} 0 & t < 0 \\ 1 & t \in [0, t_1] \\ 0 & t > t_1 \end{cases}$$

and graph its modulus for $t_1 = 10$ for $\omega' > 0$. The Fourier transform of a square pulse is singular at the origin. Therefore, you should avoid including that point and its immediate neighborhood of points when plotting it.

4. The continuous version of the Bartlett window is the following piece-wise defined function:

$$w(t) = \begin{cases} 0 & t < 0 \\ 1 - \frac{|t - t_1/2|}{t_1/2} & t \in [0, t_1] \\ 0 & t > t_1. \end{cases}$$

   (a) Compute its Fourier transform, then show that

$$\Re[\tilde{w}(\omega')] = -\frac{2}{t_1(\omega')^2}\cos\omega' t_1 - \frac{2}{t_1(\omega')^2} + \frac{4}{t_1(\omega')^2}\cos\frac{\omega' t_1}{2} + \frac{1}{\omega'}\sin\frac{\omega' t_1}{2},$$

$$\Im[\tilde{w}(\omega')] = \frac{2}{t_1(\omega')^2}\sin\omega' t_1 - \frac{4}{t_1(\omega')^2}\sin\frac{\omega' t_1}{2} + \frac{1}{\omega'}\cos\frac{\omega' t_1}{2}.$$

   (b) Therefore, the Bartlett window also causes spurious side peaks. Write a small python program to compute and graph of $|\tilde{w}(\omega')|^2$. The Fourier transform of the Bartlett window is also singular at the origin. A graph of $|\tilde{w}(\omega')|^2$ is shown in Figure 14.5 for comparison.

5. Start with the solution of the Heisenberg equation for operator $\hat{A}$,

$$\hat{A}(t) = e^{iHt/\hbar}\hat{A}(0)e^{-iHt/\hbar},$$

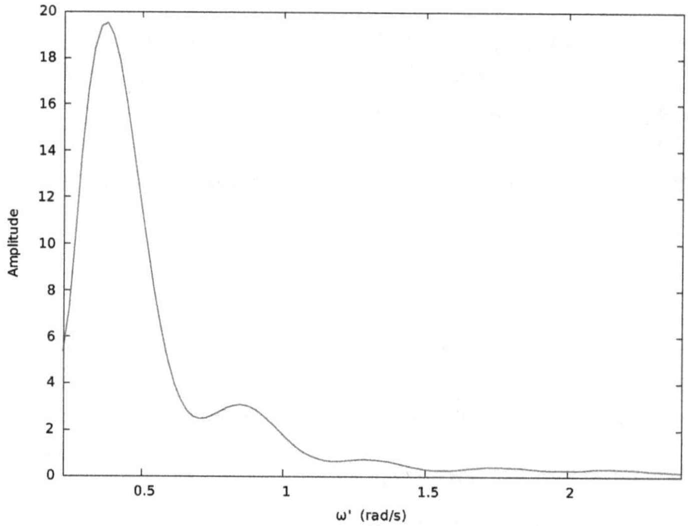

Figure 14.5: The square modulus of the Fourier transform of the Bartlett window function for $t_1 = 10$ units.

switch to imaginary time $t \to -i\hbar\lambda$ to find $\hat{A}(\lambda)$, insert your result into the definition of $\tilde{C}_{AB}(t)$ below:

$$\tilde{C}_{AB}(t) = \frac{1}{\beta Q}\int_0^\beta d\lambda[e^{-\beta H}\hat{A}(\lambda)\hat{B}(t)]$$

and derive equation (14.13).

6. In this exercise, you are guided to develop equation (14.16) from equation (14.13).
   (a) Begin with Heisenberg's equation (14.14), insert equation (14.15) into it,

$$\frac{d\hat{x}}{dt} = \frac{1}{i\hbar}[e^{iHt/\hbar}\hat{x}(0)e^{-iHt/\hbar}, \hat{H}]$$

   and show that it becomes equivalent to

$$\frac{d\hat{x}}{dt} = \frac{1}{i\hbar}e^{iHt/\hbar}[\hat{x}(0), \hat{H}]e^{-iHt/\hbar}.$$

   (b) Switch to imaginary time $t = -i\lambda\hbar$ and derive

$$\frac{d\hat{x}}{d\lambda} = e^{H\lambda}[\hat{x}(0), \hat{H}]e^{-H\lambda}$$

   from the result in part (a).
   (c) Plug the result of part (b) into the following expression:

$$C_{xx}(t) = \frac{1}{\beta Q}\int_0^\beta \mathrm{Tr}\left\{e^{-\beta H}\frac{d\hat{x}}{d\lambda}\hat{x}(t)\right\}d\lambda$$

   and prove the following result:

$$C_{xx}(t) = \frac{1}{\beta Q}\int_0^\beta \mathrm{Tr}\{e^{-(\beta-\lambda)H}[\hat{x}(0), \hat{H}]e^{-H\lambda}\, e^{iHt/\hbar}\hat{x}(0)e^{-iHt/\hbar}\}\, d\lambda.$$

   (d) Use the eigenvectors of $\hat{H}$,

$$\hat{H}|k\rangle = E_k|k\rangle, \quad \langle j|\hat{H} = \langle j|E_j$$

   to show that

$$\begin{aligned}\langle j|[\hat{x}, \hat{H}]|k\rangle &= \int_{\mathbb{R}} dx\, \psi_j^*[\hat{x}, \hat{H}]\psi_k \\ &= (E_k - E_j)\langle j|\hat{x}|k\rangle.\end{aligned}$$

(e) Use the definition of the $N \times N$ matrix product to show that, if A, B, and C are $N \times N$ matrices, and both A and C are diagonal, then

$$(ABC)_{jk} = A_j B_{jk} C_k.$$

(f) Apply the result in part (e) twice to prove that, if A, B, C, D, E, and F are $N \times N$ matrices, and A, C, D, F are diagonal, then

$$(ABCDEF)_{jk} = \sum_{k'=0}^{N-1} A_j B_{jk'} C_{k'} D_{k'} E_{k'k} F_k.$$

(g) Insert the result in parts (d) and (f) into the expression in part (c) to derive the following equation:

$$\begin{aligned} C_{xx}(t) \\ &= \frac{1}{\beta Q} \int_0^\beta \sum_{k,k'} \{ e^{-(\beta-\lambda)E_j} (E_{k'} - E_j) \langle j|\hat{x}|k'\rangle \\ &\quad \times e^{-E_{k'}\lambda} e^{iE_{k'}t/\hbar} \langle k'|\hat{x}|k\rangle e^{-iE_k t/\hbar} \} \, d\lambda \end{aligned}$$

(h) Integrate the expression in (g) to derive (14.16).

7. Find an analytical expression for the Fourier transform of equation (14.16).
8. Equation (14.12) can be extended to the product of two operators, $\hat{A}$ and $\hat{B}$,

$$C_{AB}(t) \approx \frac{1}{(2\pi\hbar)^n Q_n} \int dp_1 \cdots \int dx_n \, e^{-\beta_n H_n(p,x)} \hat{A}_n(x) \hat{B}(x,t)$$

where B is propagated in real time. Ring polymer molecular dynamics simulations approximate the following quantum expression related to the Kubo transform of $\hat{A}$ and $\hat{B}$:

$$C_{AB}(t) = \frac{1}{\beta Q} \mathrm{Tr}\left[e^{-\beta\hat{H}} \hat{A} \, e^{i\hat{H}t/\hbar} \hat{B} \, e^{-i\hat{H}t/\hbar}\right].$$

Repeat the analysis of the previous exercise and show that this autocorrelation function contains information similar to the exact Kubo-transformed real-time correlation function of the two operators. For the purpose of this exercise, assume both $\hat{A}$ and $\hat{B}$ are the position operator.

9. Reproduce the classical molecular dynamics simulations in the chapter by following the steps listed below.
   (a) Modify the leap-frog and Lennard–Jones `pot(x)` codes from Chapter 12 to compute the classical position-position autocorrelation function. Make a plot of $C_{xx}(t)$ and use it to estimate the visible frequencies.

(b) Show that the position x_{min} where the Lennard–Jones potential is at its minimum value of $-\epsilon$ is $2^{1/6}\sigma$, and that at that value

$$\frac{\partial^2 U}{\partial x^2} = \frac{72\epsilon}{2^{1/3}\sigma^2} = K_{eff},$$

where ϵ and σ are the Lennard–Jones parameters.

(c) The result in part (b) suggests that we could approximate U near x_{min} with an harmonic potential $U_h = k_{eff}(x - x_{min})^2/2$. Compute the natural frequency associated with the frequency peak reported in the chapter and the estimate in party (a).

(d) Insert the code that computes the Fourier transform via the trapezoid rule and the Bartlett window and plot the resulting frequency spectrum of the systems.

(e) There should be some small satellite peaks visible near the main feature. A quick way to determine if these are spurious is to cut Δt in half and rerun the simulation with exactly the same initial values of p. If the peak moves, it is spurious. Are the satellite peaks spurious?

(f) Switch the window to a Gaussian, and repeat the exercise in part (b). Comment on your observations.

(g) The fast Fourier transform for the test case can be implemented using the following code:

```
sf = 1/deltat
fstep = sf/steps
f = np.linspace(0,(steps-1)*fstep,num=steps) # frequency space
fs = np.fft.fft(ax)
fs_mag = np.abs(fs)/steps
tpi = 2.0*np.pi
i=1
while i < steps/2:
    print("%12.7f %12.7f" %(tpi*f[i],fs_mag[i]))
    i += 1
```

Test this code, but recall that the units used by the `np.fft.fft(ax)` numpy function returns frequencies in units of inverse time. To correct for it, we have multiplied the abscissa by 2π in the first argument of the print function.

10. The first step in carrying out ring polymer molecular dynamics simulations is to create a function that returns U_{eff} and its gradient.

(a) The gradient of U_{eff} is understood to have as components the partial derivatives with respect to $x_0, x_1, \ldots, x_{n-1}$. Obtain an expression for the gradient of U_{eff} given in equation (14.11).

(b) A Python code for the calculation of U_{eff} and its gradient given $x_0, x_1, \ldots, x_{n-1}$, m, n, and $\beta_n = \beta/n$ is below:

```
def RPMDpot(n,betan,mass,x):
    drpmdu = np.zeros(n+1)
    cn = mass/(2.0*betan*betan)
    x[n] = x[0]
    rpmdu = 0.0
    i = 1
    while i < n+1:
        dx = x[i] - x[i-1]
        mt = cn*dx*dx                          # Matzubara term
        xin = x[i]
        u,du = pot(xin)                        # U(x[i])
        rpmdu +=  mt + u
        if i == n:
            drpmdu[0] = du + 2.0*cn*(2.0*x[0] - x[n-1] - x[1])
        else:
            drpmdu[i] = 2.0*cn*(2*x[i] - x[i-1] - x[i+1]) + du
        i += 1
    return rpmdu,drpmdu
```

Note that the function `pot(x)` has to be defined and inserted into the code file before testing. Write a Python program that initializes the array `x[k]` for $k = 0, n$ with random numbers distributed uniformly around $x_{min} = 1.414$, and calls `RPMDpot()` to test the code above. Use $n = 5$ and check the outcome of each step with a spreadsheet. Just check the steps for the calculation of U_{eff} (returned in `rpmdu`).

(c) Check that the $n = 1$ case is the classical limit by testing the code once more.

(d) After parts (c) and (d) are completed, check the gradient of U_{eff} against the numerical estimate by adding the following snippet at the end of the code in part (c):

```
hh = 1.e-5
k=0
while k < n:
  x[k] += hh
  up,du = RPMDpot(n,betan,mass,x)
  x[k] -= 2.0*hh
  um,du = RPMDpot(n,betan,mass,x)
  x[k] += hh
  u,du = RPMDpot(n,betan,mass,x)
  dn = (up - um)/(2.0*hh)
  print("%5i, %12.7f, %12.7f, %12.7f" %(k,dn,du[k],(du[k] - dn)))
  k += 1
```

(e) The first term of U_{eff} in equation (14.11) is called the Matzubara potential U_M. This term corrects the classical dynamics, but it may introduce additional spurious features in the frequency spectrum of the position autocorrelation function. Therefore, it is important to identify them for the obvious reason of interpreting the spectrum correctly, but also to decide on the best time interval to run the dynamics with to avoid aliasing effects:

$$U_M = \frac{1}{2} k_M \sum_{i=1}^{n} (x_i - x_{i-1})^2.$$

Prove that the Hessian matrix for U_M is of the following form:

$$\partial_\mu \partial_\nu U_M = k_M \begin{bmatrix} 2 & -1 & 0 & \cdots & 0 & -1 \\ -1 & 2 & -1 & \cdots & 0 & 0 \\ \vdots & & & & & \vdots \\ -1 & 0 & 0 & \cdots & -1 & 2 \end{bmatrix},$$

where $\mu, \nu \in \{0, 1, \ldots, n-1\}$.

(f) Write a Python code that computes the Matzubara frequencies by diagonalizing $\partial_\mu \partial_\nu U_M / m$ where m is the mass of the particle. The following relationship $k_M/m = n^2/(\beta^2 \hbar^2)$ shows that the Matzubara frequencies are mass independent and grow linearly with n. Run your code with $n = 5$, $\beta = 20$ and verify the following frequencies 0.226127, 0.086373, 0 with the first two being doubly degenerate.

11. The code below uses the function `RPMDpot()` in part (b) of the previous exercise to compute U_{eff} and runs the ring polymer dynamics:

```
import numpy as np
import random as rnd
mass,beta,xmin,deltat,n,steps = 919.0,20.0,1.414,0.05,1,16384
betan = beta/n
x = np.full(n+1,xmin) #  0.1 + xmin    # n beads  np.full(n+1,xmin) #
x[n] = x[0]
x0 = x
#p = np.random.normal(0,np.sqrt(mass/beta),n)
p = np.zeros(n+1)
ax = np.zeros(steps+1,np.float64)
                          # array for the pos. autocorrelation function
u,du = RPMDpot(n,betan,mass,x)
k = 0
while k < steps:
    j = 0
    while j < n:   # update all the positions
        x[j] += deltat*(p[j] - deltat*du[j]/2.0)/(mass)
        j += 1
```

```
    up1,dup1 = RPMDpot(n,betan,mass,x)
    j = 0
    while j < n:   # update all the momenta
        p[j] += -deltat*(dup1[j] + du[j])/2.0
        ax[k] += x0[j]*x[j]
        du[j] = dup1[j]
        j += 1
    time = deltat*k
    k += 1
```

(a) Test the code and run it. Use it to plot `x[0]` as a function of time and compare your trajectory with one computed using the same parameters, including Δt from the leap-frog algorithm in Chapter 12.

(b) Modify the code so that the ring polymer energy is computed, then check that it is conserved for `n=1,2,4`.

(c) Before we can extract the spectrum from the autocorrelation function stored in the `ax[k]` array we have to do some preprocessing. Generally, $x(t)x(0)$ does not fluctuate about zero, and transforming it directly leads to a large spurious peak at zero frequency whose properties depend on the choice of window used. Add a loop that computes the time average value of $x(t)x(0)$. Add another loop to subtract such value from it, and to multiply the window function $w_p(t)$ onto the difference:

$$C_{xx}(t) \rightarrow w_p(t)\left[x(t)x(0) - \frac{1}{t_1}\int_0^{t_1} x(t')x(0)dt'\right].$$

(d) Use the code to explore its behavior at various values of n, and β. You will need to increase the simulation time t_1 to resolve the frequencies, and you may want to use the Python fast Fourier transform algorithm:

$$w_h(t) = \begin{cases} 0 & t < 0 \\ 1 - \left(\frac{|t - t_1/2|}{t_1/2}\right)^2 & t \in [0, t_1] \\ 0 & t > t_1. \end{cases}$$

The spectra graphed in Figure 14.6 is an example of the type of data that can be obtained from a single, sufficiently long trajectory. We did not compute the transform of the position autocorrelation function using an equilibrated set of starting conditions.

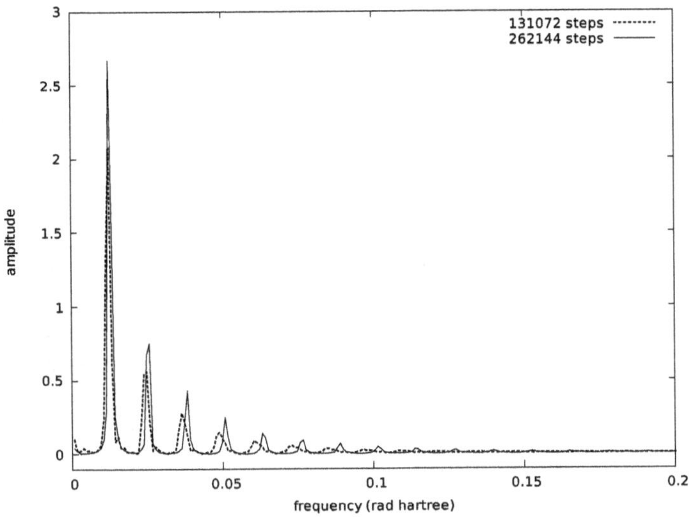

Figure 14.6: Frequency spectrum for the test case from a single trajectory $\beta = 20$, $n = 5$, with $2^{17} = 131072$ and 2^{18} steps and with the initial position uniformly distributed around 2.414 bohr.

15 Essential electrodynamics

Well, gauge theory is very fundamental to our understanding of physical forces these days. But they are also dependent on a mathematical idea, which has been around for longer than gauge theory has.

Roger Penrose

The study of electrodynamics is where the ideas of invariance under coordinate transformations, the Lorentz group, space time, special relativity, gauge invariance, and gauge theory all originated from. Maxwell's equations govern the dynamic behavior of the electric and magnetic fields **E** and **B**. These are experimentally measurable vector valued functions of x, y, z, and t that are fundamental properties of electromagnetic radiation. Maxwell's equations is the following set of coupled vector differential equations:

$$\nabla \cdot \mathbf{E} = \frac{\rho}{\epsilon_0} \tag{15.1}$$

$$\nabla \cdot \mathbf{B} = 0 \tag{15.2}$$

$$\nabla \times \mathbf{E} = -\frac{\partial \mathbf{B}}{\partial t} \tag{15.3}$$

$$\nabla \times \mathbf{B} = \mu_0 \left(\mathbf{J} + \epsilon_0 \frac{\partial \mathbf{E}}{\partial t} \right). \tag{15.4}$$

The bold font notation indicates that **E**, **B**, and **J** are 3 vector-valued functions, meaning they have components along the three directions of physical space **i**, **j**, and **k**.

The purpose of this chapter is to briefly review the origin of equations (15.1) through (15.4), explore some special cases using the nineteen century notation, introduce the vector potential, and explain how its nonunique nature can be exploited to uncouple the differential equations, and finally show how Hamilton's least action principle can be used to derive the source-free version of Maxwell's equation using the tensor language of Chapter 8 through 11.

The first two equations involve the divergence of the electric and magnetic fields and these are scalar quantities, whereas the second two involve the curl of the two fields representing a total of six differential equations for the respective components. One can prove mathematically that if the gradient and the curl of a vector valued function are known, along with proper boundary conditions, one can find a unique solution for the components of the field at all space-time points. The mathematical statement goes under the name of Helmholtz theorem, typically proved in introductory electromagnetism courses. The divergence and the curl are the sources of the fields and these involve charges and currents.

The symbol ρ is the charge density, a scalar quantity that can be a function of x, y, z, and t. The vectorial quantity **J** represents the electric current, the source of magnetism. The charge densities and the electric currents can be both macroscopic, e. g., the charge

https://doi.org/10.1515/9783111610207-015

on a capacitor and the current in a wire, as well as microscopic, e. g., the charge and the motion of a set of point particles. At the macroscopic scale, or at the atomic scale using the laws of quantum mechanics, the charge and current densities are generally approximated by smooth continuous functions of space-time.

There is a relationship between the divergence of $\mathbf{J}$ and the time derivative of ρ. To develop the relationship, consider a volume V in $\mathbb{R}^3$ bound by a closed surface $\mathcal{S}$. The total charge inside V given a charge density ρ is simply the volume integral of the charge density:

$$q = \int_V \rho \, dV. \tag{15.5}$$

The rate of change of q inside V due to the current $\mathbf{J}$ is given by the charge flux through $\mathcal{S}$,

$$-\frac{dq}{dt} = \oint\!\!\oint_{\mathcal{S}} \mathbf{J} \cdot d\mathbf{a} = \int_V \nabla \cdot \mathbf{J} \, dV,$$

where $d\mathbf{a}$ is the differential area element and we have used Gauss's divergence theorem, to turn the surface integral into a volume integral of the divergence of $\mathbf{J}$. We now take the derivative of (15.5) with respect to t and combine it with the last expression to get

$$-\frac{dq}{dt} = \int_V \frac{\partial \rho}{\partial t} \, dV = \int_V \nabla \cdot \mathbf{J} \, dV.$$

Because the range of integration of the two integrals is the same, the only way they can be equal to each other is if the integrands are the same at every point in V. This yields the equation of continuity

$$\nabla \cdot \mathbf{J} = -\frac{\partial \rho}{\partial t}, \tag{15.6}$$

a formal mathematical statement of the conservation of charge.

The equation of continuity is the keystone of the bridge that unifies electricity and magnetism into a single theory for the interaction between matter particles and photons, the carriers of the electromagnetic force. Faraday was the first to observe experimentally how electricity and magnetism are coupled dynamically by measuring electric fields generated by moving magnetic fields. The result of his many experiments is (15.3), the equation that bears his name. In order to appreciate how the pieces of this puzzle come together, we briefly review electricity and magnetism, then we explore how Maxwell used the principle of conservation of charge enshrined in equation (15.6) to put the last piece of the puzzle in place. To start, let us observe that if fields are static, namely if $\partial \mathbf{E}/\partial t$ and $\partial \mathbf{B}/\partial t$ are zero, Maxwell's equations decouple, and we can separately derive a unique solution for the electric and magnetic fields. This is the task of the next two sections.

15.1 Electrostatics

Equation (15.1) is known as Gauss's law and is the differential equivalent of Coulomb's law, first published by the French physicist Charles Augustin de Coulomb. Coulomb's law states that the force on a point charge q located at $\mathbf{r}$, due to the charge q' located at $\mathbf{r}'$ is

$$\mathbf{F}'_{q'\to q} = \frac{1}{4\pi\epsilon_0} qq' \frac{\hat{\mathbf{R}}}{R^2} \tag{15.7}$$

where ϵ_0 is the permittivity of free space and has been measured experimentally to have a value of $8.85419 \times 10^{-12}\,\mathrm{C^2N^{-1}m^{-2}}$ in SI units. The vector $\hat{\mathbf{R}}$ is the unit vector $\mathbf{R}/R$, where $\mathbf{R} = \mathbf{r} - \mathbf{r}'$ is the relative vector that points from q' to q and $R = |\mathbf{r} - \mathbf{r}'|$ is its size. In electrostatics, it is assumed that the source vectors $\mathbf{r}'$ are time independent. In Cartesian coordinates, the relative vector is

$$\mathbf{R} = (x - x')\mathbf{i} + (y - y')\mathbf{j} + (z - z')\mathbf{k}.$$

There is also a force on q' equal in magnitude but oriented in the opposite direction. The point charge q is called the test charge whereas the charge q' is called the source charge. The electric field $\mathbf{E}$ is defined as the ratio of the Coulomb force over the test charge $\mathbf{F}'_{q'\to q}/q$. For a single source charge, it is

$$\mathbf{E}(r) = \frac{1}{4\pi\epsilon_0} q' \frac{\hat{\mathbf{R}}}{R^2}. \tag{15.8}$$

For a system of static source particles, the field at point r and time t becomes

$$\mathbf{E} = \frac{1}{4\pi\epsilon_0} \sum_{i=1}^{n} q_i \frac{\hat{R}_i}{R_i^2} = \frac{1}{4\pi\epsilon_0} \sum_{i=1}^{n} q_i \frac{\mathbf{R}_i}{R_i^3}. \tag{15.9}$$

Gauss's law is a mathematical statement derived by considering the electric field flux over the surface of a sphere or radius r containing the source charge,

$$\oint \mathbf{E} \cdot d\mathbf{a} = \frac{q'}{\epsilon_0}$$

where $d\mathbf{a}$ is an infinitesimal element of the area multiplied by the unit vector normal to the surface that points outward. For simplicity, we have placed the source charge at the origin and, therefore, $\mathbf{R} = \mathbf{r}$. This is always possible when we have a single point charge, but it is not so if the charge density is spread out over a volume. Using the spherical symmetry of the problem, $d\mathbf{a} = \hat{\mathbf{r}}\, r^2 \sin\theta d\theta d\phi$,

$$\oint \mathbf{E} \cdot \hat{\mathbf{r}} r^2 \sin\theta d\theta d\phi = \frac{q'}{\epsilon_0},$$

but **E** and $\hat{r}$ do not depend on the angles. Therefore,

$$\mathbf{E} \cdot \hat{r} r^2 \int_0^{2\pi} d\phi \int_0^{\pi} \sin\theta d\theta = \frac{q'}{\epsilon_0},$$

the integral is just the solid angle 4π, and we derive a special case for the electric field when the source charge is at the origin:

$$\mathbf{E} = \frac{q'}{4\pi\epsilon_0 r^2}\hat{\mathbf{r}}.$$

The electric flux yields the correct expression for the electric field in this case. The connection between the flux and the divergence follows from Gauss's divergence theorem:

$$\oiint_S \mathbf{E} \cdot d\mathbf{a} = \int_V \nabla \cdot \mathbf{E} \, dV.$$

Note that we used spherical symmetry for this simple case to derive equation (15.8), however, it is possible to show that the electric field flux over a closed surface of arbitrary shape is equal to the total charge inside the volume it encloses divided by ϵ_0.

An important property is the scalar potential Φ defined for a set of point particles as

$$\Phi = \sum_{i=1}^{n} \frac{q_i}{4\pi\epsilon_0 R_i}. \tag{15.10}$$

It is then possible to show that $\mathbf{E} = -\nabla\Phi$ and that $\nabla \times \mathbf{E} = 0$. Electrostatics is a useful tool to study the fields when charges are pinned in place by, e. g., chemical bonds as in semiconducting or insulating materials, or accumulated as excess charges on the surface of a conductor. Equation (15.3) is more general and indicates a relationship between the electric and magnetic fields when charges are free to move around. Moving charges as in, e. g., a conducting wires are electric currents, the sources of magnetic fields. These are the subjects of the next section.

15.2 Magnetostatics

The magnetic field **B** is derived from an experimentally determined law that relates the total force on a complete circuit C carrying a steady filamentary current I by a complete circuit C' carrying a steady filamentary current I'. The expression for the force is written as a double line integral,

$$\mathbf{F}'_{C'\to C} = \frac{\mu_0}{4\pi} \oint_C \oint_{C'} \frac{I\, d\mathbf{s} \times (I'\, d\mathbf{s}' \times \hat{\mathbf{R}})}{R^2} \tag{15.11}$$

where $\mu_0 = 4\pi \times 10^{-7}\ \mathrm{NA}^{-2}$ is the permeability of free space, $d\mathbf{s}$ and $d\mathbf{s}'$ are infinitesimal path elements, and $\mathbf{R} = \mathbf{r} - \mathbf{r}'$ is the relative vector from the source current $I' d\mathbf{s}'$ to the test field current $I d\mathbf{s}$. The magnetic field $\mathbf{B}$ is defined as the Ampère force per unit test field current as follows, by rearranging (15.11) slightly:

$$\mathbf{F}'_{C'\to C} = \frac{\mu_0}{4\pi} \oint_C I\, d\mathbf{s} \times \left(\oint_{C'} \frac{I'\, d\mathbf{s}' \times \hat{\mathbf{R}}}{R^2} \right),$$

$$\mathbf{F}'_{C'\to C} = \frac{\mu_0}{4\pi} \oint_C I\, d\mathbf{s} \times \mathbf{B},$$

where clearly,

$$\mathbf{B} = \oint_{C'} \frac{I'\, d\mathbf{s}' \times \hat{\mathbf{R}}}{R^2}. \tag{15.12}$$

Equation (15.12) is Biot–Savart law.

If the source and test currents are steady but spread out over a volume V' and V, respectively, then the expression for the magnetostatic force becomes

$$\mathbf{F}'_{C'\to C} = \frac{\mu_0}{4\pi} \int_V dV \int_{V'} \frac{\mathbf{J} \times (\mathbf{J}' \times \hat{\mathbf{R}})}{R^2}\, dV \tag{15.13}$$

from which one derives equation (15.2) and the static version of equation (15.4)

$$\nabla \times \mathbf{B} = \mu_0 \mathbf{J}. \tag{15.14}$$

Equations (15.11), (15.13), and (15.14) are different forms of Ampère law. This last equation should clarify what is meant by "steady currents," such as, e. g., in permanent magnets, these are such that $\partial\mathbf{E}/\partial t = 0$.

15.3 The displacement current

The term proportional to $\partial\mathbf{E}/\partial t$ of equation (15.4) is deduced theoretically by J. C. Maxwell when he realizes that without the time derivative of the electric field the equations are inconsistent. To see how this is the case, we take the divergence of (15.14) and arrive at

$$\nabla \cdot \nabla \times \mathbf{B} = \mu_0\, \nabla \cdot \mathbf{J} = -\mu_0 \frac{\partial \rho}{\partial t},$$

where in the last step we have used (15.6). It can be demonstrated that the divergence of a curl of any vector is zero. But $\partial\rho/\partial t$ need not be zero. Therefore, Maxwell concludes that (15.14) is inconsistent with the conservation of charge. To fix the problem, he introduces a so-called displacement current $\mathbf{J}_d$ so that the curl of $\mathbf{B}$ becomes

$$\nabla \times \mathbf{B} = \mu_0(\mathbf{J} + \mathbf{J}_d), \tag{15.15}$$

then the divergence of this result becomes

$$\nabla \cdot \nabla \times \mathbf{B} = \mu_0(\nabla \cdot \mathbf{J} + \nabla \cdot \mathbf{J}_d)$$
$$\nabla \cdot \nabla \times \mathbf{B} = \mu_0\left(-\frac{\partial\rho}{\partial t} + \nabla \cdot \mathbf{J}_d\right).$$

Now we use equation (15.1) rewrite the first term in the parenthesis

$$-\frac{\partial\rho}{\partial t} = -\epsilon_0\frac{\partial}{\partial t}\nabla \cdot \mathbf{E}$$

and since the sum of the two terms in the parentheses have to add up to zero, i. e.,

$$-\frac{\partial\rho}{\partial t} + \nabla \cdot \mathbf{J}_d = 0,$$
$$-\epsilon_0\frac{\partial}{\partial t}\nabla \cdot \mathbf{E} + \nabla \cdot \mathbf{J}_d = 0$$

or,

$$\nabla \cdot \mathbf{J}_d = \epsilon_0\frac{\partial}{\partial t}\nabla \cdot \mathbf{E},$$

which can only hold if the two vector whose divergence is evaluated are the same,

$$\mathbf{J}_d = \epsilon_0\frac{\partial}{\partial t}\mathbf{E}.$$

Equation (15.4) follows after this expression for $\mathbf{J}_d$ is substituted in (15.15). The displacement current is perhaps the best example of how one derives laws of physics by adhering to conservation principles. The developers of gauge theory have followed the same principle to derive the standard model of the interactions among elementary particles, and to propose the existence of a particle that had yet to be discovered, the Higgs boson [7].

The recent discovery of such particle by the Large Hadron Collider at CERN has once more confirmed the validity of gauge theory and its principles, the same way that the experiments conducted by the German physicist Heinrich Hertz in the late 1880s proved the existence of electromagnetic radiation and verified it travels at the speed of light, validating Maxwell's displacement current hypothesis. As the next section clearly demonstrates, it is the displacement current that establishes the propagation of electromagnetic waves in space.

15.4 Vacuum solutions

As discussed earlier, Faraday's law of induction, namely equation (15.3), was discovered by observing how a varying magnetic field flux through the surface bounded by an electric circuit induces an electric field in the circuit wire without a battery. The change in flux is proportional to the rate of change of **B** with respect to time, and Stoke's theorem is used to derive the law in the form of equation (15.3) from the surface integral form. We now consider the Maxwell equations when the field is sufficiently far from the sources. Then equations (15.1) through (15.4) become much more symmetric, and it is possible to decouple them,

$$\nabla \cdot \mathbf{E} = 0 \tag{15.16}$$

$$\nabla \cdot \mathbf{B} = 0 \tag{15.17}$$

$$\nabla \times \mathbf{E} = -\frac{\partial \mathbf{B}}{\partial t} \tag{15.18}$$

$$\nabla \times \mathbf{B} = \mu_0 \epsilon_0 \frac{\partial \mathbf{E}}{\partial t}. \tag{15.19}$$

By taking the curl of (15.17) and using the identity,

$$\nabla \times (\nabla \times \mathbf{E}) = \nabla(\nabla \cdot \mathbf{E}) - \nabla^2 \mathbf{E}$$

we obtain

$$\nabla(\nabla \cdot \mathbf{E}) - \nabla^2 \mathbf{E} = -\frac{\partial}{\partial t} \nabla \times \mathbf{B}$$

where on the right-hand side we have exchanged the order of operation from the curl of a derivative to the derivative of a curl. Now using equations (15.16) and (15.19), the last equation simplifies to

$$\nabla^2 \mathbf{E} = \mu_0 \epsilon_0 \frac{\partial^2 \mathbf{E}}{\partial t^2}. \tag{15.20}$$

This expression represents three independent second-order differential equations, one for each of the three components, e. g.,

$$\nabla^2 E_x = \mu_0 \epsilon_0 \frac{\partial^2 E_x}{\partial t^2}, \tag{15.21}$$

etc. This differential equation is isomorphic to the wave equation $f(x, y, z, t)$ propagating through space and time with a speed v,

$$\nabla^2 f = \frac{1}{v^2}\frac{\partial^2 f}{\partial t^2}. \tag{15.22}$$

Therefore, in free space the speed of the electromagnetic wave $c = 2.99792 \times 10^8\ \mathrm{m\,s^{-1}}$ is a fundamental constant,

$$c = \frac{1}{\sqrt{\mu_0 \epsilon_0}}. \tag{15.23}$$

Finally, by evaluating the curl of (15.19) and following the same procedure used earlier to derive the wave equation for **E** we can show that **B** satisfies exactly the same wave equation.

$$\nabla^2 \mathbf{B} = \mu_0 \epsilon_0 \frac{\partial^2 \mathbf{B}}{\partial t^2}. \tag{15.24}$$

It is possible to show by inserting the solutions back into (15.16) through (15.19) that the electric and magnetic fields propagate in phase, with amplitudes in planes perpendicular to each other.

15.5 The vector potential

It is very convenient, when working from the theoretical perspective to define a potential from which the magnetic field can be derived. From equation (15.2) and the fact that the divergence of a curl is zero one defines a vector potential **A**,

$$\mathbf{B} = \nabla \times \mathbf{A}. \tag{15.25}$$

From equation (15.13), we can deduce

$$\mathbf{B} = \int_{V'} \frac{\mathbf{J}' \times \hat{\mathbf{R}}}{R^2}\, dV' \tag{15.26}$$

and we can show that

$$\mathbf{A} = \frac{\mu_0}{4\pi}\int_{V'} \frac{\mathbf{J}'}{R}\, dV',$$

by evaluating the curl of this expression using the following vector identity $\nabla(u\mathbf{F}) = -\mathbf{F} \times (\nabla u) + u(\nabla \times \mathbf{F})$ with $u = 1/R$, a function, and $\mathbf{F} = \mathbf{J}'$. Then

$$\nabla \times \mathbf{A} = \frac{\mu_0}{4\pi}\int_{V'} \left[\nabla \times \left(\frac{\mathbf{J}'}{R}\right)\right] dV'$$

where we exchange the order integral-curl to curl-integral. Using the identity, we get

$$\mathbf{B} = \nabla \times \mathbf{A} = \frac{\mu_0}{4\pi} \int_{V'} \left[-\mathbf{J}' \times \left(\nabla \frac{1}{R} \right) + \frac{1}{R} \nabla \times \mathbf{J}' \right] dV'.$$

The cross-product $\nabla \times \mathbf{J}'$ is zero because the curl is evaluated with respect to the field coordinates x, y, z whereas the current density is a function of the source coordinates x', y', z'. The results follows from

$$\nabla \frac{1}{R} = -\frac{\hat{\mathbf{R}}}{R^2},$$

which is left to the reader to prove. When we insert (15.25) into Faraday's induction law (15.3), we obtain

$$\nabla \times \mathbf{E} = -\frac{\partial}{\partial t} (\nabla \times \mathbf{A}).$$

Upon exchanging the order of operations on the right and comparing the vectors whose curl, we derive

$$\mathbf{E} = -\frac{\partial \mathbf{A}}{\partial t}.$$

But there is a problem with this result. The electrostatic term $\mathbf{E} = -\nabla \Phi$ is missing. So, let us add it in:

$$\mathbf{E} = -\nabla \Phi - \frac{\partial \mathbf{A}}{\partial t}.$$

What follows is now critical. How can we, without direction or planning, haphazardly add terms to our equation? It is definitely bad mathematical practice, but our intuition here turns out to be correct. We now show that the choice of the vector potential $\mathbf{A}$ is not unique, meaning one can add to $\mathbf{A}$ the gradient of any well behaved function χ and obtain the same magnetic field $\mathbf{B}$,

$$\begin{aligned} \mathbf{A}' &= \mathbf{A} + \nabla \chi, \\ \mathbf{B}' &= -\nabla \times \mathbf{A}' = -\nabla \times \mathbf{A} - \nabla \times \nabla \chi \end{aligned} \tag{15.27}$$

and since the curl of a gradient is zero for "well-behaved" functions $\nabla \times \nabla \chi = 0$, this gives immediately

$$\mathbf{B}' = -\nabla \times \mathbf{A} = \mathbf{B}' = \mathbf{B}.$$

Additionally, if we let

$$\Phi' = \Phi - \frac{\partial \chi}{\partial t} \tag{15.28}$$

for the same function χ, we recover the desired electric field, meaning

$$\mathbf{E} = -\nabla\Phi - \frac{\partial \mathbf{A}}{\partial t} = -\nabla\Phi' - \frac{\partial \mathbf{A}'}{\partial t}. \tag{15.29}$$

The reader is asked to prove this result as an exercise. So, the magnetic field is okay, but to fix the electric field we have made a specific choice for $\nabla\chi$. Deriving the same fields from "similar" vector potentials is reminiscent of the identical Lagrangians we explore in Chapter 12.

Adding terms to the vector potential to simplify expressions or to force additional constraints such as, e. g., setting $\nabla \cdot \mathbf{A}' = 0$ is an example of choosing a particular gauge. Note that Maxwell's equations are gauge invariant, because the fields do not change under a gauge transformation. Choosing $\nabla \cdot \mathbf{A}'$ to vanish amounts to choosing χ so that the following equation is satisfied:

$$\nabla \cdot \mathbf{A}' = \nabla \cdot (\mathbf{A} + \nabla\chi) = 0,$$
$$\nabla^2\chi = -\nabla \cdot \mathbf{A}.$$

This is the so-called Coulomb gauge. The choice of χ such that

$$\nabla \cdot \mathbf{A} = -\frac{1}{c^2}\frac{\partial \Phi}{\partial t} \tag{15.30}$$

holds is the Lorentz gauge. Depending on the conditions, these choices can simplify the analysis tremendously. To see how this works, consider inserting (15.25) and (15.29) into (15.4), the curl of **B**,

$$\nabla(\nabla \cdot \mathbf{A}) - \nabla^2\mathbf{A} = \mu_0\mathbf{J} - \frac{1}{c^2}\left(\nabla\frac{\partial \Phi}{\partial t} + \frac{\partial^2 \mathbf{A}}{\partial t^2}\right) \tag{15.31}$$

where the identity $\nabla \times (\nabla \times \mathbf{A}) = \nabla(\nabla \cdot \mathbf{A}) - \nabla^2\mathbf{A}$ was used on the left- hand side. Then, consider inserting (15.29) into the divergence of **E**, equation (15.1),

$$\nabla \cdot \mathbf{E} = -\nabla^2\Phi - \nabla \cdot \frac{\partial \mathbf{A}}{\partial t} = \frac{1}{\epsilon_0}\rho \tag{15.32}$$

this is a pair of coupled second-order vector differential equations. Now consider inserting the Lorentz gauge, equation (15.30) in the last two expressions. Equation (15.31) becomes

$$-\frac{1}{c^2}\nabla\left(\frac{\partial \Phi}{\partial t}\right) - \nabla^2\mathbf{A} = \mu_0\mathbf{J} - \frac{1}{c^2}\left(\nabla\frac{\partial \Phi}{\partial t} + \frac{\partial^2 \mathbf{A}}{\partial t^2}\right)$$

which after canceling the same terms on both sides and rearranging yields

$$-\nabla^2\mathbf{A} + \frac{1}{c^2}\frac{\partial^2 \mathbf{A}}{\partial t^2} = \mu_0\mathbf{J}. \tag{15.33}$$

The divergence of $\partial\mathbf{A}/\partial t$ in (15.32) becomes

$$\nabla\cdot\frac{\partial\mathbf{A}}{\partial t}=\frac{\partial}{\partial t}(\nabla\cdot\mathbf{A})=-\frac{1}{c^2}\frac{\partial^2\Phi}{\partial t^2}.$$

Therefore,

$$-\nabla^2\Phi+\frac{1}{c^2}\frac{\partial^2\Phi}{\partial t^2}=\frac{1}{\epsilon_0}\rho. \tag{15.34}$$

Now equations (15.33) and (15.34) are decoupled, and in free space (away from sources) are isomorphic to those for the components of **E** and **B**.

15.6 Maxwell's tensor

The isomorphism between equations (15.34) and (15.33) is striking and it is suggesting that it is time to make use of the concise tensor notation we have learned in Chapters 8 through 10. The differential expression of equations (15.34) can be interpreted as a kind of Laplace–Beltrami operator in space time if we make the following identifications:

$$x^\mu=(x^0,x^1,x^2,x^3)=(ct,x,y,z)$$

where absorbing the speed of light c into t cleans up the notation, promotes time into an extra dimension of space, albeit with a different signature (see below), and removes the constant from the definition of the differential operator we are about to discuss. It is possible to show that, with this convention,

$$\partial_\mu\partial^\mu=\partial_\lambda g^{\lambda\nu}\partial_\nu\equiv\frac{\partial^2}{\partial t^2}-\nabla^2 \tag{15.35}$$

where

$$g^{\mu\nu}=g_{\mu\nu}=\begin{pmatrix}1&0&0&0\\0&-1&0&0\\0&0&-1&0\\0&0&0&-1\end{pmatrix}, \tag{15.36}$$

is the metric tensor of flat space-time in its canonical form. The choice of −1 for the signature of space-like coordinates is dictated by the negative sign in front of the Laplacian. However, this is just a convention; there are many authors that reverse the signature making time-like negative and space-like positive. The expression in (15.35) goes under the name of the D' Alembertian and in some books it is notated with the symbol $\square^2$. We now want to show a few important results. First, note that the Lorentz gauge now reads

$$\partial_\mu A^\mu=0$$

where the vector $A^\mu = (A^0, A^1, A^2, A^3) = (\Phi, A_x, A_y, A_z)$. To prove the statement, consider (15.30) slightly rearranged and with the time in meters,

$$\nabla \cdot \mathbf{A} + \frac{\partial \Phi}{\partial t}.$$

The expression $\partial_\mu A^\mu = 0$ follows from

$$\partial_\mu = \left(\frac{\partial}{\partial t}, \frac{\partial}{\partial x}, \frac{\partial}{\partial y}, \frac{\partial}{\partial z} \right).$$

Second, we note that defining the Maxwell tensor $F^{\mu\nu}$ in terms of the 4-vector potential derivatives,

$$F^{\mu\nu} = \partial^\mu A^\nu - \partial^\nu A^\mu \tag{15.37}$$

allows us to write both equations (15.34) and (15.33) in a single expression,

$$\partial_\mu F^{\mu\nu} = \mu_0 j^\nu \tag{15.38}$$

where $j^\mu = (\rho, j_x, j_y, j_z)$, the zero component is the charge density, the remaining three components are those of the electric current **J**. Let us demonstrate how this works, and how crucial it is to work in the Lorentz gauge. Taking the derivative of equation (15.37) and using the right-hand side of (15.38) yield

$$\partial_\mu \partial^\mu A^\nu - \partial_\mu \partial^\nu A^\mu = \mu_0 j^\nu.$$

We note that we can interchange order of differentiation, so by the Lorentz gauge condition $\partial_\mu \partial^\nu A^\mu = \partial^\nu (\partial_\mu A^\mu) = 0$. Then the last equation simplifies to

$$\partial_\mu \partial^\mu A^\nu = \mu_0 j^\nu.$$

The $\nu = 0$ case

$$\partial_\mu \partial^\mu A^0 = \mu_0 j^0,$$

expands to

$$\frac{\partial^2}{\partial t^2}\Phi - \nabla \Phi = \mu_0 \rho,$$

which is (15.34) since $\mu_0 = c^{-2} \epsilon_0^{-1}$ in these units.

It is not difficult to write out the components of Maxwell's tensor in terms of the components of **E** and **B** by using

$$\partial^{\mu} = g^{\mu\lambda}\partial_{\lambda} = \left(\frac{\partial}{\partial t}, -\frac{\partial}{\partial x} - \frac{\partial}{\partial y} - \frac{\partial}{\partial z}\right),$$

and equation (15.37). For example, the x component of (15.29) reads

$$E_x = -\frac{\partial}{\partial x}\Phi - \frac{\partial}{\partial t}A_x = -F^{01},$$

whereas the z component of (15.25) reads

$$B_z = \frac{\partial}{\partial x}A_y - \frac{\partial}{\partial y}A_x = F^{21}.$$

Continuing this way, we derive

$$F^{\mu\nu} = \begin{pmatrix} 0 & -E_x & -E_y & -E_z \\ E_x & 0 & -B_z & B_y \\ E_y & B_z & 0 & -B_x \\ E_z & -B_y & B_x & 0 \end{pmatrix}. \tag{15.39}$$

Only two additional principles need to be added to our present discussion before gauge theory can be understood. First, we need to see how Hamilton's principle can be used to arrive at Maxwell's equations as written in (15.38). This is achieved by the following Lagrangian:

$$\mathscr{L}^{(A)} = -\frac{1}{4\mu_0}F_{\mu\nu}F^{\mu\nu} = -\frac{1}{4\mu_0}(\partial_\mu A_\nu - \partial_\nu A_\mu)(\partial^\mu A^\nu - \partial^\nu A^\mu). \tag{15.40}$$

The Euler–Lagrange equations are obtained by considering the derivatives terms $\partial_\mu A_\nu$ as the analog of $\dot{x}$ in Chapter 12, where t, x, y, z play the role of t in the mechanical analog.

$$\partial_\mu\left[\frac{\partial\mathscr{L}^{(A)}}{\partial(\partial_\mu A_\nu)}\right] = \frac{\partial\mathscr{L}^{(A)}}{\partial A_\nu} \tag{15.41}$$

and far away from sources one arrives at

$$\partial_\mu F^{\mu\nu} = 0.$$

This is the $j^\mu = 0$ case of equation (15.38).

Second, we need to have some rudimentary notion of invariance, and that is the content of the next section. It is very important to note that the Lagrangian in (15.40) cannot produce the current vector j^μ because there is no interaction term for particles in $\mathscr{L}^{(A)}$, it contains only electromagnetic radiation. The details of the derivations of the last equations are left and an exercise.

15.7 The Kline–Gordon equation

Here, we introduce a field Lagrangian for a spinless particle with charge e and (rest) mass m so that we may inch our way forward toward a complete matter-radiation interaction,

$$\mathscr{L} = (\partial_\mu \varphi^*)(\partial^\mu \varphi) - m^2 \varphi^* \varphi \tag{15.42}$$

where, as in the previous section we are working in t, x, y, z space endowed with the metric tensor in equation (15.36) and t has units of length. Just as in the Schrödinger–Lagrangian in Chapter 12, φ^* and φ are scalar field variables independent of each other representing a massive boson field. The Euler–Lagrange equations,

$$\partial_\nu \left[\frac{\partial \mathscr{L}}{\partial(\partial_\mu \varphi^*)} \right] = \frac{\partial \mathscr{L}}{\partial(\varphi^*)}$$

yield the partial differential equations for the scalar φ,

$$\partial_\nu \partial^\mu \varphi = -m^2 \varphi, \tag{15.43}$$

the Kline–Gordon equation for a massive particle.

Equation (15.43) is the relativistic version of the Schrödinger equation for a spinless particle. Note that $\hbar$ is also set to one by choice of units. At this stage, one might be tempted to simply add $\mathscr{L}$ to $\mathscr{L}^{(A)}$ and have a complete theory of particle and radiation combined, but such an attempt will fail, unless of course, we know the answer and haphazardly, without any direction, we add it in. There is a better way, and that requires the last piece of the puzzle to complete the machinery of gauge theory: Noether's theorem.

For the ensuing discussion, we need two additional quantities, the conjugate field momentum,

$$p^\mu = \frac{\partial \mathscr{L}}{\partial(\partial_\mu \varphi^*)}$$

and the Hamiltonian density tensor,

$$\mathscr{H}_\nu^\mu = p^\mu \partial_\nu \varphi + (p^\mu)^* \partial_\nu \varphi^* - \delta_\nu^\mu \mathscr{L}. \tag{15.44}$$

With these quantities, we can state (without proof) Noether's theorem [99]. If we transform the coordinates and the field variables using a member of the Lorentz group of transformations, e. g., a rotation of the space axis, and we consider the infinitesimal changes to the space and field variables,

$$x^\mu \to x^\mu + \tau^\mu, \quad \varphi \to \varphi + \xi,$$

where τ^μ and ξ are the generators of the Lie algebra connected to the group of transformations, then the so-called Noether's currents

$$p^\mu \xi + (p^\mu)^* \xi^* - \mathcal{H}^\mu_\nu \tau^\nu \tag{15.45}$$

are constants, i. e., conserved quantities. To see how this works, consider a scalar transformation of φ and φ^* as performed by the member of the $U(1)$ group,

$$\varphi' = e^{ie\epsilon}\varphi, \quad (\varphi^*)' = e^{-ie\epsilon}\varphi^*,$$

where e is the fundamental charge, and ϵ is a constant parameter. Then the infinitesimal generator ξ can be obtained by expanding the last equation up to the linear term

$$\varphi' \approx \varphi + ie\epsilon\varphi, \quad (\varphi^*)' \approx \varphi^* - ie\epsilon\varphi^*,$$

so that ξ is just the constant $ie\epsilon$ and $\xi^* = -\xi$. Since we do not change the space variables and $p^\mu = \partial^\mu \varphi$, then Noether's theorem states that

$$j^\mu = ie\epsilon(\varphi\partial^\mu\varphi - \varphi^*\partial^\mu\varphi^*)$$

are constant, namely

$$\partial_\mu j^\mu = 0.$$

This is the relativistic version of the equation of continuity written in tensor notation. To derive equation (15.6), we would need to carry out the same analysis with the Schrödinger–Lagrangian introduced in Chapter 12. The essence of gauge theory is to make the change in the field variables nonuniform. In the next chapter, we show how these types of local transformations destroy the conservation laws of the Noether currents. The restoration of them is accomplished by redefining the derivative, and this creates the correct interaction terms. We are presently ready to put all of these tools to work in our exploration.

15.8 Exercises

1. For our purposes, we do not need to prove Gauss's divergence theorem or Stoke's integral theorem. In $\mathbb{R}^3$, these proofs can be found in most engineering and mathematical physics textbooks. It takes some additional modern mathematics (which is within your grasp), to extend these tools to manifolds. This exercise is designed to explain the surface portion of the theorem. The divergence theorem regarding the flux through a piecewise smooth surface $\mathcal{S}$ enclosing a volume V of a vector field $\mathbf{F}$ in $\mathbb{R}^3$ is

$$\oint\!\!\!\oint_{S} \mathbf{F} \cdot d\mathbf{a} = \iiint_{V} (\nabla \cdot \mathbf{F})\, dV.$$

(a) A surface in $\mathbb{R}^3$ can be represented using two independent parameters u, v as a set of three functions

$$x = x(u, v), \quad y = y(u, v), \quad z = z(u, v)$$

and the infinitesimal surface vector $d\mathbf{a}$ becomes $d\mathbf{a} = \hat{\mathbf{N}} du dv$. At a point p on the surface, $p = u_0, v_0$, we can compute the two tangent vectors $\mathbf{T}_u, \mathbf{T}_v$ pointing in the direction of u and v, respectively, using the following expressions:

$$\mathbf{T}_u = \frac{\partial x}{\partial v}\mathbf{i} + \frac{\partial y}{\partial v}\mathbf{j} + \frac{\partial z}{\partial v}\mathbf{k}$$

and

$$\mathbf{T}_v = \frac{\partial x}{\partial u}\mathbf{i} + \frac{\partial y}{\partial u}\mathbf{j} + \frac{\partial z}{\partial u}\mathbf{k}.$$

The cross-product $\mathbf{T}_u \times \mathbf{T}_v$ is a vector $\mathbf{N}$ normal to the plane tangent to $\mathcal{S}$ at point p. Find an expression for $\mathbf{N}$ in terms of the partial derivatives in $\mathbf{T}_u$ and $\mathbf{T}_v$.

(b) It is possible for some simple cases to characterize the surface by a simple parametrization $z = \mathcal{S}(x,y)$, where in this case, $x = u, y = v$. Use the result in part (a) to show that

$$\mathbf{N} = -\frac{\partial \mathcal{S}}{\partial x}\mathbf{i} - \frac{\partial \mathcal{S}}{\partial y}\mathbf{j} + \mathbf{k}.$$

2. While the expression in part (b) of the last exercise is useful in some cases, when a more general parametric form for the surface is available, one needs to employ some of the machinery of Chapter 8 to evaluate the surface integral. This exercise will show you one of the reasons multiple integrals in non-Cartesian coordinates require a Jacobian factor. Let A be the area of $\mathcal{S}$, then it can be shown that

$$A = \oint\!\!\!\oint_{\mathcal{S}} |\mathbf{N}| du dv.$$

In this example, $\mathcal{S}$ is the surface of a sphere of radius r, so $A = 4\pi r^2$.

(a) Consider x, y, and z in terms of spherical polar coordinates:

$$x = r\cos\phi\sin\theta, \quad y = r\sin\phi\sin\theta, \quad z = r\cos\theta.$$

Follow the procedure in part (a) of the last exercise to prove that

$$\mathbf{N} = r^2 \cos\phi \sin^2\theta\, \mathbf{i} + r^2 \sin\phi \sin^2\theta\, \mathbf{j} + r^2 \cos\theta \sin\theta\, \mathbf{k}.$$

(b) The result in part (a) is just an intermediate step. You now need to show that **N** is perpendicular to any point θ, ϕ of the surface. To do so, you need to transform the unit basis vectors according to

$$e_\mu = \frac{\partial x^{\mu'}}{\partial x^\mu} e_{\mu'}$$

where $\{e_{\mu'}\} = (\mathbf{i}, \mathbf{j}, \mathbf{k})$. For example,

$$e_\theta = \frac{\partial x}{\partial \theta}\mathbf{i} + \frac{\partial y}{\partial \theta}\mathbf{j} + \frac{\partial z}{\partial \theta}\mathbf{k},$$

etc. In this part, just find the expressions for $e_r, e_\theta, e_\phi = \hat{\mathbf{r}}, \hat{\boldsymbol{\theta}}, \hat{\boldsymbol{\phi}}$ in terms of $\mathbf{i}, \mathbf{j}, \mathbf{k}$.

(c) Combine the results in part (a) and (b) to show that

$$\mathbf{N} = r^2 \sin\theta\, \hat{\mathbf{r}}.$$

Then insert $|\mathbf{N}|$ and $dudv = d\theta d\phi$ into the double integral. The rest should be trivial.

(d) Let $\mathcal{S}$ be the same as in parts (a) through (c) and let $\mathbf{F} = x\mathbf{i} + y\mathbf{j} - z\mathbf{k}$. Compute the two integrals in Gauss's divergence theorem and show that they are equal to each other. The resulting trigonometric integrals are well known. The volume integral is trivial, and it should make you appreciate the power of Gauss's divergence theorem.

3. Show that

$$\nabla \cdot \frac{\mathbf{R}}{R^3} = 0.$$

4. Use the result from the previous exercise to prove that for the field produced by n point particles at point $r \neq r_i$,

$$\nabla \cdot \mathbf{E} = 0.$$

5. Prove the following relationship:

$$\nabla \frac{1}{R} = -\frac{\mathbf{R}}{R^3}.$$

6. Let Φ be the potential in equation (15.10). Find an expression for $\nabla \times \mathbf{E}$.
7. Consider the scalar quantity χ, and assume the following generic form for the vector potential:

$$\mathbf{A} = A_x\mathbf{i} + A_y\mathbf{j} + A_z\mathbf{k}$$

where A_x, A_y, A_z, and χ are known and supposedly "well-behaved" functions of x, y, z, and t.

(a) Find a generic expression for $\mathbf{B}$.
(b) Find an expression for $\nabla \times \mathbf{B}$.
(c) Show that $\nabla \cdot \nabla \times \mathbf{A} = 0$.
(d) Show that $\nabla \times \nabla \chi = 0$.
(e) Find an expression for $\nabla \times \mathbf{E}$.
(f) Find an expression for $\nabla \times \nabla \times \mathbf{A}$ and compare it with $\nabla(\nabla \cdot \mathbf{A}) - \nabla^2 \mathbf{A}$.

8. This exercise demonstrates how one can derive equation (15.2) from (15.26).
   (a) Prove the following vector identity by expanding both sides for two arbitrary vectors $\mathbf{A} = A_x\mathbf{i} + A_y\mathbf{j} + A_z\mathbf{k}$ and $\mathbf{B} = B_x\mathbf{i} + B_y\mathbf{j} + B_z\mathbf{k}$,

$$\nabla \cdot (\mathbf{A} \times \mathbf{B}) = \mathbf{B} \cdot (\nabla \times \mathbf{A}) - \mathbf{A} \cdot (\nabla \times \mathbf{B}).$$

   (b) Prove that

$$\nabla \times \frac{\hat{\mathbf{R}}}{R^2} = \nabla \times \frac{\mathbf{R}}{R^3} = 0$$

   by direct computation, where $\mathbf{R} = (x - x')\mathbf{i} + (y - y')\mathbf{j} + (z - z')\mathbf{k}$, the derivatives are with respect to the unprimed (field) coordinates and zero in this equation means the zero vector. Note that we have used the first equality on the left of the last expression to define $\hat{\mathbf{R}}$ as the unit vector along $\mathbf{R}$.
   (c) Now apply the results in part (a) and (b) to show that

$$\nabla \cdot \mathbf{B} = \nabla \cdot \int_{V'} \frac{\mathbf{J}' \times \hat{\mathbf{R}}}{R^2}\, dV' = 0.$$

   You may begin by moving the divergence operator inside the integral. Then recall that $\mathbf{J}'$ does not depend on the field coordinates; rather, it is a function of the primes (source) coordinates.
9. Derive equation (15.24).
10. Show that equation (15.29) holds for a function χ that satisfies (15.27) and (15.28).
11. Prove the following vector identity by expanding both sides for an arbitrary vector $\mathbf{F} = F_x\mathbf{i} + F_y\mathbf{j} + F_z\mathbf{k}$:

$$\nabla \times (\nabla \times \mathbf{F}) = \nabla(\nabla \cdot \mathbf{F}) - \nabla^2 \mathbf{F}.$$

12. Show that $f = f_0 \cos(\mathbf{K} \cdot \mathbf{r} - \omega t)$ satisfies (15.22) if it satisfies the dispersion condition. The wave-vector $\mathbf{K} = k_x\mathbf{i} + k_y\mathbf{j} + k_z\mathbf{k}$ points in the direction of the propagation of the wave, and $\mathbf{r} = x\,\mathbf{i} + y\,\mathbf{j} + z\,\mathbf{k}$. The dispersion condition is a relationship between the components of $\mathbf{K}$ and the speed of the wave v.

13. Show that the dispersion conditions for electromagnetic waves become $\lambda\nu = c$, by using the solution of the wave equation in the previous exercise, with $\nu = c/\lambda$, and $\omega = 2\pi\nu$. You will need to find a relationship between the wavelength λ and $\mathbf{K}\cdot\mathbf{r}$
14. Derive the remaining components of equation (15.39).
15. Derive the 2-form version of the Maxwell tensor

$$F_{\mu\nu} = \begin{pmatrix} 0 & E_x & E_y & E_z \\ -E_x & 0 & -B_z & B_y \\ -E_y & B_z & 0 & -B_x \\ -E_z & -B_y & B_x & 0 \end{pmatrix}$$

in two ways:
(a) directly by using

$$F_{\mu\nu} = \partial_\mu A_\nu - \partial_\nu A_\mu$$

(b) by using the metric tensor to lower the indices

$$F_{\mu\nu} = g_{\mu\mu'} g_{\nu\nu'} F^{\mu'\nu'}.$$

It is perhaps best to observe that $g_{\mu\mu'} g_{\nu\nu'} = \delta_{\mu\mu'}\delta_{\nu\nu'}$ if $\mu, \nu > 0$ and $g_{\mu\mu'} g_{\nu\nu'} = -\delta_{\mu\mu'}\delta_{\nu\nu'}$ if either μ or ν equal zero. The rest should be trivial.
16. Show that

$$\mathscr{L}^{(A)} = \frac{1}{2\mu_0}\left(|\mathbf{E}|^2 - |\mathbf{B}|^2\right).$$

Begin by writing out explicitly all sixteen terms of the double sum $F_{\mu\nu}F^{\mu\nu} = F_{00}F^{00} + F_{01}F^{01} + \cdots + F_{33}F^{33}$. Then considering that $F^{\mu\nu}$ and $F_{\mu\nu}$ are both antisymmetric, you should end up with only six surviving terms. Complete your proof by using the result in the previous exercise and equation (15.39).
17. Evaluating the derivatives of $\mathscr{L}^{(A)}$ with respect to $\partial_\mu A_\nu$ is not exactly trivial. It helps to write out the surviving terms of the Lagrangian explicitly by expanding the sum:

$$\begin{aligned} F_{\mu\nu}F^{\mu\nu} &= 2(\partial_0 A_1 - \partial_1 A_0)(\partial^0 A^1 - \partial^1 A^0) + 2(\partial_0 A_2 - \partial_2 A_0)(\partial^0 A^2 - \partial^2 A^0) + \cdots \\ &\quad + 2(\partial_2 A_3 - \partial_3 A_2)(\partial^2 A^3 - \partial^3 A^2). \end{aligned}$$

(a) Consider evaluating

$$\frac{\partial F_{\mu\nu}F^{\mu\nu}}{\partial(\partial_0 A_1)}$$

and note that $\partial^0 A^1 = -\partial_0 A_1$. Using the product rule, show that

$$\frac{\partial F_{\mu\nu}F^{\mu\nu}}{\partial(\partial_0 A_1)} = 2(\partial^0 A^1 - \partial^1 A^0) + 2(\partial_0 A_1 - \partial_1 A_0)(-1) = 4F^{01}.$$

(b) Use this strategy to evaluate

$$\frac{\partial F_{\mu\nu}F^{\mu\nu}}{\partial(\partial_2 A_3)}.$$

18. In this exercise, we explore how Noether's theorem is implemented in practice.
    (a) Derive an expression for the Hamiltonian tensor density in equation (15.44) using the Lagrangian in (15.40).
    (b) The three-dimensional version of the Schrödinger–Lagrangian of Chapter 12 is

$$\begin{aligned}\mathcal{L} = &-\frac{\hbar^2}{2m}\left[(\partial_x\psi)\partial_x\psi^* + (\partial_y\psi)\partial_y\psi^* + (\partial_z\psi)\partial_z\psi^*\right]\\ &+ i\frac{\hbar}{2}(\psi^*\partial_t\psi - \psi\partial_t\psi^*) - U\psi\psi^*.\end{aligned}$$

    Find the expressions for p^μ and its complex conjugate. Remember that the configuration space is $\mathbb{R}^3$, t is just a parameter. However, it is helpful to let p^μ be a four-vector, with

$$p^0 = \frac{\partial\mathcal{L}}{\partial(\partial_t\psi)}$$

    as in the relativistic case.
    (c) Use Noether's theorem to derive the conserved currents when the field variables are gauged uniformly using a member of SU(1). Show that the result is equation (15.6).

16 Nonrelativistic gauge theory

> No one undertakes research in physics with the intention of winning a prize. It is the joy of discovering something no one knew before.
>
> Stephen Hawking

We begin this chapter by explaining how the traditional, fully relativistic scalar gauge theory is developed and interpreted. In the last chapter, we have worked with two Lagrangian functionals, one for the derivation of Maxwell's equation and another for the derivation of the Kline–Gordon equation for a spinless particle. As we mention there, we cannot simply add the two together, because the electromagnetic radiation-matter coupling terms would be missing. Here, we demonstrate how such interaction is derived by a type of change of dependent variables that goes under the name of a "local" $U(1)$ gauge. This process requires the introduction of the covariant derivative so that the Lagrangian remains invariant under the transformation. The local phase change is interpreted as the result of the particle absorbing a photon. A phase change is indeed a change in the state of the particle, because the current is impacted, i. e., the particle has gained some momentum. Pure relativistic gauge theory is a scattering theory, the phase change is in a continuous range of energies. This means that absorbing a photon of any wavelength changes the quantum state of a particle from a scattering state with energy E to another "excited" scattering state with energy $E + \hbar\omega$ for any frequency ω of the photon.

Our interest in gauge theory is to explore the possibility of constructing accurate wave functions for bound states of a nonrelativistic quantum system. There are some significant differences between the scattering relativistic and the bound nonrelativistic scenarios. For bound states, an arbitrary photon frequency will create a complicated mixture of ground and excited states, which in a conservative system oscillates in real time from one state to another. In laboratory settings, at finite temperatures these virtual states are short lived, as the system relaxes back to its equilibrium mixture. More specifically, our goal is to construct accurate approximations for the wave functions of excited states from the ground state, since we can find a relatively accurate approximation for it with stochastic methods. To begin the exploration, we derive an expression for the covariant derivative for $U(1)$ field gauges applied to the Schrödinger–Lagrangian. It is straightforward to demonstrate that the change of dependent variable can be interpreted as a special type of guiding function. However, unlike the scattering state case, the scalar gauge theory does not have sufficient structure to construct approximate representations of excited states. Therefore, we derive the interaction terms form two types of non-Abelian gauge theories, then we derive the corresponding Green functions that may be used to propagate a number of states simultaneously.

Before we embark on this laborious task, it is reasonable to ask why that is even necessary in the first place. After all, the nonrelativistic matter-electromagnetic radia-

https://doi.org/10.1515/9783111610207-016

tion interaction Hamiltonian can be found in textbooks [92–94, 100] typically written for a single particle with mass m and charge q:

$$H = \frac{1}{2m}[\mathbf{P} - q\mathbf{A}]^2 + U - \frac{q}{m}\mathbf{S} \cdot \mathbf{B} \tag{16.1}$$

where $\mathbf{A}$ is the 3-vector potential, $\mathbf{S}$ is the spin, and $\mathbf{B}$ the magnetic field. Expanding the square in the last equation, and being careful with commutations one can derive the Schrödinger equation in the form $H = H_0 + W(t)$ where H_0 is the radiation-free Hamiltonian. These expressions are useful when H_0 is sufficiently simple that we can compute eigenvalues and eigenvectors and use perturbation theory to derive a number of insightful and physically meaningful results. Some of these results are discussed in Chapter 17 when we interpret the results that we obtain in this chapter. However, one would like to know how the states of the interaction Hamiltonian can be simulated, if at all, by diffusion Monte Carlo. Or to phrase the question another way, what is the Green function that propagates the population foreward when the radiation terms are present, and what is the equilibrium state that is reached? We have found the journey to be a fruitful one, though a number of interesting questions remain to be explored.

16.1 Scalar relativistic gauge theory

Let us again consider the Lagrangian for a spinless particle

$$\mathscr{L} = \frac{1}{2}(\partial_\mu \varphi^*)(\partial^\mu \varphi) - \frac{1}{2}m^2\varphi^2, \tag{16.2}$$

along with the following change in the wave function:

$$\varphi' = \exp(ig\epsilon)\varphi,$$

where $i = \sqrt{-1}$, g is the coupling constant for the electromagnetic interaction and ϵ is a scalar function. In this case, g is the charge on the particle. It follows that $(\varphi')^* = \exp(-ig\epsilon)\varphi^*$.

The first goal is to use these modified wave functions to transform the Lagrangian in equation (16.2). We can immediately note that the term containing $\varphi^2 = \varphi^*\varphi$ is unaffected, since the phases simply cancel. Next, let us consider the special case where ϵ is uniform. Then it is straightforward to check that the derivative term of the Lagrangian is also unaffected since

$$\partial^\mu \varphi' = \partial^\mu e^{ig\epsilon}\varphi = e^{ig\epsilon}\partial^\mu \varphi \tag{16.3}$$

and

$$\partial_\mu (\varphi')^* = \partial_\mu e^{-ig\epsilon}\varphi^* = e^{-ig\epsilon}\partial_\mu \varphi^*. \tag{16.4}$$

We can clearly see that the phase cancels as well. Such a transformation goes under the name of "global gauge." We have already shown in the previous chapter that this invariance leads to the conservation of charge and currents, but it is important to repeat it here to help us understand the next steps.

For the ensuing discussion, it is sufficient (but not necessary) to consider a rather stringent definition of invariance. If upon changing the dependent variable $\varphi \rightarrow \varphi'$ using a so-called "local gauge," we define a covariant derivative D_μ that transforms the same way as the common derivative does in equation (16.3), then the modified Lagrangian

$$\mathscr{L}^{(g)} = \frac{1}{2}(D_\mu\varphi)^*(D^\mu\varphi) - \frac{1}{2}m^2\varphi^2$$

is also "invariant." Therefore, a local gauge transformation is nonuniform, meaning it is represented by an operator-valued function of configuration space. The covariant derivative is constructed as follows: $D_\mu \rightarrow \partial_\mu + igA_\mu$ where A_μ is just the vector potential of the last chapter. Note that the covariant derivative has a complex conjugate as well $D^*_\mu \rightarrow \partial_\mu - igA^*_\mu$.

We now demonstrate that $A_\mu = (A_0, \ldots, A_3)$ must transform as follows:

$$A'_\mu = A_\mu - \partial_\mu\epsilon.$$

This expression is equation (15.27), hence the name gauge theory. To prove this expression holds, we require that D_μ transforms the same way as ∂_μ for the global gauge, i. e.,

$$D'_\mu\varphi' = \exp(ig\epsilon)D_\mu\varphi. \tag{16.5}$$

Inserting the expressions for $D_\mu, D'_\mu = \partial_\mu + igA'_\mu$, and φ'_μ into equation (16.5) gives

$$(\partial_\mu + igA'_\mu)\exp(ig\epsilon)\varphi = \exp(ig\epsilon)(\partial_\mu + igA_\mu)\varphi.$$

The terms on the left after the product rule is used become

$$\exp(ig\epsilon)\partial_\mu\varphi + ig\exp(ig\epsilon)\varphi\partial_\mu\epsilon + igA'_\mu\exp(ig\epsilon)\varphi,$$

whereas the terms on the right are

$$\exp(ig\epsilon)\partial_\mu\varphi + ig\exp(ig\epsilon)A_\mu\varphi.$$

Putting the two together gives

$$\begin{aligned} & ig\exp(ig\epsilon)\varphi\partial_\mu\epsilon + igA'_\mu\exp(ig\epsilon)\varphi \\ & = ig\exp(ig\epsilon)A_\mu\varphi, \end{aligned}$$

where $\exp(ig\epsilon)\partial_\mu\varphi$ is dropped on both sides. After removing the common factor $ig\exp(ig\epsilon)\varphi$ from the remaining terms, one arrives at $\partial_\mu\epsilon + A'_\mu = A_\mu$, the intended result.

To derive an expression for the interactions, one expands the derivative terms in $\mathscr{L}^{(g)}$ and compares the resulting expression with $\mathscr{L}$ the original Lagrangian. The result is

$$\mathscr{L}^{(g)} = \mathscr{L} + j^\mu A_\mu + g^2 A_\mu A^\mu \varphi^* \varphi$$

where j^μ are Noether's currents,

$$j^\mu = ig(\varphi\partial^\mu\varphi^* - \varphi^*\partial^\mu\varphi).$$

Note that the units here are such that both $\hbar$ and c equal one. If we now combine $\mathscr{L}^{(A)}$ from Chapter 15 with $\mathscr{L}^{(g)}$,

$$\mathscr{L}^{(g)} + \mathscr{L}^{(A)} = \mathscr{L} + j^\mu A_\mu + g^2 A_\mu A^\mu \varphi^* \varphi - \frac{1}{4} F_{\mu\nu} F^{\mu\nu}$$

and derive the corresponding Euler–Lagrange equations we are presented with a coupled set of differential equations that describe the dynamics of the interaction between the particle and the electromagnetic field. As an important aside, these equations are considered "classical" by theoretical physicists. Their derivations are the typical starting point for a quantum field theory course, where the next item would be to turn position and momentum into operators for the matter particle and the carrier boson, and then these are used to generate the respective creation and annihilation operators. The coupling terms $j^\mu A_\mu + g^2 A_\mu A^\mu \varphi^* \varphi$ are treated perturbatively by computing Feynman diagrams of various order. This has been the tried and true recipe for modeling successfully high energy events in particle accelerators for more than half a century.

16.1.1 Schrödinger field Lagrangian in $\mathbb{R}^3$

The derivation of the imaginary time diffusion equation from a field Lagrangian can be generalized to a multidimensional space. For now, let that space be $\mathbb{R}^3$. Then the Greek indices $\mu = 1, 2, 3$ label the components of vectorial quantities such as, e. g., the coordinates, $x^\mu = x^1, x^2, x^3 \equiv x, y, z$, and tensor quantities such as the metric, $\delta_{\mu\nu}$, a 3×3 unit matrix. The field Lagrangian is

$$\mathscr{L} = -\frac{\hbar^2}{2m}\psi^*_\mu \delta^{\mu\nu} \psi_\nu + \frac{1}{2} i\hbar(\psi^* \psi_t - \psi^*_t \psi) - \psi^* U \psi \tag{16.6}$$

where we are adopting a shorthand to clean up some of the notation we use in Chapter 15,

$$\partial_\mu \psi \equiv \psi_\mu \equiv \frac{\partial \psi}{\partial x^\mu} \quad \partial_\mu \psi^* \equiv \psi^*_\mu \equiv \frac{\partial \psi^*}{\partial x^\mu},$$
$$\partial_t \psi \equiv \psi_t \equiv \frac{\partial \psi}{\partial t} \quad \partial_t \psi^*_t \equiv \psi^*_t \equiv \frac{\partial \psi^*}{\partial t},$$

mainly to eliminate the cumbersome looking partial derivatives of $\mathscr{L}$ with respect to $\partial_\mu \psi^*$. The notation also takes care of the ambiguity of expressions like $\partial_\mu \psi^* \delta^{\mu\nu} \partial_\nu \psi$ where it is not clear, without proper use of parentheses, what the argument of the leftmost derivative is. When there is no such ambiguity and we are not deriving equations of motions from Lagrangians, we switch to the ∂_μ notation for gradients of the field variables ψ, φ, ρ, etc. Unlike the Lagrangian in the previous section, here we are working in a three-dimensional space with a unit 3×3 matrix representing the metric tensor. Time is just a parameter, and we have used the $\mathbb{R}^3$ metric tensor inverse $\delta^{\mu\nu}$ to raise the index on the gradient of ψ,

$$\delta^{\mu\nu} \partial_\nu \psi = \partial^\mu \psi.$$

The position of Greek indices is not critically important as long as we continue to use orthogonal Cartesian coordinates.

The three-dimensional Schrödinger equation for ψ, can be derived from eq. (16.6) the same way that we do in Chapter 12 for the monodimensional case:

$$\frac{\partial \mathscr{L}}{\partial \psi^*} = \frac{1}{2} i\hbar \psi_t - U\psi, \quad \frac{\partial \mathscr{L}}{\partial \psi^*_t} = -i\frac{\hbar}{2}\psi$$
$$\frac{\partial \mathscr{L}}{\partial \psi^*_\mu} = -\frac{\hbar^2}{2m} \delta^{\mu\nu} \psi_\nu, \quad \partial_t\left(\frac{\partial \mathscr{L}}{\partial \psi^*_t}\right) = -i\frac{\hbar}{2}\psi_t$$
$$\partial_\mu\left(\frac{\partial \mathscr{L}}{\partial \psi^*_\mu}\right) = -\frac{\hbar^2}{2m} \partial_\mu \partial^\mu \psi.$$

Substituting these results into equation (12.6) yields

$$i\frac{\hbar}{2}\psi_t - U\psi = -\frac{\hbar^2}{2m}\partial_\mu \partial^\mu \psi - i\frac{\hbar}{2}\psi_t,$$

from which

$$i\hbar \psi_t = -\frac{\hbar^2}{2m} \partial_\mu \partial^\mu \psi + U\psi \tag{16.7}$$

follows trivially.

16.2 Local nonrelativistic $U(1)$ gauge theory

Following the earlier example, we transform the wave function,

$$\psi' = \exp(ig\epsilon)\psi,$$

along with its complex conjugate,

$$\psi'^* = \exp(-ig\epsilon)\psi^*.$$

The first goal is to transform the Lagrangian in equation (16.6). We note that the terms $i(\psi^*\psi_t - \psi_t^*\psi)/2$ and $\psi^* U\psi$ are unaffected by the transformation if $\partial\epsilon/\partial t = \epsilon_t = 0$. The spatial derivatives, however, are affected if we assume ϵ is a function of x^μ. We now use the same procedure explained earlier to define a covariant derivative, by requiring that D_μ transforms the same way as ψ',

$$D'_\mu\psi' = \exp(ig\epsilon)D_\mu\psi \tag{16.8}$$

where A_μ can be shown to transform according to

$$A'_\mu = A_\mu - \partial_\mu\epsilon,$$

by following exactly the same steps as in the previous example. However, note that we intentionally left the vector potential time independent. We consider the time dependent, spatially uniform case separately. The interpretation of A_μ as the vector potential is not rigorous in the nonrelativistic limit, since Maxwell's equations are Lorentz invariant, where the Schrödinger equation is not. Nevertheless, we should be able to perform such transformation on the field variables in the low velocity limit to derive physically meaningful theories and perhaps novel numerical methods. Additionally, we will not merge $\mathscr{L}^{(A)}$ with equation (16.6), rather we want to consider interpreting A_μ as an external vector potential and neglect the particle generated fields. This is commonly done when simulating spectroscopic experiments. The external field is orders of magnitude larger than that generated by the particles; the latter is simply too weak to impact the source and the particle-to-particle interactions in typical experiments.

Transforming the wave function is the approach used to derive importance sampling in diffusion Monte Carlo, were $\exp(ig\epsilon)$ plays the role of a special type of guiding wave function. In practice, importance sampling diffusion Monte Carlo does not use complex guiding functions. To make any progress, we need the expression for the gauged Schrödinger–Lagrangian, from which we derive the corresponding diffusion equation. The transformation of equation (16.6) is carried out by introducing the covariant derivatives into the kinetic energy terms of the Lagrangian,

$$\psi_\mu^*\delta^{\mu\nu}\psi_\nu \to (D_\mu\psi^*)\delta^{\mu\nu}(D_\nu\psi).$$

We can now expand the term containing the covariant derivatives. Without loss of generality, we let the vector potential be a real quantity. Therefore,

$$(D_\mu\psi^*)(D^\mu\psi) = (\psi_\mu^* - igA_\mu\psi^*)(\psi^\mu + igA^\mu\psi).$$

Note that the superscripts and subscripts on the wave functions should never be interpreted as the covariant derivative: $\psi_\mu \neq D_\mu\psi$, etc., they are only meant to represent $\partial_\mu\psi$. We now expand the product into four terms,

$$(D_\mu\psi^*)(D^\mu\psi) = \psi_\mu^*\psi^\mu + ig(A^\mu\psi_\mu^*\psi - A_\mu\psi^*\psi^\mu) + g^2A_\mu A^\mu\psi^*\psi,$$

and insert them into equation (16.6):

$$\begin{aligned}\mathscr{L} &\longrightarrow \mathscr{L}^{(g)}\\ &= -\frac{\hbar^2}{2m}\psi_\mu^*\psi^\mu - i\frac{g\hbar^2}{2m}(A^\mu\psi_\mu^*\psi - A_\mu\psi^*\psi^\mu)\\ &\quad - \frac{g^2\hbar^2}{2m}g^2A_\mu A^\mu\psi^*\psi + i\frac{\hbar}{2}(\psi^*\psi_t - \psi_t^*\psi) - \psi^*U\psi + \cdots\end{aligned} \tag{16.9}$$

or

$$\begin{aligned}\mathscr{L}^{(g)} &= \mathscr{L} - i\frac{g\hbar^2}{2m}(A^\mu\psi_\mu^*\psi - A_\mu\psi^*\psi^\mu)\\ &\quad - \frac{g^2\hbar^2}{2m}g^2A_\mu A^\mu\psi^*\psi.\end{aligned} \tag{16.10}$$

The interaction terms are very similar to those derived earlier for the relativistic case. The Euler–Lagrange equation is a modified Schrödinger equation. The relevant derivatives are

$$\begin{aligned}\frac{\partial\mathscr{L}^{(g)}}{\partial\psi^*} &= i\frac{g\hbar^2}{2m}A_\mu\psi^\mu - \frac{g^2\hbar^2}{2m}A_\mu A^\mu\psi + i\frac{\hbar}{2}\psi_t - U\psi\\ \frac{\partial\mathscr{L}^{(g)}}{\partial\psi_\mu^*} &= -\frac{\hbar^2}{2m}\psi^\mu - i\frac{g\hbar^2}{2m}A^\mu\psi\\ \frac{\partial\mathscr{L}^{(g)}}{\partial\psi_t^*} &= -i\frac{\hbar}{2}\psi,\end{aligned}$$

from which it follows that

$$\begin{aligned}\partial_\mu\left(\frac{\partial\mathscr{L}^{(g)}}{\partial\psi_\mu^*}\right) &= -\frac{\hbar^2}{2m}\partial_\mu\partial^\mu\psi - i\frac{g\hbar^2}{2m}A^\mu\psi_\mu - i\frac{g\hbar^2}{2m}(\partial_\mu A^\mu)\psi\\ \partial_t\left(\frac{\partial\mathscr{L}^{(g)}}{\partial\psi_t^*}\right) &= -i\frac{\hbar}{2}\psi_t.\end{aligned}$$

Upon combining all the terms and rearranging, we obtain the modified Schrödinger equation,

$$-i\hbar\psi_t = \frac{\hbar^2}{2m}\partial_\mu\partial^\mu\psi - U\psi$$
$$-\frac{g^2\hbar^2}{2m}A_\mu A^\mu\psi + i\frac{g\hbar^2}{m}A_\mu\partial^\mu\psi + i\frac{g\hbar^2}{2m}(\partial_\mu A^\mu)\psi, \quad (16.11)$$

where the identity $A^\mu\partial_\mu\psi = A_\mu\partial^\mu\psi$ has been used, and for ease of comparison we have changed the sign on both sides. The second line contains the interaction terms. The first is a real term, the second and the third are imaginary. Moreover, the first and the third are "potential energy-like" terms because they multiply ψ, while the second seems to be advection-like because it multiplies the gradient of ψ. To interpret the interaction terms further and to obtain the correct energy and advection expressions, we must compare this equation with the corresponding Smoluchowski operator expression. It is to this task that we now turn our attention.

16.2.1 Comparing the Smoluchowski operator with the $U(1)$ gauged ELEs

If we use twentieth century notation, we can write the Smoluchowski operator as follows:

$$\hbar\rho_\tau = \frac{\hbar^2}{2m}\partial_\mu\partial^\mu\rho - \frac{\hbar^2}{m}\partial_\mu(\rho\varphi^{-1}\partial^\mu\varphi) - E_l\rho, \quad (16.12)$$

where

$$E_l = -\varphi^{-1}\frac{\hbar^2}{2m}\partial_\mu\partial^\mu\varphi + U.$$

One can show that equation (16.11) is a special case of equation (16.12) with the following result.

Theorem 42. *If we let $\varphi = \exp(ig\epsilon)$, $\rho = \psi$, and $A_\mu = -\partial_\mu\epsilon$, equation* (16.11) *can be written as*

$$-i\hbar\psi_t = \frac{\hbar^2}{2m}\partial_\mu\partial^\mu\psi + ig\frac{\hbar^2}{m}\partial_\mu(A^\mu\psi) - E_l\psi \quad (16.13)$$

where the second term on the right is a complex drift-like term and

$$E_l = \frac{\hbar^2}{2m}(ig\partial_\mu A^\mu + g^2A_\mu A^\mu) + U$$

are the local energy terms.

Proof. First, let us note that the choice $A_\mu = -\partial_\mu \epsilon$ satisfies the symmetry requirements for the covariant derivative D_μ. With this choice,

$$\varphi^{-1}\partial_\mu \varphi = -igA_\mu$$

and

$$\varphi^{-1}\partial_\mu \partial^\mu \varphi = -ig\partial_\mu A^\mu - g^2 A_\mu A^\mu.$$

Let us consider the drift expression in equation (16.12) separately into three terms by means of the product rule along with the kinetic term of E_l,

$$-\frac{\hbar^2}{m}\partial_\mu(\rho\varphi^{-1}\partial^\mu\varphi) + \left(\varphi^{-1}\frac{\hbar^2}{2m}\partial_\mu\partial^\mu\varphi\right)\rho.$$

We need to examine each term in detail. Term 1:

$$-\frac{\hbar^2}{m}(\varphi^{-1}\partial^\mu\varphi)\partial_\mu\rho = i\frac{g\hbar^2}{m}A_\mu\partial^\mu\psi.$$

Term 2:

$$-\frac{\hbar^2}{m}(\partial_\mu\varphi^{-1})(\partial^\mu\varphi)\rho = -\frac{g^2\hbar^2}{m}A_\mu A^\mu\psi.$$

Term 3:

$$-\frac{\hbar^2}{m}(\varphi^{-1}\partial_\mu\partial^\mu\varphi)\rho = i\frac{g\hbar^2}{m}(\partial_\mu A^\mu)\psi + \frac{g^2\hbar^2}{m}A_\mu A^\mu\psi.$$

Term 4:

$$\left(\varphi^{-1}\frac{\hbar^2}{2m}\partial_\mu\partial^\mu\varphi\right)\rho = -i\frac{g\hbar^2}{2m}(\partial_\mu A^\mu)\psi - \frac{g^2\hbar^2}{2m}A_\mu A^\mu\psi.$$

Adding these four terms to $(\hbar^2/2m)\partial_\mu\partial^\mu\psi$ yields equation (16.11) after some cancellations. □

Therefore, in nonrelativistic $U(1)$ gauge theory, we can interpret A_μ as the logarithmic derivative of a guiding function rather than the gauge boson vector field. However, note that if we interpret $A_\mu = -\partial_\mu\epsilon$, as some external vector potential of electrodynamics, because its curl vanishes, the magnetic field is zero. While the comparison with the Smoluchowski operator is insightful, the drift and one term of the local energy are imaginary. How one could implement these in a method like the diffusion Monte Carlo is not obvious unless we examine the Green function propagator. That is the subject of the next section.

16.2.2 The Fourier transform of the $U(1)$ gauged Schrödinger equation

In imaginary time the $U(1)$-gauged Schrödinger equation becomes

$$\psi_t = D\partial_\mu\partial^\mu\psi + iv^\mu\partial_\mu\psi - c\psi - if\psi$$

where we have introduced the following shorthand:

$$D = \frac{\hbar}{2m}, \quad v^\mu = g\frac{\hbar}{m}A^\mu \tag{16.14}$$

and

$$c = g^2\frac{\hbar}{2m}A_\mu A^\mu + \hbar^{-1}U, \quad f = g\frac{\hbar}{2m}\partial_\mu A^\mu. \tag{16.15}$$

Assuming Cartesian coordinates, the solution in the frequency domain is

$$\widetilde{\psi}(\alpha, t) = e^{-Dt\alpha_\mu\delta^{\mu\nu}\alpha_\nu - v^\mu\alpha_\mu t - ct - ift}.$$

The inverse integral is obtained using the following multidimensional Gaussian integral theorem:

$$I(A, b) = \int d^n x \, \exp\left(-\frac{1}{2}\sum_{i,j=1}^{n} A_{ij}x_i x_j + \sum_{i=1}^{n} b_i x_i\right)$$

where A is nonsingular and symmetric:

$$I(A, b) = (2\pi)^{n/2}[\det(A)]^{-1/2}\exp\left(\frac{1}{2}\sum_{i,j=1}^{n} A_{ij}^{-1}b_i b_j\right), \tag{16.16}$$

by identifying

$$A^{\mu\nu} = 2Dt\delta^{\mu\nu}, \tag{16.17}$$

as the coupling matrix, which is diagonal in this case, and

$$b^\mu = -v^\mu t + ix^\mu \tag{16.18}$$

the linear term vector. Then the propagator becomes

$$\mathcal{K} \propto e^{-ift}e^{-ct}\exp\left(-\frac{\delta_{\mu\nu}x^\mu x^\nu}{4Dt}\right)\exp\left(-i\frac{v_\mu x^\mu}{2D}\right)\exp\left(\frac{v_\mu v^\mu}{4D}\right). \tag{16.19}$$

Notably, the imaginary drift vector gave rise to an energy-like term, in the last exponential on the right. When we put together the real part of the branching pieces of

Green's function, namely those that have an explicit linear time dependence, we observe that all the gauge interactions, i. e., terms proportional to g cancel

$$\frac{v^{\mu}v_{\mu}t}{4D} - ct = g^2 A_{\mu}A^{\mu}\frac{\hbar t}{2m} - g^2 A_{\mu}A^{\mu}\frac{\hbar t}{2m} - \frac{1}{\hbar}U = -\hbar^{-1}U$$

leaving the original potential energy to govern the branching part, and a couple of phase factors:

$$\mathscr{K} \propto e^{-Ut/\hbar}\exp\left(-\frac{\delta_{\mu\nu}x^{\mu}x^{\nu}}{4Dt}\right)\exp\left(-i\frac{gA_{\mu}x^{\mu}}{2} - ig\frac{\hbar}{2m}\partial_{\mu}A^{\mu}t\right). \tag{16.20}$$

From left to right, the first exponential function is the usual branching operator, the second is the diffusion generator we have derived in Chapter 5. There are no actual drift terms. Rather, the third exponential is a pure phase change $e^{-i\phi}$. When evaluating the ground state energy, the phase factor vanishes. We can show this by using the method in Chapter 5, using $w_i = e^{-i\phi/2}e^{-Ut/2\hbar}$. Then $w_i^2 = w_i^* w_i = e^{-Ut/\hbar}$. Therefore, a local $U(1)$ gauge transformation leaves the ground state energy and the probability density unchanged. This perhaps should not come as a surprise. After all, the scalar transformations, albeit local, are phases. What this result ultimately tells us is that the solution of the imaginary time version of equation (16.1), in the static field case for a spinless system is just the ground state of H_0; a rather disappointing result.

16.2.3 The time dependent nonrelativistic $U(1)$ gauge theory

Before leaving the spin zero-$U(1)$ gauge theory there is one last case worth exploring. Let the gauge function ϵ be a function of time only:

$$\psi' = e^{ig\epsilon}\psi, \quad \partial_{\mu}\epsilon = 0, \quad \partial_t\epsilon = A.$$

Then, inspecting the terms of the Lagrangian as in equation (16.6), we observe that both

$$\partial_{\mu}\psi^*\partial^{\mu}\psi$$

and

$$\psi^* U\psi$$

are unchanged, but the imaginary term becomes

$$i\frac{\hbar}{2}(\psi^*\partial_t\psi - \psi\partial_t\psi^*) \to i\frac{\hbar}{2}(e^{-ig\epsilon}\psi^*\partial_t e^{ig\epsilon}\psi - e^{ig\epsilon}\psi\partial_t e^{-ig\epsilon}\psi^*),$$

which unravels as follows:

$$e^{-ig\epsilon}\psi^*\partial_t e^{ig\epsilon}\psi - e^{ig\epsilon}\psi\partial_t e^{-ig\epsilon}\psi^* = \psi^*\partial_t\psi + ig\psi^*\psi\,\partial_t\epsilon - \psi\partial_t\psi^* + ig\psi^*\psi\,\partial_t\epsilon.$$

Combining terms gives

$$\mathscr{L}^{(g)} = -\frac{\hbar^2}{2m}\partial_\mu\psi^*\partial^\mu\psi + i\frac{\hbar}{2}(\psi^*\partial_t\psi - \psi\partial_t\psi^*) - \hbar g\psi^*\psi\partial_t\epsilon - \psi^* U\psi.$$

The Euler–Lagrange equation now becomes

$$i\hbar\partial_t\psi = -\frac{\hbar^2}{2m}\partial_\mu\partial^\mu\psi + (\hbar g\partial_t\epsilon + U)\psi.$$

The added branching term $\hbar g\partial_t\epsilon$ is simply an additional potential energy function. When ϵ is a function of both space and time, therefore we derive a phase shift as well as a new "potential-like" term. While the new potential term does impact the ground state energy and wave functions, it is not immediately clear how one can formulate approximations to the excited state wave functions without some additional guidance. Ideally, we would like to combine the vector potential of the electromagnetic radiation in vacuum to provide the structure for the excited state guiding function.

The analysis in the previous section has revealed that the scalar version of the theory does not have the sufficient structure to support such interpretation for A_μ. The scalar nature of the $U(1)$ gauge does not permit us to directly couple the states of the unperturbed system. Relativistic nonscalar gauge theories have been developed to connect the "internal states" of the particle. By analogy with relativistic non-Abelian gauge theories, one can imagine the bound states of a Schrödinger particle to be its "internal" states. Because significant progress has been made with the static non-Abelian version of the theory, the rest of the chapter introduces it. The time dependent version of the scalar theory is quite insightful and requires further investigation.

16.3 Non-Abelian, nonrelativistic Gauge theories

There are several possibilities to generalize gauge theories when one considers quantized internal degrees of freedom. The subject of non-Abelian gauge theory is introduced using the SU(2) example, representing isospin as the internal degree of freedom of the neutron-proton particle. The process by which the interactions between gauge bosons and matter are derived mimics the $U(1)$ gauge theory of electrodynamics. Our goal is one of generating the equivalent of a Smoluchowski operator with the symmetries that a particular Lie group imposes on the system. Therefore, we develop the nonrelativistic non-Abelian gauge theories by following the same steps used for the $U(1)$ equivalent.

There are two classes of nonrelativistic field theories when quantized internal degrees of freedom are considered. The first is derived from the Schrödinger–Lagrangian, the second from its imaginary time counterpart. The Schrödinger–Lagrangian written for a multiplet is

$$\mathscr{L} = -\frac{\hbar^2}{2m}(\partial_\mu \psi^\dagger)\partial^\mu \psi + i\frac{\hbar}{2}[\psi^\dagger \partial_t \psi - (\partial_t \psi^\dagger)\psi] - \psi^\dagger U \psi, \tag{16.21}$$

where $\psi^\dagger$ is the transpose complex-conjugate of the vector ψ. Changing $t \to -i\tau$ and $\psi^\dagger \to \psi^T$ (the transpose of a real valued vector over the internal states), one obtains

$$\mathscr{L}^{(i)} = -\frac{\hbar^2}{2m}(\partial_\mu \psi^T)\partial^\mu \psi + \frac{\hbar}{2}[(\partial_\tau \psi^T)\psi - \psi^T \partial_\tau \psi] - \psi^T U \psi. \tag{16.22}$$

We derive the Euler–Lagrange equations for the second field Lagrangian. Those of the real time Lagrangian follow a similar set of steps, and are left for practice:

$$\begin{aligned} \frac{\partial \mathscr{L}^{(i)}}{\partial \psi^T} &= -\frac{\hbar}{2}\partial_\tau \psi - U\psi, \\ \partial_\mu \left(\frac{\partial \mathscr{L}^{(i)}}{\partial \psi^T_\mu} \right) &= -\frac{\hbar^2}{2m}\partial_\mu \partial^\mu \psi, \\ \partial_\tau \left(\frac{\partial \mathscr{L}^{(i)}}{\partial \psi^T_\tau} \right) &= \frac{\hbar}{2}\partial_\tau \psi. \end{aligned}$$

After some trivial rearrangements, this yields

$$\hbar \partial_\tau \psi = \frac{\hbar^2}{2m}\partial_\mu \partial^\mu \psi - U\psi$$

which looks like the regular diffusion equation, but in vector form. Namely, if we let Roman indices $a, b, c, \ldots$ represent the components of the multiplet ψ,

$$\hbar \partial_\tau \psi^a = \frac{\hbar^2}{2m}\partial_\mu \partial^\mu \psi^a - U\psi^a, \tag{16.23}$$

where the multiplet ψ spans a vector space.

The expression in equation (16.23) demonstrates that the diffusion algorithm for each component is the same when there are no interactions.

16.3.1 Two sets of Lie groups for Gauge theory

When we perform a gauge transformation of the Schrödinger–Lagrangian, and the wave function is a multiplet, the transformation $\psi' = G\psi$ is carried out by a matrix in the internal degrees of freedom, namely

$$(\psi^a)' = G^a_b \, \psi^b$$

with the Einstein sum implied over the Roman index b. Only a selected set of matrices will fit the first symmetry requirement, meaning G must be such that

$$\psi^{\dagger\prime}\psi' = \psi^\dagger\psi.$$

When we spell this out in terms of G, we get

$$\psi^\dagger G^\dagger G\psi = \psi^\dagger\psi,$$

this means that

$$G^\dagger G = \mathbb{I}$$

where $\mathbb{I}$ is the unit matrix in the internal states space,

$$(G^\dagger)^a_b G^b_c = \delta^a_c.$$

Therefore, $G \in \mathrm{SU}(N)$.

When we perform a global or local gauge transformation of the diffusion field Lagrangian, the symmetry requirement for G is

$$\psi^{T\prime}\psi' = \psi^T\psi$$

or

$$\psi^T G^T G\psi = \psi^T\psi,$$

this means that G must be orthogonal

$$G^T G = \mathbb{I}.$$

Therefore, $G \in \mathrm{SO}(N)$. Each one of these groups can be generated by matrix exponentiation as we have seen in Chapter 4.

We use both classes of Lie groups and corresponding Lie algebras in our exploration of nonrelativistic gauge theory applied to the two field Lagrangians introduced earlier. As explained in Chapter 4, every Lie algebra is a vector space endowed with a basis set E^c,

$$A = \epsilon_c E^c$$

as the generators of the corresponding Lie groups. The structure constants for the SU(N) family are imaginary,

$$[E_a, E_b] = iC^c_{ab}E_c,$$

and are a particularly important property of the Lie algebras in local gauge theory. Those for the SO(N) are real valued tensors,

$$[E_a, E_b] = C^c_{ab}E_c.$$

16.3.2 SU(*N*) nonrelativistic gauge theory

Here, we apply the field transformation $\psi' = G\psi$ where $G = e^{ig\epsilon_a E^a}$ to the Schrödinger–Lagrangian, and we assume that $\partial_t \epsilon_a = 0$. We note that three terms, namely $(\psi^\dagger \partial_t \psi - \psi \partial_t \psi^\dagger)' = \psi^\dagger \partial_t \psi - \psi \partial_t \psi^\dagger$ and $(\psi^\dagger U \psi)' = \psi^\dagger U \psi$ remain unchanged. However, the derivatives terms are affected if $\partial_\mu \epsilon_a \neq 0$. The covariant derivative operator in now a $N \times N$ matrix (we are omitting the Roman indices for their components for clarity):

$$(\partial_\mu \psi^\dagger)\partial^\mu \psi \to (D_\mu^\dagger \psi^\dagger)(D^\mu \psi^\mu)$$

with the auxiliary field W_μ,

$$D_\mu = \partial_\mu + igW_\mu.$$

The ∂_μ term multiplies the identity matrix in the internal state space, whereas each component of the vector W_μ is a $N \times N$ matrix. The first task is to determine how W_μ changes under a gauge transformation. By imposing the symmetry requirement,

$$D'_\mu \psi' = G D_\mu \psi$$

we arrive at the following rule.

Theorem 43. *The auxiliary field W_μ transforms as follows:*

$$W'_\mu = G W_\mu G^{-1} + \frac{i}{g}(\partial_\mu G)G^{-1}.$$

Proof. Beginning with

$$D'_\mu \psi' = G D_\mu \psi$$

and inserting the definition of D_μ and $\psi^\dagger$,

$$(\partial_\mu + igW'_\mu)G\psi = G(\partial_\mu + igW_\mu)\psi$$

distributing but being careful not to permute the order of multiplication among the various matrix-like terms,

$$\partial_\mu G\psi + igW'_\mu G\psi = G\partial_\mu \psi + igGW_\mu \psi$$

making use of the product rule and canceling terms that appear on both sides gives

$$(\partial_\mu G)\psi + igW'_\mu G\psi = igGW_\mu \psi.$$

Dropping ψ from all terms and multiplying by the matrix G^{-1} from the right finally yields

$$(\partial_\mu G)G^{-1} + igW'_\mu = igGW_\mu G^{-1}.$$

The theorem follows after some trivial rearrangements. □

If we rewrite the group operation G as an exponential over the generators of the Lie group, we get

$$W'_\mu = e^{ig\epsilon^a E_a} W_\mu e^{-ig\epsilon^a E_a} + \frac{i}{g}(\partial_\mu e^{ig\epsilon^a E_a})e^{-ig\epsilon^a E_a}.$$

The derivative of an exponential map such as $\partial_\mu \exp(A)$ is quite messy because the matrix-valued function A typically does not commute with its derivative, $[A, \partial_\mu A] \neq 0$. To make progress, one expands the exponential map into a power series and keeps only the linear terms,

$$\begin{aligned}
W'_\mu &= (\mathbb{I} + ig\epsilon^a E_a)W_\mu(\mathbb{I} - ig\epsilon^a E_a) - (\partial_\mu \epsilon^a)E_a(\mathbb{I} - ig\epsilon^a E_a)\\
W'_\mu &= W_\mu + ig\epsilon^a E_a W_\mu - W_\mu ig\epsilon^a E_a - E_a\partial_\mu\epsilon^a\\
W'_\mu &= W_\mu + ig\epsilon^a[E_a, W_\mu] - E_a\partial_\mu\epsilon^a.
\end{aligned}$$

This equation suggests that W_μ is in $\mathscr{A}$, and as such, it has its own expansion in terms of the generators,

$$W_\mu = w^b_\mu E_b.$$

Then

$$w^{a\prime}_\mu E_a = w^a_\mu E_a + ig\epsilon^a w^b_\mu [E_a, E_b] - E_a\partial_\mu\epsilon^a,$$

or using the structure constants in place of the commutator,

$$w^{a\prime}_\mu E_a = w^a_\mu E_a - g\epsilon^a w^b_\mu C^c_{ab}E_c - E_a\partial_\mu\epsilon^a. \tag{16.24}$$

The gauged Lagrangian

$$\mathscr{L}^{(g)} = -\frac{\hbar^2}{2m}(D^\dagger_\mu\psi^\dagger)D^\mu\psi + i\frac{\hbar}{2}(\psi^\dagger\partial_t\psi - \psi\partial_t\psi^\dagger) - \psi^\dagger U\psi$$

is expanded, and the Euler–Lagrange equations are derived the same way as the $U(1)$ case. The corresponding diffusion equation derived by changing to imaginary time is

$$\hbar\partial_t\psi = \frac{\hbar^2}{2m}\partial_\mu\partial^\mu\psi + i\frac{\hbar^2}{m}g(W^\mu\partial_\mu\psi) - E_l\psi$$

where

$$E_l = U + \frac{\hbar^2}{2m} g^2 W^\mu W_\mu + i\frac{\hbar^2}{2m} g(\partial_\mu W^\mu).$$

These equations are similar to those for the $U(1)$ nonrelativistic gauge theory; once more we are presented with an imaginary drift and an imaginary potential energy. However, there are some key differences. In particular, each term of the Schrödinger equation is a scalar in the configuration space as evidenced by the contractions over the Greek indices, but each term is a vector in the internal quantized space. Moreover, each vectorial quantity in the drift and branching terms is the result of a product of a matrix with the vector ψ^a in the internal quantized space. To better evidence this, we introduce the Roman indices in the notation,

$$\hbar\partial_t \psi^a = \frac{\hbar^2}{2m}\partial_\mu\partial^\mu \delta^a_b \psi^b + i\frac{\hbar^2}{m} g(\partial_\mu W^\mu)^a_b \psi^b - (E_l)^a_b \psi^b.$$

At this point, we investigate if the results obtained for the nonrelativistic $U(1)$ gauge theory apply for the SU(N) cases. To continue, we transform the coupled set of differential equation.

16.3.3 Fourier transform of the SU(N) nonrelativistic gauge theory

Let us begin by writing the generic coupled diffusion equations of the form,

$$\partial_t \psi^a = D\partial_\mu\partial^\mu \delta^a_b \psi^b + i(v^\mu)^a_b \partial_\mu \psi^b - (c)^a_b \psi^b$$

where $D = \hbar(2m)^{-1}$, $(v^\mu)^q_b = \hbar g(W^\mu)^a_b/m$, and $c = E_l/\hbar$ for the case in hand. The Fourier transform of the coupled system is

$$\partial_t \tilde{\psi}^a = -D\alpha_\mu\alpha^\mu \delta^a_b \tilde{\psi}^b - \alpha_\mu (v^\mu)^a_b \tilde{\psi}^b - (c)^a_b \tilde{\psi}^b.$$

The propagator for a multiplet is clearly a matrix in the internal space,

$$\psi_a(x^\mu_{k+1}) = \int d^n x_k \mathcal{K}^a_b\, \psi^b(x^\mu_k)$$

where

$$\mathcal{K}^a_b \propto \int d^n\alpha \left\{\exp(-D\alpha_\mu\alpha^\mu t - \alpha_\mu v^\mu t + i\alpha_\mu x^\mu - ct)\right\}^c_b.$$

The exponent in the integrand requires a closer look. We can write it as

$$-Dt\alpha_\mu\alpha_\nu\delta^{\mu\nu}\delta^c_b + (b^\mu)^c_b \alpha_\mu$$

where $(b^\mu)^c_b = -(v^\mu t)^c_b + ix^\mu \delta^c_b$. The $-ct$ term in the exponent can be split in to e^{-ct} and taken out of the integral. To further inspect the structure of the matrix in the quadratic form, we consider the following notation:

$$-Dt\alpha^\mu \alpha_\mu \,\delta^b_a + (b^\mu)^b_a \alpha_\mu.$$

In other words, the size of the matrix in the exponent is the same as the size of the matrix representation of the su(N) Lie algebra. The inverse Fourier transform integral can be evaluated even for this case by a change of coordinates similar to the approach needed to prove the quadratic form Gaussian integral in Chapter 11. Unlike that case, the exponent is a $N \times N$ matrix, not a scalar quantity. Therefore, it is important to walk through the proof in detail. Care must be used to ensure the order of multiplication between matrices is handled properly.

Theorem 44. *The following result holds,*

$$\int d^n\alpha \exp(-D\alpha_\mu \alpha^\mu t - \alpha_\mu v^\mu t + i\alpha_\mu x^\mu - ct)^b_a$$
$$\propto \left(\frac{\pi}{Dt}\right)^{m/2} \{e^{-x_\mu x^\mu/(4Dt)} \times e^{-iv_\mu x^\mu/(2D) - ct + v_\mu v^\mu t/(4D)}\}^b_a. \tag{16.25}$$

Proof. The process of simplifying the quadratic form begins with finding the values of α that minimizes it, namely

$$\frac{\partial}{\partial \alpha_\sigma}[-Dt\alpha^\mu \alpha_\mu \,\delta^b_a + (b^\mu)^b_a \alpha_\mu] = 0,$$

with $b^\mu = v^\mu t + ix^\mu$. Using the resulting equation for α to remap the frequency space with the proper change of variables, i. e., the derivative of the quadratic form set to zero,

$$-2Dt\alpha^\sigma \,\delta^b_a + (b^\sigma)^b_a = 0$$

yields an equation for α,

$$\alpha^\sigma \,\delta^b_a = \frac{1}{2Dt}(b^\sigma)^b_a$$

to which we add a new set of variables β^μ,

$$\alpha^\mu \,\delta^b_a = \frac{1}{2Dt}(b^\mu)^b_a + \beta^\mu \,\delta^b_a.$$

We can lower the Greek index in the usual way on both sides of this equation using the Euclidean metric tensor, $\alpha_\mu = \delta_{\mu\nu}\alpha^\nu$. Note that the integral measure does not change, $d^n\alpha = d^n\beta$. When the two expressions for α are substituted into the quadratic form, we get

$$
\begin{aligned}
&-Dt\alpha^\mu\alpha_\mu\,\delta_a^b + (b^\mu)_a^b\alpha_\mu \\
&\quad = -Dt\left(\frac{1}{2Dt}(b_\mu)_a^b + \beta_\mu\,\delta_a^b\right)\left(\frac{1}{2Dt}(b^\mu)_a^b + \beta^\mu\,\delta_a^b\right) \\
&\qquad + \frac{1}{2Dt}(b^\mu b_\mu)_a^b + (b^\mu)_a^b\beta_\mu.
\end{aligned}
$$

After some multiplications and cancellations, we arrive at the following expression:

$$
-Dt\alpha^\mu\alpha_\mu\,\delta_a^b + (b^\mu)_a^b\alpha_\mu = \frac{1}{4Dt}(b^\mu b_\mu)_a^b - Dt\beta_\mu\beta^\mu\,\delta_a^b.
$$

When b^μ is expressed in terms of the drift and x^μ,

$$
(b_\mu b^\mu)_a^b = -x_\mu x^\mu\,\delta_a^b - 2i(v_\mu)_a^b x^\mu t + (v_\mu v^\mu)_a^b t^2
$$

and inserted into the last expression for the exponent we are presented with

$$
\begin{aligned}
&-Dt\alpha^\mu\alpha_\mu\,\delta_a^b + (b^\mu)_a^b\alpha_\mu \\
&\quad = -\frac{1}{4Dt}x_\mu x^\mu\,\delta_a^b - i\frac{1}{2D}(v_\mu)_a^b x^\mu + \frac{1}{4D}(v_\mu v^\mu)_a^b t - Dt\beta_\mu\beta^\mu\,\delta_a^b.
\end{aligned}
$$

The theorem follows by inserting this expression into the exponent of the integrand, using the fact that the matrices $\beta_\mu\beta^\mu\,\delta_a^b$ and $x_\mu x^\mu\,\delta_a^b$ commute, in the $\Delta t \to 0$ limit, with all the other matrices, allowing us to break up the exponential into separate exponential functions, and finally moving all the terms that do not depend on β out of the integral sign.

The remaining integral over β^μ can be evaluated analytically,

$$
\begin{aligned}
\int d^n\beta\, e^{(-Dt\beta_\mu\beta^\mu\,\delta_a^b)} &= \int d^n\beta \begin{pmatrix} e^{(-Dt\beta_\mu\beta^\mu)} & 0 & \cdots \\ 0 & e^{(-Dt\beta_\mu\beta^\mu)} & 0 \\ \vdots & \vdots & \vdots \end{pmatrix}_a^b \\
&= \begin{pmatrix} \int d^n\beta\, e^{(-Dt\beta_\mu\beta^\mu)} & 0 & \cdots \\ 0 & \int d^n\beta\, e^{(-Dt\beta_\mu\beta^\mu)} & 0 \\ \vdots & \vdots & \vdots \end{pmatrix}_a^b = \delta_a^b \int d^n\beta\, e^{(-Dt\beta_\mu\beta^\mu)}
\end{aligned}
$$

where the last is a standard multidimensional Gaussian integral with a scalar argument. When this is evaluated, a proportionality constant for $\mathcal{K}$ is the result

$$
\int d^n\beta\, e^{(-Dt\beta_\mu\beta^\mu\,\delta_a^b)} = \delta_a^b\left(\frac{\pi}{Dt}\right)^{n/2}.
$$

However, its value does not impact the random walk algorithm and can be set to one in our implementation without loss of generality. □

The business end of the propagator is in the exponent outside of the integral in equation (16.25). The diffusion part can be written as follows:

$$\begin{pmatrix} e^{-x_\mu x^\mu/(4Dt)} & 0 & \cdots \\ 0 & e^{-x_\mu x^\mu/(4Dt)} & 0 \\ \vdots & \vdots & \vdots \end{pmatrix}_a^b ,$$

indicating that the diffusion for each component of ψ^a can be carried out separately. The rest of the exponent simplifies further when we replace the expressions for D, v^μ, and c,

$$\begin{aligned} &-iv_\mu x^\mu/(2D) - ct + v_\mu v^\mu t/(4D) \\ &= -igW_\mu x^\mu - \frac{U}{\hbar}t - \frac{\hbar g^2}{2m}W_\mu W^\mu t - i\frac{\hbar g}{2m}\partial_\mu W^\mu t + \frac{\hbar g^2}{2m}W_\mu W^\mu t \\ &= -igW_\mu x^\mu - \frac{U}{\hbar}t - i\frac{\hbar g}{2m}\partial_\mu W^\mu t. \end{aligned}$$

As in the $U(1)$ theory, the quadratic terms cancel leaving the potential U and two terms explicitly proportional to i. However, unlike the $U(1)$ case, these remaining linear perturbations are not purely imaginary, rather they are complex in general, and are not necessarily diagonal.

The form of the propagator for a static SU(N) gauge theory is

$$\mathscr{K} \propto e^{-x_\mu x^\mu/(4Dt)} e^{-Ut/\hbar} \exp\left[-igW_\mu x^\mu - ig\frac{\hbar}{2m}(\partial_\mu W^\mu)t\right]. \tag{16.26}$$

It is clear from this expression that when the coupling constant g is small enough to permit the perturbation to be ignored, one recovers the traditional diffusion Monte Carlo algorithm for each of the components ψ^a. In fact, ψ can be written as a product of a function of space and the internal state vector. The quantization of each part can then be carried out separately. When g is such that it cannot be ignored, the operator on the right of equation (16.26) rotates the state vector into a mixture of ground and excited states as we demonstrate in the next chapter. If we gauge the Lagrangian with a time dependent phase, it can be shown that the time derivative terms of $\mathscr{L}^{(g)}$ are

$$\psi^\dagger\partial_t\psi - (\partial_t\psi^\dagger)\psi \to \psi^\dagger(\partial_t\psi) - (\partial_t\psi^\dagger)\psi + 2ig\psi^\dagger(\partial_t\epsilon^a)E_a\psi \tag{16.27}$$

by following the same procedure we use for the scalar gauge theory case.

16.4 The general SO(N) gauge theory

There are seemingly small differences between the SU(N) and the SO(N) gauge theories, but as we work through the field transformations

$$\psi' = G\psi,$$

we discover that the outcomes are perhaps simpler to implement into numerical strategies for excited states. Unlike the SU(N) case, $G = e^{g\epsilon_a E^a}$. If $\partial_t \epsilon_a = 0$, the covariant derivative operator is $D_\mu = \partial_\mu + gW_\mu$. Theorem 43 becomes

$$W'_\mu = GW_\mu G^{-1} - \frac{1}{g}(\partial_\mu G)G^{-1}. \tag{16.28}$$

If we rewrite the group operation G as an exponential over the generators of the Lie group once more, we get

$$W'_\mu = e^{g\epsilon^a E_a} W_\mu e^{-g\epsilon^a E_a} - \frac{1}{g}(\partial_\mu e^{g\epsilon^a E_a})e^{-g\epsilon^a E_a}.$$

Expanding the exponential functions up to linear terms, using $W_\mu = w_\mu^b E_b$, and performing the same steps as in the SU(N) case, we obtain

$$w_\mu^{a\prime} E_a = w_\mu^a E_a + g\epsilon^a w_\mu^b C_{ab}^c E_c - \partial_\mu \epsilon^a E_a, \tag{16.29}$$

which is similar to equation (16.24).

The gauged Lagrangian

$$\mathscr{L}^{(g)} = -\frac{\hbar^2}{2m}(D_\mu\psi)^\dagger D^\mu\psi + i\frac{\hbar}{2}[\psi^\dagger\partial_t\psi - (\partial_t\psi^\dagger)\psi] - \psi^\dagger U\psi \tag{16.30}$$

expands to

$$\begin{aligned}\mathscr{L}^{(g)} = &-\frac{\hbar^2}{2m}[(\partial_\mu\psi^\dagger)\partial^\mu\psi + g(\partial_\mu\psi^\dagger)W^\mu\psi - g\psi^\dagger W_\mu\partial^\mu\psi - g^2\psi^\dagger W_\mu W^\mu\psi] \\ &+ i\frac{\hbar}{2}[\psi^\dagger\partial_t\psi - (\partial_t\psi^\dagger)\psi] - \psi^\dagger U\psi,\end{aligned} \tag{16.31}$$

where we use

$$D^\mu\psi = \partial^\mu\psi + gW^\mu\psi \tag{16.32}$$

and

$$(D_\mu\psi)^\dagger = \partial_\mu\psi^\dagger - g\psi^\dagger W_\mu. \tag{16.33}$$

The Schrödinger equation is

$$\hbar\partial_t\psi = \frac{\hbar^2}{2m}\partial_\mu\partial^\mu\psi + \frac{\hbar^2}{m}gW^\mu\partial_\mu\psi - E_l\psi \tag{16.34}$$

where

$$E_l = U - \frac{\hbar^2}{2m} g(\partial_\mu W^\mu) - \frac{\hbar^2}{2m} g^2 W_\mu W^\mu. \tag{16.35}$$

The steps to arrive at the Green function are similar to those followed for the SU(N) case and are left to the reader. The end result is

$$\mathcal{K}_b^a \propto e^{-Ut/\hbar} \exp\left(-\frac{m}{2\hbar t} x^\mu x_\mu\right) \exp\left[-gW_\mu x^\mu + \frac{\hbar}{2m} g(\partial_\mu W^\mu) t\right]. \tag{16.36}$$

This expression resembles the operator we derive for the SU(N) gauge theory, but there are no imaginary terms. Instead, the anti-Hermitian matrices of equation (16.26) are replaced with real skew-symmetric ones of the same order. These two results suggest strategies to approximate excites states when an accurate ground state wave function is available as we demonstrate in the subject of the next chapter.

16.5 Exercises

1. Use equations (16.16), (16.17), and (16.18) to derive (16.19).
2. Derive equation (16.20) from (16.19) using (16.14) and (16.15).
3. Derive equation (16.28) by using the same strategy used to prove Theorem 43.
4. Derive equation (16.31) starting from (16.30) and using equations (16.32) and (16.33). Note that the † operation is a transpose-complex conjugate transformation, and generally, for a matrix A and a vector b, $(Ab)^T = b^T A^T$.
5. Derive equations (16.34) and (16.35) by following the steps below:
   (a) Show that the relevant derivatives of the Lagrangian are

$$\frac{\partial \mathcal{L}^{(g)}}{\partial \psi_\mu^\dagger} = -\frac{\hbar^2}{2m}\partial^\mu \psi - \frac{\hbar^2}{2m} g W^\mu \psi$$

$$\frac{\partial \mathcal{L}^{(g)}}{\partial \psi_t^\dagger} = -i\frac{\hbar}{2}\psi$$

$$\frac{\partial \mathcal{L}^{(g)}}{\partial \psi^\dagger} = \frac{\hbar^2}{2m} g W_\mu \partial^\mu \psi + \frac{\hbar^2}{2m} g^2 W_\mu W^\mu \psi + i\frac{\hbar}{2}\partial_t \psi - U\psi.$$

   (b) Use the Euler–Lagrange equations

$$\partial_\mu \left(\frac{\partial \mathcal{L}^{(g)}}{\partial \psi_\mu^\dagger}\right) + \partial_t \left(\frac{\partial \mathcal{L}^{(g)}}{\partial \psi_t^\dagger}\right) = \frac{\partial \mathcal{L}^{(g)}}{\partial \psi_t^\dagger}$$

   to arrive at

$$i\hbar\partial_t \psi = -\frac{\hbar^2}{2m}\partial_\mu \partial^\mu \psi + U\psi - \frac{\hbar^2}{2m} g(\partial_\mu W^\mu)\psi - \frac{\hbar^2}{m} g W^\mu \partial_\mu \psi - \frac{\hbar^2}{2m} g^2 W_\mu W^\mu \psi.$$

(c) Switch to imaginary time, $t = -it$, and change signs on both sides to arrive at the desired result.

6. Derive equation (16.27). The proper way to express the transformation of the time derivative term of the Lagrangian for the nonrelativistic SU(N) theory is below:

$$\psi^\dagger \partial_t \psi - (\partial_t \psi^\dagger)\psi \to (e^{ige^a E_a}\psi)^\dagger \partial_t e^{ige^a E_a}\psi - [\partial_t (e^{ige^a E_a}\psi)^\dagger] e^{ige^a E_a}\psi.$$

Terms such as $(e^{ige^a E_a}\psi)^\dagger$ develop into $\psi^\dagger e^{-ige^a E_a}$ and terms like $\partial_t(e^{ige^a E_a}\psi)^\dagger$ become $\partial_t(\psi^\dagger e^{-ige^a E_a})$. Use these expressions, expand all the exponential functions to first order, and distribute all the terms, being careful to respect the order of multiplication.

17 Excited states from the Stark effect

It's unbelievable how much you don't know about the game you've been playing all your life.

Mickey Mantle

In the last chapter, we derive the modified Schrödinger equations for several types of nonrelativistic gauge transformations, and we obtain the corresponding expressions for imaginary time propagators. We observe that the scalar gauge theory does not have the proper structure to construct excited states. The SU(N) and SO(N) gauge theories provide us with a coupling matrix for the mixing of a state vector ψ^a. However, we never explicitly choose a vector space to represent ψ^a, nor do we attempt to introduce an expression for the vector potential A_μ or W_μ. In this chapter, we suggest two possible, physically meaningful vector potentials, and we explore possible interpretations of the propagators we derived in the last chapter. We first explore a static external electric field, and we demonstrate that the interaction terms simply create mixtures of ground and excited states. But we can untangle these mixtures, at least approximately, by orthogonalizing. This seems a natural choice given what the Sturm–Liouville theorem tells us about nondegenerate excited states. The result are the approaches of Chapter 7. In fact, it is simple to see that the linear expansion of the coupling terms results in the functional forms adopted in Chapter 7 for the first excited states as a precise mathematical limit. This is the most important result from this chapter; a direct connection between the first principles explored in Chapters 3, 4, 12, 15, and 16 and the state of the art excited bosonic state strategies in diffusion Monte Carlo, such as the ground state probability amplitude approach, when problems with more than one variable are considered.

We next propose a possible way to improve on the methods of Chapter 7 by expanding the size of the orthogonal basis set and show that it is possible to create a convergent representation of the excited state wave function. When systems with more than one degree of freedom are considered, the matter is complicated by the ordering of polynomials. We demonstrate how product polynomials can be used to find excited states, and we demonstrate how one can achieve this by the generalized eigenvalue problem. Then we discuss a particular form of orthogonal polynomials of many variables that can be used to ease the automation of the method to $\mathbb{R}^d$ so that second excitations or higher if needed for greater accuracy can be easily included. Finally, the dynamic version of the Stark effect and the difficulties of its implementation in a DMC simulation are briefly discussed.

17.1 The linear terms of the SO(N) theory

Perhaps the most straightforward way is to begin with the SO(N) propagator since there are no complex quantities for us to handle,

https://doi.org/10.1515/9783111610207-017

$$\mathcal{K}_b^a \propto e^{-Ut} \exp\left(-\frac{m}{2\hbar t} x^\mu x_\mu\right) \exp\left[-gW_\mu x^\mu + \frac{\hbar}{2m} g(\partial_\mu W^\mu)t + (\hbar g J_a \partial_t \epsilon^a)t\right]. \tag{17.1}$$

Here, J_a is one of the generators of the SO(N) Lie group. There are well- known isomorphisms between the SO(N') and the SU(N) groups and their corresponding Lie algebras; therefore, the choice among the two possibilities is immaterial, as is the size N for that matter. The SO(2) is sufficient to derive the results we need for one dimension as we explain later. We have included the most general case where the gauge transformation is a function of both x^μ and t. The quantum theory of matter-radiation interactions can be found in a number of excellent textbooks, and its presentation normally begins with Plank's radiation law and the phenomenological representation of the absorption, emission, and stimulated emission processes in terms of the Einstein coefficients. It is then typical to use time dependent quantum mechanics to find expressions for the absorption and emission coefficients in terms of the square of the dipole matrix element between two stationary states of the unperturbed system. Such calculations are very insightful, and for those cases where the states of the unperturbed system can be found, they provide the way of computing the coefficients for simple optical processes directly.

Presently, we seek to achieve the converse, i. e., we assume only the ground state is known to us, we introduce a physically meaningful perturbing vector potential and determine the form of the excited state wave functions that follow from mathematical and established physical principles. At this point, we have accumulated a considerable amount of machinery to proceed with the endeavor. We have a way of computing an excellent representation of the ground state wave function ψ_0, we have at our disposal a basis set we can use to compute matrix elements in terms of monomial integrals, and we have the necessary group theory knowledge to construct viable solutions. As we have argued in Chapter 16, the vector potential W_μ that best represents a typical spectroscopy experiment is the solution of the source-free Maxwell equations. Namely, we can simply consider

$$\epsilon^a = A \sin(k_\mu x^\mu - \omega t).$$

The k_μ components are those of the propagation vector, and $x^\mu = x, y, z$, and its size equals $2\pi/\lambda$ where λ is the wavelength of the external radiation. Since we are not interested in scattering states, or x-ray experiments, the wavelength is almost always many orders of magnitude larger than the size of a typical system; namely the size of a typical atomic or molecular system is a few bohr, whereas the wavelengths of light that excites its states is generally on the order of hundreds of nm. At the system scale, the electromagnetic radiation is essentially uniform in all three directions. Therefore, $\partial_\mu W^\mu \approx 0$ is a very good approximation, and consequently, we can drop the term containing the gradient of W^μ in $\mathcal{K}_b^a$,

$$\mathcal{K}_b^a \propto e^{-Ut} \exp\left(-\frac{m}{2t} x^\mu x_\mu\right) \exp[(gJ_a \partial_t \epsilon^a)t] \exp[-gW_\mu x^\mu]. \tag{17.2}$$

The vector gx^μ in the argument of the last exponential on the right is simply the system's dipole moment. It is best to consider first what the static Stark effect yields. It is a straightforward exercise to show that even a static perturbation is sufficient to produce mixtures of ground and excited states. Therefore, we set $\exp[(gJ_a \partial_t \epsilon^a)t] \approx 1$, and we expand $\exp[-gW_\mu x^\mu]$ into a power series

$$\mathscr{K}_b^a \propto e^{-Ut} \exp\left(-\frac{m}{2t} x^\mu x_\mu\right) \left\{\mathbb{I} - gW_\mu x^\mu + \frac{1}{2} g^2 W_\mu W_\nu x^\mu x^\nu + \cdots\right\}, \tag{17.3}$$

where the first two factors from the left propagate the ground state. These are diagonal matrices that simply create n replicas of the ground state when acting on a n state vector:

$$e^{-Ut} \exp\left(-\frac{m}{2t} x^\mu x_\mu\right) \mathbb{I} \to \psi_0 \begin{pmatrix} 1 \\ 1 \\ \vdots \end{pmatrix}.$$

Note that we have used atomic units starting with equation (17.2) so that $\hbar = 1$ for notational convenience, and these are adapted for the rest of the chapter. The second term inside the curly brackets multiplied by the first two exponential functions produces a n vector with different mixtures of ground and excited states,

$$gW_\mu x^\mu e^{-Ut} \exp\left(-\frac{m}{2t} x^\mu x_\mu\right) \mathbb{I} \to \begin{pmatrix} \varphi_1 \\ \varphi_2 \\ \vdots \end{pmatrix}.$$

Since we have assumed that gW_μ is a constant, all mixtures produced by a $n \times n$ operator are of the general form (in $\mathbb{R}$),

$$\varphi_i = \varphi_0 \left\{1 - \sum_{k=1}^{n} c_k \frac{(W_\mu x^\mu g)^k}{k!}\right\}.$$

We can now discuss what group order we use for the implementation. For simplicity, we only discuss applications to systems in $\mathbb{R}$, where we have a ground state, a first excited state, etc. Then the order N of the SO(N) theory we need is the number of excited states desired, with $N \geq 2$. Choosing $N = 1$ returns us to the scalar theory. If we choose to expand up the propagator in (17.1) the linear term, then we are targeting just the first excited state and we use $N = 2$. For systems of particles with d ro-vibrational degrees of freedom (see Chapter 10), then $N = d + 1$ should be chosen if only first excitations are desired, and as we discuss later

$$N = \binom{2+n}{2}$$

if first and second excitations is the goal. Finally, we argued that the time dependent terms $(\hbar g J_a \partial_t \epsilon^a)t$ can be ignored as well. It turns out that eliminating that term amounts to simulating a physically meaningful process as we explain in the next section.

17.2 The static Stark effect

Perturbing a system in the ground state with a uniform static field creates mixtures of excited states that evolve in real time. We now want to demonstrate how this mixture takes place using a two state case. For the present discussion, we assume that the eigenstates of a system, ψ_n and E_n are known. Then, in the eigenbasis $\langle x|n\rangle = \psi_n$ the Hamiltonian matrix

$$\langle n'|\hat{H}|n\rangle = \delta_{n'n} E_n$$

is diagonal. We now introduce the linear term in (17.3) to produce

$$\langle n'|\hat{H} + \hat{W}|n\rangle = \langle n'|\hat{H}|n\rangle + \mathscr{W}_{n'n}$$

where the second term is a matrix proportional to $\langle n'|x|n\rangle$ as the perturbation where the proportionality constant is gW.

The two-state case is sufficient to demonstrate how such mixtures are produced,

$$\langle n'|\hat{H}|n\rangle = \begin{pmatrix} E_0 + \mathscr{W}_{00} & \mathscr{W}_{01} \\ \mathscr{W}_{10} & E_1 + \mathscr{W}_{11} \end{pmatrix}. \tag{17.4}$$

Note that $\mathscr{W}_{01}^* = \mathscr{W}_{10}$. Because of the off diagonal terms, the state vector evolves in time even though the perturbation is static,

$$\psi(t) = c_0(t)\psi_0 + c_1(t)\psi_1. \tag{17.5}$$

The evolution of $\psi(t)$ is governed by the time dependent Schrödinger equation,

$$i\frac{\partial}{\partial t}\psi(t) = (\hat{H}_{(1)})\psi(t).$$

Inserting (17.5) into the last expression yields a set of coupled first-order differential equations for the two coefficients,

$$i\frac{\partial c_0}{\partial t} = c_0(E_0 + \mathscr{W}_{00}) + c_1\mathscr{W}_{01},$$
$$i\frac{\partial c_1}{\partial t} = c_1(E_1 + \mathscr{W}_{11}) + c_0\mathscr{W}_{10}.$$

Using the methods of Chapters 3 and 4, we can write the solution formally as follows:

$$\begin{pmatrix} c_0(t) \\ c_1(t) \end{pmatrix} = \exp(-\hat{H}_{(1)}t) \begin{pmatrix} c_0(0) \\ c_1(0) \end{pmatrix} \tag{17.6}$$

and we take $c_0(0) = 1$, $c_1(0) = 0$, i. e., we start in the ground state. The expression for the time evolution operator in the last equation is simple to implement into code, but it is a standard exercise in introductory quantum mechanics to explicitly diagonalize the 2×2 $\hat{H}_{(1)}$ matrix, by using a rotation matrix (cf. example 9 in Chapter 3),

$$R = \begin{pmatrix} \cos\theta & -\sin\theta \\ \sin\theta & \cos\theta \end{pmatrix},$$

where

$$\theta = \frac{1}{2} \tan^{-1} \frac{2|\mathscr{W}_{10}|}{E_0 + \mathscr{W}_{00} - E_1 - \mathscr{W}_{11}}.$$

Consequently,

$$R^T \begin{pmatrix} c_0(t) \\ c_1(t) \end{pmatrix} = R^T \exp(-\hat{H}_{(1)}t) R R^T \begin{pmatrix} c_0(0) \\ c_1(0) \end{pmatrix},$$

and using Taylor's theorem we find

$$R^T \begin{pmatrix} c_0(t) \\ c_1(t) \end{pmatrix} = \begin{pmatrix} e^{-E_+ t} & 0 \\ 0 & e^{-E_- t} \end{pmatrix} \begin{pmatrix} \cos\theta \\ -\sin\theta \end{pmatrix} \tag{17.7}$$

where the eigenvalues are

$$\begin{aligned} E_+ &= \frac{1}{2}(E_0 + E_1 + \mathscr{W}_{00} + \mathscr{W}_{11}) \\ &\quad + \frac{1}{2}\sqrt{(E_0 + \mathscr{W}_{00} - E_1 - \mathscr{W}_{11})^2 + 4|\mathscr{W}_{10}|^2}, \end{aligned} \tag{17.8}$$

$$\begin{aligned} E_- &= \frac{1}{2}(E_0 + E_1 + \mathscr{W}_{00} + \mathscr{W}_{11}) \\ &\quad - \frac{1}{2}\sqrt{(E_0 + \mathscr{W}_{00} - E_1 - \mathscr{W}_{11})^2 + 4|\mathscr{W}_{10}|^2}. \end{aligned} \tag{17.9}$$

We now seek an expression for $|c_1(t)|^2$, i. e., the probability of finding the system in the first excited state. By multiplying both sides of (17.7) by R, we can isolate the state vector

$$\begin{pmatrix} c_0(t) \\ c_1(t) \end{pmatrix} = \begin{pmatrix} \cos\theta & -\sin\theta \\ \sin\theta & \cos\theta \end{pmatrix} \begin{pmatrix} \cos\theta e^{-iE_+ t} \\ -\sin\theta e^{-iE_- t} \end{pmatrix},$$

$$\begin{pmatrix} c_0(t) \\ c_1(t) \end{pmatrix} = \begin{pmatrix} \cos^2\theta e^{-iE_+ t} + \sin^2\theta e^{-iE_- t} \\ \cos\theta \sin\theta (e^{-iE_+ t} - e^{-iE_- t}) \end{pmatrix},$$

we get

$$|c_1(t)|^2 = \frac{1}{4}\sin^2(2\theta)|e^{-iE_+t} - e^{-iE_-t}|^2$$
$$|c_1(t)|^2 = \frac{1}{2}\sin^2(2\theta)\{1 - \cos(E_+ - E_-)t\}$$
$$|c_1(t)|^2 = \frac{4|\mathscr{W}_{10}|^2}{(E_0 + \mathscr{W}_{00} - E_1 - \mathscr{W}_{11})^2 + 4|\mathscr{W}_{10}|^2}\{1 - \cos(E_+ - E_-)t\}$$

or

$$|c_1(t)|^2 = \frac{4|\mathscr{W}_{10}|^2}{(E_0 + \mathscr{W}_{00} - E_1 - \mathscr{W}_{11})^2 + 4|\mathscr{W}_{10}|^2} \times \{1 - \cos(\sqrt{(E_0 + \mathscr{W}_{00} - E_1 - \mathscr{W}_{11})^2 + 4\mathscr{W}_{01}^2}\, t)\}. \tag{17.10}$$

This is Rabi's formula for a generic 2×2 case. For even parity potential energies such as the harmonic oscillator, $\mathscr{W}_{00} = \mathscr{W}_{11} = 0, E_0 - E_1 = \omega_0$, we obtain a slightly simpler expression,

$$|c_1(t)|^2 = \frac{4|\mathscr{W}_{10}|^2}{(\omega_0)^2 + 4|\mathscr{W}_{10}|^2}\{1 - \cos(\sqrt{(\omega_0)^2 + 4\mathscr{W}_{10}^2}\, t)\}. \tag{17.11}$$

We note that the system never reaches a pure excited state. Having shown that the static field creates mixtures of states, it is important to keep in mind that when more than one excited state exists, all such states are part of the mixture that evolves in time even when we restrict the size of the vector ψ^a to two mixtures as we do here.

17.3 A configuration interaction approach

One approach that has shown some success is to consider expanding the Hamiltonian H in a subspace of size N of the vector space spanned by

$$\psi_n = \varphi_0 \sum_{k=0}^{N} c_k^{(N)} x^k. \tag{17.12}$$

From this general expansion, we can consider some cases. For example, if we let $c_k^{(n)} = \delta_{k,n}$ we are expanding ψ_n in the canonical vector space $\{\varphi_0, x\varphi_0, x^2\varphi_0, \ldots\}$. One can construct a mathematically convergent expansion of the Hamiltonian matrix $\langle n'|\hat{H}|n\rangle$ in this basis set; the orthogonality of the basis set is not needed for such purpose. The convergence with respect to the subspace size N, $\{\varphi_0, x\varphi_0, x^2\varphi_0, \ldots, x^N\varphi_0\}$ is expected to be very slow in general. A much better choice is to build a representation using $c_k^{(n)} = w_k^{(n)}$, i. e., the orthogonal polynomials introduced in Chapters 1 and 7. We now use equation (17.12) to derive the following expression:

$$\begin{aligned}\langle n'|\hat{H}|n\rangle &= -\frac{1}{2m}\left\{\sum_{l=0}^{n'}\sum_{k=0}^{n} w_l^{(n')} w_k^{(n)}\,\tilde{\mathscr{P}}_{k+l}\right.\\ &\quad + 2\sum_{l=0}^{n'}\sum_{k=1}^{n} k\, w_l^{(n')} w_k^{(n)}\,\tilde{\mathscr{Q}}_{k+l-1}\\ &\quad \left.+ \sum_{l=0}^{n'}\sum_{k=2}^{n} k(k-1)\, w_l^{(n')} w_k^{(n)}\,\tilde{\mathscr{I}}_{k+l-2}\right\} + \sum_{l=0}^{n'}\sum_{k=0}^{n} w_l^{(n')} w_k^{(n)}\,\tilde{\mathscr{U}}_{k+l}. \end{aligned} \tag{17.13}$$

Note that if the coefficients the $\langle 0|H|0\rangle$ and $\langle 1|H|1\rangle$ matrix elements are the estimates of E_0 and E_1 in Chapter 7. This particular Hamiltonian matrix is not symmetric as the result of statistical fluctuations, but the eigenvalues can be shown to converge to their exact values as N increases. To produce a proof of concept, we used the system in Chapter 7, and we compute the eigenvalues and eigenvectors using the discrete variable representation. Then we compute the monomial integrals and the polynomial coefficients using the trapezoid rule and the algorithm in Chapter 2. The Singular Value Decomposition (SVD) is used to find the eigenvalues of the Hamiltonian matrix, though one can use regular diagonalizing routines if H is symmetrized first. In Tables 17.1 and 17.2, we compare the energies for a subspace of size 4, and 6, respectively (in the column labeled SVD with the discrete variable representation results. The column labeled with the symbol $\langle n|H|n\rangle$ contains the values of the energies estimated with the approaches

Table 17.1: Energies of the ground and first excited states for the system in Chapter 7. A subspace of size $N = 4$ is used to produce these results.

n	$\langle n\|H\|n\rangle$	E_n (SVD)	E_n (DVR)
0	0.00313606	0.00313606	0.00313613
1	0.01130537	0.01124006	0.01123793
2	0.02239396	0.02207031	0.02205101
3	0.03528789	0.03535321	0.03444055
4	0.04960539	0.04992904	0.04809605

Table 17.2: Energies of the ground and first excited states for the system in Chapter 7. A subspace of size $N = 6$ is used to produce these results.

n	$\langle n\|H\|n\rangle$	E_n (SVD)	E_n (DVR)
0	0.00313606	0.00313606	0.00313613
1	0.01130537	0.01123767	0.01123793
2	0.02239396	0.02205188	0.02205101
3	0.03528789	0.03451501	0.03444055
4	0.04960539	0.04829506	0.04809605
5	0.06511216	0.06595276	0.06281471

in Chapter 7. Note how, e. g., the second excitation energy is correct up to three figures, and it becomes accurate to five figures with the SVD on a subspace of size six. While the results in Tables 17.1 and 17.2 are encouraging, the time has come to consider how this machinery can be implemented in multidimensional spaces.

17.4 The product polynomial coefficients for two variables

Our discussion of the strategies implemented so far would be of little use if one could not extend it to problems with more than one degree of freedom. This section is dedicated to the discussion of how excited state approaches that build states from the Diffusion Monte Carlo ground state like those in Chapter 7 and in the present chapter may be implemented. We begin by building expressions for the polynomial coefficients of the two-dimensional case using the coefficients of the orthogonal polynomials in one variable. Because, in general, the dot product weight is not separable, i. e.,

$$\omega(x,y) = |\varphi_0(x,y)|^2 \neq \omega_x(x)\omega_y(y)$$

the polynomials constructed by multiplication are not orthonormal. However, if a "good choice of coordinates" is made (e. g., normal modes), then the couplings may be small. The GSPA method [42] accounts for the couplings that remain for systems with d degrees of freedom by building the $d \times d$ sub-block of the Hamiltonian for single and double excitations along with the corresponding overlap matrix and submitting the resulting two objects to a generalized eigenvalue subroutine. The procedure is the equivalent of evaluating the $\langle 1|H|1\rangle$ matrix element in the monodimensional case as we do in Chapter 7. We should mention here that GSPA yields sufficient accuracy to reproduce the experimental infrared spectrum of complicated aggregates of molecules and ions, and could, in principle, be extended to include double or higher order excitations. The resulting eigenvectors are transformed into carefully selected internal coordinates GSPA can be used to assign the various features [42].

In unpublished work, we have shown that the accuracy of the method in Chapter 7, and by extension the GSPA method, become increasingly more accurate as the effective masses increase. For cases when the accuracy is not sufficient, it is possible to extend the configuration interaction strategy discussed earlier in this chapter for problems in $\mathbb{R}^b$ by augmenting the basis set, e. g., by adding all the second excitations

$$\binom{d+1}{2} \times \binom{d+1}{2}$$

subblock, etc.

To represent a two-dimensional Hamiltonian in terms of monomial integrals and product polynomial coefficients for the $d = 2$ case, we consider the following expansion built on the ground state wave function:

$$\psi_{n,l} = \sum_{k=0}^{n} \sum_{k'=0}^{l} \tilde{w}_k^{(n,x)} \tilde{w}_{k'}^{(l,y)} x^k y^{k'} \varphi_0. \tag{17.14}$$

With this expression, we can formulate the Hamiltonian matrix and the overlap matrix in terms of monomial integrals and polynomial coefficients.

17.4.1 The Hamiltonian elements

Let

$$\hat{H} = -\frac{\hbar^2}{2m}\left(\frac{\partial^2}{\partial x^2} + \frac{\partial^2}{\partial y^2}\right) + U(x,y) \tag{17.15}$$

and suppose φ_0 is the ground state wave function of $\hat{H}$ obtained by DMC and fitted to a Fourier series with a Hartree product. Then it is possible to obtain analytical expressions for the derivatives of φ_0. We find an expression for the second partial derivatives for $\psi_{n,l}$, using equation (17.14),

$$\begin{aligned}
\frac{\partial}{\partial x}\psi_{n,l} &= \sum_{k=1}^{n} \sum_{k'=0}^{l} \tilde{w}_k^{(n,x)} \tilde{w}_{k'}^{(l,y)} k x^{k-1} y^{k'} \varphi_0 \\
&\quad + \sum_{k=0}^{n} \sum_{k'=0}^{l} \tilde{w}_k^{(n,x)} \tilde{w}_{k'}^{(l,y)} x^k y^{k'} \frac{\partial \varphi_0}{\partial x} \\
\frac{\partial^2}{\partial x^2}\psi_{n,l} &= \sum_{k=2}^{n} \sum_{k'=0}^{l} \tilde{w}_k^{(n,x)} \tilde{w}_{k'}^{(l,y)} k(k-1) x^{k-2} y^{k'} \varphi_0 \\
&\quad \times 2 \sum_{k=1}^{n} \sum_{k'=0}^{l} \tilde{w}_k^{(n,x)} \tilde{w}_{k'}^{(l,y)} k x^{k-1} y^{k'} \frac{\partial \varphi_0}{\partial x} \\
&\quad + \sum_{k=0}^{n} \sum_{k'=0}^{l} \tilde{w}_k^{(n,x)} \tilde{w}_{k'}^{(l,y)} x^k y^{k'} \frac{\partial^2 \varphi_0}{\partial x^2}.
\end{aligned} \tag{17.16}$$

A similar equation applies for the partial w. r. t. y,

$$\begin{aligned}
\frac{\partial^2}{\partial y^2}\psi_{n,l} &= \sum_{k=0}^{n} \sum_{k'=2}^{l} \tilde{w}_k^{(n,x)} \tilde{w}_{k'}^{(l,y)} k'(k'-1) x^{k} y^{k'-2} \varphi_0 \\
&\quad \times 2 \sum_{k=1}^{n} \sum_{k'=0}^{l} \tilde{w}_k^{(n,x)} \tilde{w}_{k'}^{(l,y)} k' x^{k} y^{k'-1} \frac{\partial \varphi_0}{\partial y} \\
&\quad + \sum_{k=0}^{n} \sum_{k'=0}^{l} \tilde{w}_k^{(n,x)} \tilde{w}_{k'}^{(l,y)} x^k y^{k'} \frac{\partial^2 \varphi_0}{\partial y^2}.
\end{aligned} \tag{17.17}$$

Now let us express the following matrix element:

$$\int_{\mathcal{D}} \{\psi^*_{n',l'} \hat{H}\, \psi_{n,l}\}\, dA = \int_{\mathcal{D}} \{\psi^*_{n',l'} \hat{K}\, \psi_{n,l}\}\, dA + \int_{\mathcal{D}} \{\psi^*_{n',l'} U\, \psi_{n,l}\}\, dA$$

where $dA = dxdy$. For convenience, we split the Hamiltonian into the kinetic and the potential operators. Let us address the potential energy integral first assuming the most general form, which contains coupling terms,

$$\int_{\mathcal{D}} \{\psi^*_{n',l'} U\, \psi_{n,l}\}\, dA$$
$$= \sum_{j=0}^{n'} \sum_{j'=0}^{l'} \sum_{k=0}^{n} \sum_{k'=0}^{l} \tilde{w}_j^{(n',x)} \tilde{w}_{j'}^{(l',y)} \tilde{w}_k^{(n,x)} \tilde{w}_k^{(l,y)} \tilde{\mathcal{U}}_{k+j,k'+j'},$$

where we have used (17.14) twice with appropriate indices,

$$\tilde{\mathcal{U}}_{k+j,k'+j'} = \int_{\mathcal{D}} x^{k+j} y^{k'+j'}\, U \varphi_0^2\, dA$$

and we have moved some things around. We can evaluate this matrix using the DMC population distributed according to φ_0,

$$\tilde{\mathcal{U}}_{k,k'} \approx \frac{1}{M} \sum_{i=1}^{M} x_i^k y_i^{k'}\, U(x_i, y_i) \varphi_0(x_i, y_i).$$

Next, we separate the kinetic energy operator as follows:

$$\int_{\mathcal{D}} \{\psi^*_{n',l'} \hat{K}\, \psi_{n,l}\}\, dA = \int_{\mathcal{D}} \{\psi^*_{n',l'} \hat{K}_x\, \psi_{n,l}\}\, dA + \int_{\mathcal{D}} \{\psi^*_{n',l'} \hat{K}_y\, \psi_{n,l}\}\, dA$$

where, e. g.,

$$\int_{\mathcal{D}} \{\psi^*_{n',l'} \hat{K}_x\, \psi_{n,l}\}\, dA = -\frac{\hbar^2}{2m} \int_{\mathcal{D}} \left\{\psi^*_{n',l'} \frac{\partial^2}{\partial x^2} \psi_{n,l}\right\} dA.$$

Using (17.16) and (17.14), we can unravel this as follows:

$$\int_{\mathcal{D}} \{\psi^*_{n',l'} \hat{K}_x\, \psi_{n,l}\}\, dA$$
$$= -\frac{\hbar^2}{2m} \sum_{j=0}^{n'} \sum_{j'=0}^{l'} \sum_{k=2}^{n} \sum_{k'=0}^{l} \tilde{w}_j^{(n',x)} \tilde{w}_{j'}^{(l',y)} \tilde{w}_k^{(n,x)} \tilde{w}_{k'}^{(l,y)}\, k(k-1)\, \mathcal{I}_{k+j-2,k'+j'}$$
$$- \frac{\hbar^2}{m} \sum_{j=0}^{n'} \sum_{j'=0}^{l'} \sum_{k=1}^{n} \sum_{k'=0}^{l} \tilde{w}_j^{(n',x)} \tilde{w}_{j'}^{(l',y)} \tilde{w}_k^{(n,x)} \tilde{w}_{k'}^{(l,y)}\, k\, \tilde{\mathcal{Q}}^{(x)}_{k+j-1,k'+j'}$$

$$-\frac{\hbar^2}{2m}\sum_{j=0}^{n'}\sum_{j'=0}^{l'}\sum_{k=0}^{n}\sum_{k'=0}^{l}\tilde{w}_j^{(n',x)}\tilde{w}_{j'}^{(l',y)}\tilde{w}_k^{(n,x)}\tilde{w}_{k'}^{(l,y)}\,\tilde{\mathscr{P}}^{(x)}_{k+j,k'+j'}$$

where the doubly indexed monomial integrals are

$$\begin{aligned}
\tilde{\mathscr{I}}_{k+j-2,k'+j'} &= \int_{\mathcal{D}} x^{k+j-2}y^{k'+j'}\varphi_0^2\,dA\\
\tilde{\mathscr{I}}_{k,k'} &\approx \frac{1}{M}\sum_{i=1}^{M} x_i^k y_i^{k'}\varphi_0\\
\tilde{\mathscr{Q}}^{(x)}_{k+j-1,k'+j'} &= \int_{\mathcal{D}}\left\{x^{k+j-1}y^{k'+j'}\varphi_0\frac{\partial\varphi_0}{\partial x}\right\}dA\\
\tilde{\mathscr{Q}}^{(x)}_{k,k'} &\approx \frac{1}{M}\sum_{i=1}^{M} x_i^k y_i^{k'}\frac{\partial\varphi_0}{\partial x}\\
\tilde{\mathscr{P}}^{(x)}_{k+j,k'+j'} &= \int_{\mathcal{D}}\left\{x^{k+j}y^{k'+j'}\varphi_0\frac{\partial^2\varphi_0}{\partial x^2}\right\}dA\\
\tilde{\mathscr{P}}^{(x)}_{k,k'} &\approx \frac{1}{M}\sum_{i=1}^{M} x_i^k y_i^{k'}\frac{\partial^2\varphi_0}{\partial x^2}
\end{aligned}$$

It is straightforward to show that

$$\begin{aligned}
&\int_{\mathcal{D}}\{\psi^*_{n',l'}\hat{K}_y\,\psi_{n,l}\}\,dA\\
&= -\frac{\hbar^2}{2m}\sum_{j=0}^{n'}\sum_{j'=0}^{l'}\sum_{k=0}^{n}\sum_{k'=2}^{l}\tilde{w}_j^{(n',x)}\tilde{w}_{j'}^{(l',y)}\tilde{w}_k^{(n,x)}\tilde{w}_{k'}^{(l,y)}\,k'(k'-1)\,\tilde{\mathscr{I}}_{k+j,k'+j'-2}\\
&\quad-\frac{\hbar^2}{m}\sum_{j=0}^{n'}\sum_{j'=0}^{l'}\sum_{k=0}^{n}\sum_{k'=1}^{l}\tilde{w}_j^{(n',x)}\tilde{w}_{j'}^{(l',y)}\tilde{w}_k^{(n,x)}\tilde{w}_{k'}^{(l,y)}\,k'\,\tilde{\mathscr{Q}}^{(y)}_{k+j,k'+j'-1}\\
&\quad-\frac{\hbar^2}{2m}\sum_{j=0}^{n'}\sum_{j'=0}^{l'}\sum_{k=0}^{n}\sum_{k'=0}^{l}\tilde{w}_j^{(n',x)}\tilde{w}_{j'}^{(l',y)}\tilde{w}_k^{(n,x)}\tilde{w}_{k'}^{(l,y)}\,\tilde{\mathscr{P}}^{(y)}_{k+j,k'+j'}
\end{aligned}$$

where

$$\begin{aligned}
\tilde{\mathscr{Q}}^{(y)}_{k+j,k'+j'} &= \int_{\mathcal{D}}\left\{x^{k+j}y^{k'+j'-1}\varphi_0\frac{\partial\varphi_0}{\partial y}\right\}dA\\
\tilde{\mathscr{P}}^{(y)}_{k+j,k'+j'} &= \int_{\mathcal{D}}\left\{x^{k+j}y^{k'+j'}\varphi_0\frac{\partial^2\varphi_0}{\partial y^2}\right\}dA.
\end{aligned}$$

Finally, we need to consider the overlap matrix

$$S_{n'l',nl} = \langle n', l' | n, l \rangle = \int_{\mathcal{D}} \{\psi^*_{n',l'} \psi_{n,l}\} \, dA$$
$$= \sum_{j=0}^{n'} \sum_{j'=0}^{l'} \sum_{k=0}^{n} \sum_{k'=0}^{l} \tilde{w}_j^{(n',x)} \tilde{w}_{j'}^{(l',y)} \tilde{w}_k^{(n,x)} \tilde{w}_{k'}^{(l,y)} \tilde{\mathcal{I}}_{k+j,k'+j'}.$$

These two matrices can be submitted to a generalized eigenvalue method to produce the desired eigenvalues and eigenvectors. In Python, the generalized eigenvalue problem,

$$\mathbf{H}v = \lambda \mathbf{S}v,$$

is solved using a `scipy.linalg.eigh()` call.

17.5 Orthogonal polynomials of multiple variables

In this section, we explore possibilities to create polynomial representations of the solutions of the Schrödinger equation using the ground state probability distribution to define the dot product in the polynomial vector space. The subject of orthogonal polynomials in multiple variables is much more involved than its monodimensional counterpart; it has been thoroughly investigated, and it turns out that it is possible to generate orthogonal polynomials in general for an arbitrary number of dimensions. The orthogonality conditions can always be enforced between pairs of polynomial of different *overall power*, where by overall power we mean the sum of the powers of all variables. Polynomials with the same overall powers are called *Homogeneous*, and homogeneous multivariable polynomials are *not orthogonal.*

We have chosen to follow the notation in Dunkl and Xu, which seems to be clear and easy to follow. We let $\mathbf{x}$ represent the column vector $(x^{\mu})_{\mu=1}^{d}$ where μ is the same Greek index we introduce in Chapter 8 and d is the dimension, and we let $\mathbf{x}^n$ represent a column vector containing the monomials of power n. For example, for $n = 1, 2$, and $d = 2$,

$$\mathbf{x} = \begin{pmatrix} x \\ y \end{pmatrix} \quad \mathbf{x}^2 = \begin{pmatrix} x^2 \\ xy \\ y^2 \end{pmatrix}.$$

Next, we define a set of d indices for a monomial of power n defined as $x^{\alpha} = [x^{(1)}]^{\alpha_1} \times [x^{(2)}]^{\alpha_2} \times \cdots \times [x^{(d)}]^{\alpha_d}$ where $\alpha = \{\alpha_1, \alpha_2, \ldots, \alpha_d\}$ is one of a number of ordered sets of positive integers $\alpha_i \in \{0, 1, \ldots n\}$ such that $\sum_{i=1}^{n} \alpha_i = n$. The order for the set α must be the same as the order in which the components of $\mathbf{x}$ are listed. For example, for $d = 2$, $n = 4$ a valid monomial $x^{\alpha} = x^3 y$ where α is the set $\{3, 1\}$. The vector space of all homogeneous polynomials of degree n in d variables is $\mathcal{V}_n^d$. For example,

$$\mathcal{V}_2^2 = \text{span}\{x^2, xy, y^2\},$$

so that $4x^2 + xy - 3y^2 \in \mathcal{V}_2^2$.

It is possible to show that there are

$$\dim \mathcal{V}_n^d = \binom{n+d-1}{n} = r_n^d$$

linearly independent monomial terms of power n in d dimensions.

For example, there are 5 basis for $\mathcal{V}_4^2$:

$$\begin{array}{llll} x^4 & \alpha = \{4,0\}, & x^3y & \alpha = \{3,1\} \\ x^2y^2 & \alpha = \{2,2\}, & xy^3 & \alpha = \{1,3\} \\ y^4 & \alpha = \{0,4\} & & \end{array}$$

and

$$r_4^2 = \binom{5}{4} = 5.$$

Here, we have used a graded lexicographic ordering, namely x comes before y by arbitrarily choosing alphabetic ordering, followed by sorting the exponents of x in descending order to list monomials. For the most general case, graded lexicographic ordering of the elements of α can be achieved by sorting in descending order the following set of integers $\{N_\alpha\}_n^d = \{\alpha_1 10^{d-1} + \alpha_2 10^{d-2} + \cdots + \alpha_d\}$ after arbitrarily choosing one of the $d!$ ordering of the components $\{x^\mu\}$. The vector space of all polynomials with degree at most n is denoted Π_n^d, so for instance, $P(x,y) = 4 + x - y^2 \in \Pi_2^2$. It can be shown that

$$\dim \Pi_n^d = \binom{n+d}{n}.$$

At this point, we can introduce a determinantal construction of orthogonal polynomials in $P_\alpha^n \in \Pi_n^d$ as follows:

$$P_\alpha^n = \frac{1}{\Delta_{n-1,d}} \det \begin{vmatrix} & & & \mathbf{s}_{\alpha,0} \\ & M_{n-1,d} & & \mathbf{s}_{\alpha,1} \\ & & & \vdots \\ 1 & \mathbf{x}^T \quad \dots & (\mathbf{x}^{n-1})^T & x^\alpha \end{vmatrix} \tag{17.18}$$

where $M_{n,d}$ is the moment matrix, similar in construction to the Hankel determinant of Chapter 2,

$$M_{n,d} = [\mathscr{I}(\mathbf{x}^k(\mathbf{x}^j)^T)]_{k,j=0}^n,$$

$\Delta_{n,d} = \det M_{n,d}$. In equation (17.18), $\mathbf{s}_{\alpha,\beta}$ is the following monomial moment integral $\mathscr{I}(x^{\alpha+\beta})$, and finally x^α is the monomial identified by a specific α set, such as, e.g., $\alpha =$

$(2,0)$, $x^\alpha = x^2$. The matrix $\mathscr{I}(\mathbf{x}^k(\mathbf{x}^j)^T)$ is $r_k^d \times r_j^d$ in size and contains the integrals over the d-dimensional space of $x^{\alpha+\beta}$ for $\dim\alpha, \dim\beta \le n$. For example, the $M_{1,2}$ matrix is built as follows:

$$M_{1,2} = \begin{bmatrix} \mathscr{I}(1) & \mathscr{I}\begin{pmatrix} x & y \end{pmatrix} \\ \mathscr{I}\begin{pmatrix} x \\ y \end{pmatrix} & \mathscr{I}\begin{pmatrix} x^2 & xy \\ xy & y^2 \end{pmatrix} \end{bmatrix}. \tag{17.19}$$

As this example shows, there is a systematic process for building P_α^n by continuing the listing of all matrices $\mathscr{I}(\mathbf{x}^k(\mathbf{x}^j)^T)$ and fit them uniquely in $M_{n,d}$.

For further clarification, the expression for $M_{1,2}$ translates into

$$M_{1,2} = \begin{bmatrix} \mathscr{I}_{0,0} & \mathscr{I}_{1,0} & \mathscr{I}_{0,1} \\ \mathscr{I}_{1,0} & \mathscr{I}_{2,0} & \mathscr{I}_{1,1} \\ \mathscr{I}_{0,1} & \mathscr{I}_{1,1} & \mathscr{I}_{0,2} \end{bmatrix}$$

when the notation for the $d = 2$ case in the previous section is used. Unlike the $d = 1$ case, $M_{n-1,d}$ is not a Hankel matrix. A specific example for $d = 2$ and $n = 2$ is

$$P_{(2,0)}^2(x,y) = \frac{1}{\Delta_{1,2}} \det \begin{vmatrix} \mathscr{I}_{0,0} & \mathscr{I}_{1,0} & \mathscr{I}_{0,1} & \mathscr{I}_{2,0} \\ \mathscr{I}_{1,0} & \mathscr{I}_{2,0} & \mathscr{I}_{1,1} & \mathscr{I}_{3,0} \\ \mathscr{I}_{0,1} & \mathscr{I}_{1,1} & \mathscr{I}_{0,2} & \mathscr{I}_{2,1} \\ 1 & x & y & x^2 \end{vmatrix}.$$

With this example, we can show that $P_{(2,0)}^2(x,y)$ is orthogonal to all polynomials of power 1 or less. For example,

$$\begin{aligned} \langle P_{(2,0)}^2 | 1 \rangle_\omega &= \frac{1}{\Delta_{1,2}} \left\langle \det \begin{vmatrix} \mathscr{I}_{0,0} & \mathscr{I}_{1,0} & \mathscr{I}_{0,1} & \mathscr{I}_{2,0} \\ \mathscr{I}_{1,0} & \mathscr{I}_{2,0} & \mathscr{I}_{1,1} & \mathscr{I}_{3,0} \\ \mathscr{I}_{0,1} & \mathscr{I}_{1,1} & \mathscr{I}_{0,2} & \mathscr{I}_{2,1} \\ 1 & x & y & x^2 \end{vmatrix} \middle| 1 \right\rangle_\omega \\ &= \frac{1}{\Delta_{1,2}} \det \begin{vmatrix} \mathscr{I}_{0,0} & \mathscr{I}_{1,0} & \mathscr{I}_{0,1} & \mathscr{I}_{2,0} \\ \mathscr{I}_{1,0} & \mathscr{I}_{2,0} & \mathscr{I}_{1,1} & \mathscr{I}_{3,0} \\ \mathscr{I}_{0,1} & \mathscr{I}_{1,1} & \mathscr{I}_{0,2} & \mathscr{I}_{2,1} \\ \langle 1|1\rangle_\omega & \langle x|1\rangle_\omega & \langle y|1\rangle_\omega & \langle x^2|1\rangle_\omega \end{vmatrix} \\ &= \frac{1}{\Delta_{1,2}} \det \begin{vmatrix} \mathscr{I}_{0,0} & \mathscr{I}_{1,0} & \mathscr{I}_{0,1} & \mathscr{I}_{2,0} \\ \mathscr{I}_{1,0} & \mathscr{I}_{2,0} & \mathscr{I}_{1,1} & \mathscr{I}_{3,0} \\ \mathscr{I}_{0,1} & \mathscr{I}_{1,1} & \mathscr{I}_{0,2} & \mathscr{I}_{2,1} \\ \mathscr{I}_{0,0} & \mathscr{I}_{1,0} & \mathscr{I}_{0,1} & \mathscr{I}_{2,0} \end{vmatrix} = 0 \end{aligned}$$

where the determinant of a matrix with two identical rows (or column) vanishes.

The factor $1/\Delta_{n-1,d}$ yields a *monic* polynomial since $\Delta_{1,2}$ is the cofactor of the last monomial term in the determinant definition of P_α^n. With their construction, Dunkl and

Xu derive expressions for orthogonal polynomials of arbitrary order and with arbitrary number of variables. However, as we mentioned earlier, homogeneous polynomials of two or more variables are not mutually orthogonal, and additional steps need to be taken to orthogonalize them. Finally, a more succinct and general expression for the

$$\binom{n+d}{n} \times \binom{n+d}{n}.$$

Hamiltonian matrix can be obtained by expanding

$$\psi_{n,a} = \varphi_0 \sum_{k=0}^{n} \sum_{|\alpha|=k} C_\alpha x^\alpha,$$

e. g., the potential energy part is

$$\langle \beta | U | \alpha \rangle_\omega = \sum_{k=0}^{n} \sum_{k'=0}^{n'} \sum_{|\alpha|=k} \sum_{|\beta|=k'} C_\beta C_\alpha \mathscr{U}_{\alpha+\beta}$$

where

$$\mathscr{U}_{\alpha+\beta} = \mathscr{I}(x^{\alpha+\beta} U)$$

are the monomial integrals. Lastly, it is possible to use Einstein's notation to express the derivatives of ψ_α and the kinetic energy matrix

$$\psi_{n,a} = \varphi_0 \sum_{k=0}^{n} \sum_{|\alpha|=k} C_\alpha \partial_\mu x^\alpha + (\partial_\mu \varphi_0) \sum_{k=0}^{n} \sum_{|\alpha|=k} C_\alpha x^\alpha,$$

etc.

One last observation worth mentioning here is the rather unfavorable exponential scaling if the power n needs to be much larger than d. Then the Hamiltonian will have on the order of n^{2d} elements. The computation of eigenpairs scales as n^{6d}. This is the worst case scenario. The best case would be when first excitations are sufficient (as is the case for the GSPA method), then the excited state computation scales as d^6 when $d \gg 1$. If double or triple excitations are needed to achieve convergence and $d \gg 3$, one arrives at the same conclusion. Can one find a better scaling method? Perhaps one is only interested in a handful (on the order of d) states.

17.6 The dynamic Stark effect

In this section, we explore the possibility of introducing the dynamic Stark effect to develop excited states methods by means Diffusion Monte Carlo simulations that have a more favorable scaling than the other approaches we have discussed thus far. Unlike the static Stark effect, which produces mixtures of all excited states, the dynamic version contains, theoretically, a detuning term in the denominator of the expansion coefficients. That term causes the contribution of excited states far from the target to become, in principle, arbitrarily small if the field is applied for a sufficiently long time. In other words, the dynamic Stark effect produces a mixture that is predominantly in the ground and a single excited state if one tunes the frequency of the external radiation to be sufficiently close to the actual energy difference. The question we discuss here is, can a DMC simulation mimic these experimental results?

First, let us explain the origin of the tuning properties of the dynamic Stark effect. An approximate, but insightful expression can be obtained in the weak field limit. Let $\hat{H}$ be the system Hamiltonian. We add to it the external time dependent electromagnetic perturbation,

$$\hat{H} + W \cos \omega t, \tag{17.20}$$

and we let $\Psi_n = \exp(-iE_n t)\psi_n$ be the eigenfunctions of $\hat{H}$. The solutions of the Schrödinger equation with the perturbation included is of the form

$$\Psi_{\text{tot}} = \sum_{j=0}^{N} c_j(t) \exp(-iE_n t)\psi_n. \tag{17.21}$$

We only need to consider two states if we assume that the energy levels are well separated. Setting $N = 2$, and inserting (17.21) we obtain

$$\begin{aligned} i\frac{d}{dt}&\{c_0(t)\exp(-iE_0 t)\psi_0 + c_1(t)\exp(-iE_1 t)\psi_1\} \\ &= (\hat{H}_0 + \mathcal{W} \cos \omega t)\{c_0(t)\exp(-iE_0 t)\psi_0 + c_1(t)\exp(-iE_1 t)\psi_1\}. \end{aligned}$$

Upon evaluating the derivative with respect to t on the left, some terms cancel since

$$i\frac{d}{dt}\exp(-iE_n t)\psi_n = \hat{H}_0 \exp(-iE_n t)\psi_n$$

where ψ_n is time independent. This leaves

$$\begin{aligned} i\exp(-iE_0 t)\psi_0 \frac{dc_0}{dt} &+ i\exp(-iE_1 t)\psi_1 \frac{dc_1}{dt} \\ &= \mathcal{W} \cos \omega t\{c_0 \exp(-iE_0 t)\psi_0 + c_1 \exp(-iE_1 t)\psi_1\}. \end{aligned} \tag{17.22}$$

Now we multiply both sides by $\exp(iE_0 t)\psi_0^*$ and integrate over all space. Then we multiply (17.22) by $\exp(iE_1 t)\psi_1^*$ and integrate over all space to get a pair of first-order coupled differential equations,

$$\frac{dc_0}{dt} = -i\cos\omega t(c_0\mathscr{W}_{00} + c_1 e^{-i\omega_0 t}\mathscr{W}_{01}) \tag{17.23}$$

$$\frac{dc_1}{dt} = -i\cos\omega t(c_1\mathscr{W}_{11} + c_0 e^{i\omega_0 t}\mathscr{W}_{10}) \tag{17.24}$$

where $\omega_0 = E_1 - E_0$ in atomic units, $\mathscr{W}_{kj} = \langle k|\mathscr{W}|j\rangle$, and $\mathscr{W}^*_{jk} = \mathscr{W}_{kj}$. In the weak field limit, the mixture is predominantly in the ground state together with excited states whose contributions drops with the amount of detuning $(\omega_0 - \omega)$ where $\omega_0 = E_j - E_0$ for state j, and ω is the frequency of the impinging field in equation (17.20). In the weak field regime, $|\mathscr{W}_{01}| \ll \omega_0$, one can make several valid approximations to solve the coupled equations in (17.23) and (17.24). The main results are

$$|c_0|^2 \approx 1 + \frac{|\mathscr{W}_{00}|^2}{\omega^2}\sin^2\omega t \tag{17.25}$$

$$|c_1|^2 \approx |\mathscr{W}_{10}|^2 \frac{\sin^2\{(\omega_0 - \omega)t/2\}}{(\omega_0 - \omega)^2}. \tag{17.26}$$

The second expression is particularly useful, demonstrating that the probability of finding the system in state 1 varies inversely with the square of the detuning factor. If the excited states are well separated, then the frequency $\omega_0' = E_2 - E_0$ of second excited state will be far from the frequency ω of the impinging field and $|c_2|^2$ should be much smaller than $|c_1|^2$. To have a better idea of the significance of these results, consider the graph of $f(\omega) = \sin^2\{(\omega_0 - \omega)t/2\}/(\omega_0 - \omega)^2$ in Figure 17.1 for a large value of t.

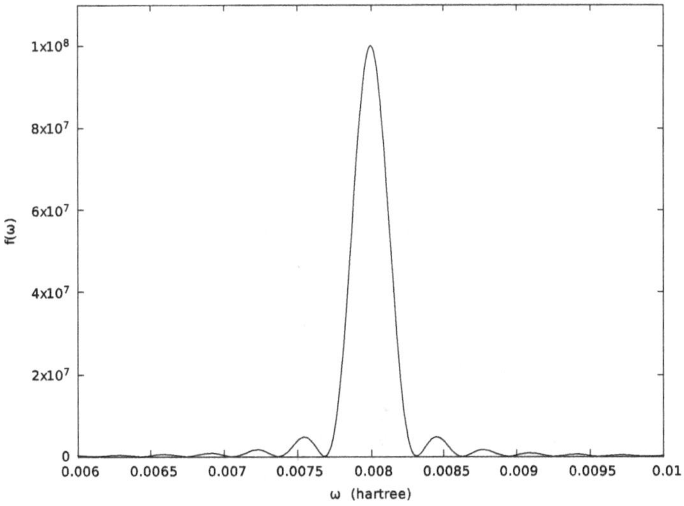

Figure 17.1: A graph of equation (17.25) with $|\mathscr{W}_{10}|^2 = 1$, $\omega_0 = 0.008$ hartree and $t = 2 \times 10^4$ hartree^{-1}.

It can be shown that

$$\lim_{\omega\to\omega_0} \frac{\sin^2\{(\omega_0-\omega)t/2\}}{(\omega_0-\omega)^2} = \frac{t^2}{4}$$

and that the full width at half-height (FWHH) is $2(1.391557\ldots)/t$. Clearly, the role played by t is one of sharpening both the main peaks as well as the satellites in a way that is similar to how collecting signal for a long time increase the resolution of features in the frequency domain of Chapter 14. Therefore, in principle, a large value of t should allow one to fine tune the frequency as much as desired while effectively separating other excited states.

While these results are encouraging, implementing a diffusion Monte Carlo strategy with the time dependent external field requires a new type of algorithm that has not yet been conceived to the best of our knowledge. We have, of course, begun exploring several possibilities, and here we discuss some of the difficulties we have thus far encountered. The branching approach of Chapter 7 entails a single random walk involving one population. This is combined with some orthogonalization procedure to construct matrix elements and estimate excited state wave functions and energies. By contrast, the implementation of a time dependent gauge theory requires the simultaneous evolution of n distinct components of the internal state vector ψ^a. Here is the rationale to motivate such need.

If we continue to assume that the wavelength is much larger than the system, and we consider an external field that oscillates with a particular frequency, e. g., $\partial_t \epsilon^a = E_{x,0} x \cos \omega t$ then equation (17.1) becomes

$$\begin{aligned}\mathscr{K}_b^a &\propto e^{-Ut} \exp\left(-\frac{m}{2\hbar t}x^2\right) \\ &\times \exp\left[-E_{x,0} x \sin \omega t \begin{pmatrix} 0 & 1 \\ -1 & 0 \end{pmatrix} + (\hbar g E_{x,0} x \cos \omega t)\begin{pmatrix} 0 & 1 \\ -1 & 0 \end{pmatrix} t\right].\end{aligned} \tag{17.27}$$

Here, we have chosen the monodimensional case and the SO(2) group for a two-state system for simplicity and $E_{x,0}$ is the external field amplitude. Let us now consider each interaction term (those proportional to $gE_{x,0}x$) individually. The one on the left has been the focus of this chapter thus far. Adding a $\sin \omega t$ is equivalent to changing the strength of the static stark effect on the state vector. Namely, changes in t or ω will simply create complicated mixtures of states of H without any obvious tuning effects.

The new interaction term on the right has a more complicated time dependence. To emphasize it better, we write the matrix elements to the left of the so(2) generator, and the t factor to the right. Because of the explicit linear t term, one obvious interpretation is as some type of potential energy not too dissimilar from the one in Chapter 16, where the time dependent $U(1)$ gauge is considered. Potential energy terms are implemented in diffusion Monte Carlo walks as branching terms. To properly implement these into the random walk, the t factor is converted into Δt as detailed in Chapter 6, since branching decisions are made at each step of the walk. However, unlike the diagonal e^{-Ut} branching step, which is implemented in the previous approaches to only evolve the ground state, the new interaction term is a skew symmetric object that in the present example couples the branching weights of two separate populations.

Martingales and branching over Markovian processes can be found in the mathematics and computer science literature. Generally, these tend to be well-defined for matrices with fixed sizes. In the implementation of DMC, we have used thus far, the size of the various populations fluctuates at every step as a result of the annihilation-growth process we use. Therefore, the details of implementing the cross-terms over populations of different sizes have to be carefully considered to avoid accidentally introducing sampling biases. Thankfully, there is a way to perform DMC walks with a fixed population [51] but several algorithmic steps are necessary to achieve equilibrium. Even the diagonal branching term is problematic when sampling multiple distributions simultaneously. When it is used to sample the ground state wave function, the feedback step is implemented to update the estimate of the reference energy, which becomes the ground state energy when the walk equilibrates. When multiple populations are sampled it is not clear if separate feedback and reference energy updates have to be implemented and if so it is not clear how it can be done.

Next, one needs to consider carefully what value for t in the argument of the sinusoidal functions should be used. It is clear, based on the theoretical results obtained earlier in this section, that Δt would be a poor choice, since fine tuning requires large values for this variable. This leads to possibly having two separate physical times, with t in the argument of the sinusoidal functions being the actual physical time the radiation has been "turned on," to be updated at each step by Δt. This sounds reasonable but not mathematically or physically sound. In one instance, t would be used as the actual Newtonian time, while Δt is a chunk of imaginary time updating it. The inconsistency is glaring, however, multiple "algorithmic" definitions of time are not new. The evolution time of the ring polymer molecular dynamics in Chapter 14 is an important multi-time scale algorithm example; the Matzubara frequencies contain the imaginary time information, while ring polymer trajectories evolve in "real" time. The preceding long discussion makes it clear that there are many difficulties and a number of unanswered questions, which will require careful exploration before the dynamic Stark effect using the diffusion Monte Carlo method can be successfully implemented. It always seems to be the case that gaining new theoretical perspectives in science is deeply humbling. It may lead to a slightly greater insight, but always presents a fresh multitude of challenges to be overcome if any new knowledge is to be gained from it. Perhaps this may be the true value of the scientific process, i. e., discovering new unknowns, the seemingly everlasting source of things yet to be explored.

17.7 Exercises

1. Derive equations (17.8) through (17.11).
2. For this exercise, consider the coupled differential equations in (17.23) and (17.24) for a special case when the parity in the potential $U(r) = U(-r)$ is such that $\mathscr{W}_{00} = \mathscr{W}_{11} = 0$, a situation common in atomic theory, where emissions, absorptions, and stimulated emission processes from a single multielectron atom centered at the origin are studied.
   (a) Prove that

$$\frac{d}{dt}(|c_0|^2 + |c_1|^2) = 0,$$

   i. e., the normalization constant for Ψ_{tot} does not change in time. Begin by taking the derivative on the left-hand side of this expression, then substitute (17.23) and (17.24) into your expressions and set $\mathscr{W}_{00}$ and $\mathscr{W}_{11}$ to zero. Recall that the coefficient c_j is a complex quantity; therefore, e. g., $|c_0|^2 = c_0c_0^*$, etc.

   (b) Consider further the case when $\omega = 0$, which corresponds to a static external filed. Show that $c_1(t)$ satisfies the following second- order differential equation:

$$\frac{d^2c_1}{dt^2} - i\omega_0\frac{dc_1}{dt} + |\mathscr{W}_{10}|^2c_1 = 0.$$

   (c) The differential equation in part (b) is linear and second-order with constant coefficients. Solve it using the traditional method, namely assume c_1 has the following general form $c_1 = e^{\lambda t}$ under the following starting conditions $c_0(0) = 1$, $c_1(0) = 0$. Then use your solution to show that

$$|c_1(t)|^2 = \frac{4|\mathscr{W}_{10}|^2}{\omega_0^2 + 4|\mathscr{W}_{10}|^2}\sin^2\{(\omega_0^2 + 4|\mathscr{W}_{10}|^2)^{1/2}t/2\}.$$

   (d) For weak fields, where $\mathscr{W}_{10} \ll \omega_0$, we can derive solutions to the differential equations in part (a) by an iterative method where, at the first iteration one inserts the initial conditions to obtain a first approximation, which is then reintroduced into them to obtain the second iteration and so forth. Insert $c_0(0) = 1$, $c_1(0) = 0$ into the coupled set of differential equations in part (a) of this exercise and show that upon integration you get

$$c_0(t) \approx 1$$

$$c_1(t) \approx \frac{\mathscr{W}_{10}}{2}\left\{\frac{1 - e^{i(\omega_0+\omega)t}}{\omega_0 + \omega} + \frac{1 - e^{i(\omega_0-\omega)t}}{\omega_0 - \omega}\right\}.$$

   (e) The second term inside the curly brackets expression for c_1 found in the previous part is much larger than the first when ω approaches ω_0. Evaluate the magnitude of both terms

$$\frac{1 - e^{i(\omega_0+\omega)t}}{\omega_0 + \omega} \quad \text{and} \quad \frac{1 - e^{i(\omega_0-\omega)t}}{\omega_0 - \omega}$$

for a typical transition where $\omega_0 = 8.1\times10^{-3}$ hartree and ω is at a 10 % detuning, $\omega = 8.8 \times 10^{-3}$ hartree, for a few values of t between 1000 and 14000 hartree^{-1}. You should find that the magnitude of the second term is about 24 times larger than the first. Neglecting the fast oscillatory terms in these expressions is named the "rotating wave approximation."

(f) Use the rotating wave approximation in the expression for c_1 in part (d) and show that, with the help of a double angle trigonometric identity you obtain

$$|c_1|^2 = |\mathscr{W}_{10}|^2 \frac{\sin^2\{(\omega_0 - \omega)t/2\}}{(\omega_0 - \omega)^2}.$$

3. For this exercise, you will carry out the analysis used in the previous one to handle the more generic case of the coupled differential equations in (17.23) and (17.24) where the parity of the potential energy cannot be assumed.

(a) Show that

$$\frac{d}{dt}(|c_0|^2 + |c_1|^2) = 0$$

still holds.

(b) Insert $c_0(0) = 1$, $c_1(0) = 0$ into the equations (17.23) and (17.24), integrate and find the first iterative solutions for c_0 and c_1. Show that the expression for $|c_1|^2$ is exactly the same as in part (f) of the previous problem.

4. In Chapter 2, we have shown that the following holds:

$$\partial_x^2 P_n \varphi_0 = (\partial_x^2 P_n)\varphi_0 + P_n \partial_x^2 \varphi_0 + 2(\partial_x P_n)\partial_x \varphi_0.$$

(a) Use $P_n = \sum_{k=0}^{n} w_k^{(n)} x^k$ to show that

$$\begin{aligned}
&-\frac{1}{2m}\partial_x^2 |n\rangle \\
&= -\frac{1}{2m}\left(\sum_{k=2}^{n} k(k-1) w_k^{(n)} x^{k-2}\, \varphi_0 + \sum_{k=0}^{n} w_k^{(n)} x^k\, \partial_x^2 \varphi_0 \right.\\
&\quad \left. + \sum_{k=1}^{n} k w_k^{(n)} x^{k-1}\, \partial_x \varphi_0 \right).
\end{aligned}$$

(b) Left multiply both sides by $\langle n'| = \sum_{l=0}^{n'} w_l^{(n')} x^l$ and derive equation (17.13), where $\hat{H}$ is a Hamiltonian in $\mathbb{R}$ with a generic potential $U(x)$. Note that the $\langle 1|H|1\rangle$ case of this equation is the estimate of E_1 in Chapter 7. Therefore, the coupling element $\langle 0|H|1\rangle$ is expected to be quite small.

(c) Show that $\langle n'|\hat{H}|n\rangle$ in part (b) is diagonal if $U = kx^2/2$ and that $\langle n|\hat{H}|n\rangle = \hbar\omega(n+1/2)$.

(d) Write the Python code that computes the matrix elements for a vector space size of 3, 4, and 5 excited states. Then submit the matrix to the singular value decomposition. The entries of Σ converge to eigenvalues of H as the number of excited states increases.

(e) With the code in the previous part, using the same systems in Chapter 7 compare your eigenvalues with those in Tables 17.1 and 17.2.

5. For this exercise, you will consider a quadratic expansion of the $\exp[-gW_\mu x^\mu]$ term in the propagator rather than the fully exponentiated rotation:

$$\begin{pmatrix} \cos wx & \sin wx \\ -\sin wx & \cos wx \end{pmatrix} \approx \begin{pmatrix} 1 & 0 \\ 0 & 1 \end{pmatrix} + \begin{pmatrix} -\frac{1}{2}(wx)^2 & x \\ -x & -\frac{1}{2}(wx)^2 \end{pmatrix} + \cdots .$$

Of course, this truncated expansion too destroys the orthogonality of the excited state, but it can once more be restored by applying the Gram–Schmidt procedure. This suggests the following form for ψ_1:

$$\psi_1 \approx [\tilde{w}_0^{(1)} + \tilde{w}_1^{(1)} x(1 - \alpha x)]\varphi_0$$

where φ_0 is the unnormalized ground state wave function, and α is an additional, presumably small, parameter to be optimized variationally while remaining orthogonal to $\psi_0 = w_0^{(0)}\varphi_0$.

Demonstrate that the resulting expressions of the two coefficients $\tilde{w}_0^{(1)}$ and $\tilde{w}_1^{(1)}$ for a given value of α are obtained using the procedure outlined in Chapters 1 and 2,

$$\tilde{w}_1^{(1)} = \tilde{w}_0^{(0)} \frac{1}{[\tilde{\mathscr{I}}_2 - 2\alpha\tilde{\mathscr{I}}_3 + \alpha^2\tilde{\mathscr{I}}_4 - (\tilde{\mathscr{I}}_1 - \alpha\tilde{\mathscr{I}}_2)^2]^{1/2}},$$

and

$$\tilde{w}_0^{(1)} = -\tilde{w}_0^{(0)} \frac{(\tilde{\mathscr{I}}_1 - \alpha\tilde{\mathscr{I}}_2)}{[\tilde{\mathscr{I}}_2 - 2\alpha\tilde{\mathscr{I}}_3 + \alpha^2\tilde{\mathscr{I}}_4 - (\tilde{\mathscr{I}}_1 - \alpha\tilde{\mathscr{I}}_2)^2]^{1/2}}.$$

6. Using the expression for ψ_1 from the previous exercise derive the following expression for $E_1(\alpha)$:

$$\begin{aligned}\langle 1|\hat{H}|1\rangle = -\frac{\hbar^2}{2m}\{&-2\alpha\tilde{w}_1^{(1)}[\tilde{w}_0^{(1)}\tilde{\mathscr{I}}_0 + \tilde{w}_1^{(1)}(\tilde{\mathscr{I}}_1 - \alpha\tilde{\mathscr{I}}_2)] \\ &+ 2\tilde{w}_1^{(1)}[\tilde{w}_0^{(1)}(\tilde{\mathscr{Q}}_0 - 2\alpha\tilde{\mathscr{Q}}_1) + \tilde{w}_1^{(1)}(\tilde{\mathscr{Q}}_1 - 3\alpha\tilde{\mathscr{Q}}_2 + 2\alpha^2\tilde{\mathscr{Q}}_3)] \\ &+ [(\tilde{w}_0^{(1)})^2\tilde{\mathscr{P}}_0 + (\tilde{w}_1^{(1)})^2(\tilde{\mathscr{P}}_2 - 2\alpha\tilde{\mathscr{P}}_3 + \alpha^2\tilde{\mathscr{P}}_4) + 2\tilde{w}_0^{(1)}\tilde{w}_1^{(1)}(\tilde{\mathscr{P}}_1 - \alpha\tilde{\mathscr{P}}_2)]\} \\ &+ (\tilde{w}_0^{(1)})^2\tilde{\mathscr{U}}_0 + (\tilde{w}_1^{(1)})^2(\tilde{\mathscr{U}}_2 - 2\alpha\tilde{\mathscr{U}}_3 + \alpha^2\tilde{\mathscr{U}}_4) + 2\tilde{w}_0^{(1)}\tilde{w}_1^{(1)}(\tilde{\mathscr{U}}_1 - \alpha\tilde{\mathscr{U}}_2).\end{aligned}$$

Here are some suggested steps:

(a) Show that

$$\partial_x^2\psi_1 = -2a\tilde{w}_1^{(1)}\varphi_0 + 2\tilde{w}_1^{(1)}(1-2ax)\partial_x\varphi_0 + [\tilde{w}_0^{(1)} + \tilde{w}_1^{(1)}x(1-ax)](\partial_x^2\varphi_0).$$

(b) Use the result in part (a) and the definition of the monomial integrals $\tilde{\mathscr{I}}_n$, $\tilde{\mathscr{Q}}_n$, $\tilde{\mathscr{P}}_n$ in Chapter 7 to show that the $\langle 1|\partial_x^2|1\rangle$ matrix elements become

$$\begin{aligned}\langle 1|\partial_x^2|1\rangle = &-2a\tilde{w}_1^{(1)}[\tilde{w}_0^{(1)}\tilde{\mathscr{I}}_0 + \tilde{w}_1^{(1)}(\tilde{\mathscr{I}}_1 - a\tilde{\mathscr{I}}_2)]\\ &+ 2\tilde{w}_1^{(1)}[\tilde{w}_0^{(1)}(\tilde{\mathscr{Q}}_0 - 2a\tilde{\mathscr{Q}}_1) + \tilde{w}_1^{(1)}(\tilde{\mathscr{Q}}_1 - 3a\tilde{\mathscr{Q}}_2 + 2a^2\tilde{\mathscr{Q}}_3)]\\ &+ [(\tilde{w}_0^{(1)})^2\tilde{\mathscr{P}}_0 + (\tilde{w}_1^{(1)})^2(\tilde{\mathscr{P}}_2 - 2a\tilde{\mathscr{P}}_3 + a^2\tilde{\mathscr{P}}_4) + 2\tilde{w}_0^{(1)}\tilde{w}_1^{(1)}(\tilde{\mathscr{P}}_1 - a\tilde{\mathscr{P}}_2)].\end{aligned}$$

(c) Use the definition of the monomial integral $\tilde{\mathscr{U}}_n$ to prove that the potential energy matrix elements are

$$\langle 1|U|1\rangle = (\tilde{w}_0^{(1)})^2\tilde{\mathscr{U}}_0 + (\tilde{w}_1^{(1)})^2(\tilde{\mathscr{U}}_2 - 2a\tilde{\mathscr{U}}_3 + a^2\tilde{\mathscr{U}}_4) + 2\tilde{w}_0^{(1)}\tilde{w}_1^{(1)}(\tilde{\mathscr{U}}_1 - a\tilde{\mathscr{U}}_2).$$

(d) Combine the results from part (b) and (c), then consider the $a \to 0$ limit and show that the result is E_1 in equation (7.11).

7. List the fifteen monomials $\mathbf{x}^a$ for the case $n = 4$, $d = 3$. Then sort them in lexicographic order by computing N_a for all of them. The ordered components of $\{x^\mu\}$ are x, y, z. Finally, check that

$$\binom{n+d-1}{n}$$

for this case evaluates to 15.

8. Find the elements of the matrix $\mathbf{x}^2(\mathbf{x}^2)^T$ for the $d = 2$ and 3 case.
9. Build the matrix $M_{2,2}$ using the same form as in equation (17.18) then translate it into the notation in equation (17.19).
10. Use the example where $P^2_{(2,0)}(x,y)$ is constructed to derive the determinant expression for $P^2_{(1,1)}(x,y)$ and $P^2_{(0,2)}(x,y)$.
11. Fill in the details of the result in the text that $\langle P^2_{(2,0)}|1\rangle_\omega = 0$. Then show that $\langle P^2_{(2,0)}|x\rangle_\omega$ and $\langle P^2_{(2,0)}|y\rangle_\omega$ are also zero.
12. Show that $\langle P^2_{(2,0)}|xy\rangle_\omega \neq 0$.
13. By expanding about the last row of $P^2_{(2,0)}(x,y)$, show that it can be written as follows:

$$P^2_{(2,0)}(x,y) = C^{(2)}_{(0,0)} + C^{(2)}_{(1,0)}x + C^{(2)}_{(0,1)}y + C^{(2)}_{(2,0)}x^2.$$

Then find expressions for the coefficients.

14. Use results from the relevant previous exercises to show that
(a) $P^2_{(2,0)}(x,y)$ is normalized;
(b) $\langle P^2_{(2,0)}|P^2_{(1,1)}\rangle_\omega$, $\langle P^2_{(2,0)}|P^2_{(0,2)}\rangle_\omega$ and $\langle P^2_{(1,1)}|P^2_{(0,2)}\rangle_\omega$ do not necessarily vanish.

(c) Prove that when $n \neq m$,

$$\langle P_\alpha^n | P_\beta^m \rangle_\omega = 0,$$

where d is arbitrary.

15. One defines the density matrix elements as the product of the coefficients $\rho_{jk} = c_j c_k^*$. The differential equations satisfied by the density matrix elements are obtained by evaluating the time derivative

$$\frac{d}{dt}\rho_{jk} = c_j \frac{dc_k^*}{dt} + c_k^* \frac{dc_j}{dt}.$$

Show that

$$\frac{d\rho_{00}}{dt} = -i \cos \omega t (\mathscr{W}_{01} e^{-i\omega_0 t} \rho_{10} - \mathscr{W}_{10} e^{i\omega_0 t} \rho_{01}),$$
$$\frac{d\rho_{11}}{dt} = -\frac{d\rho_{00}}{dt},$$

and

$$\frac{d\rho_{01}}{dt} = \frac{d\rho_{10}^*}{dt} = -i \cos \omega t \{(\mathscr{W}_{00} - \mathscr{W}_{11})\rho_{01} + \mathscr{W}_{01} e^{-i\omega_0 t}(\rho_{11} - \rho_{00})\}.$$

(a) Combine trigonometric functions using Euler's formula and some trigonometric identities, use the rotating wave approximation, and derive the following version of the differential equations:

$$\frac{d\rho_{00}}{dt} = -i(\mathscr{W}_{01} e^{-i(\omega_0-\omega)t} \rho_{10} - \mathscr{W}_{10} e^{i(\omega_0-\omega)t} \rho_{01}),$$
$$\frac{d\rho_{11}}{dt} = -\frac{d\rho_{00}}{dt},$$

and

$$\frac{d\rho_{01}}{dt} = \frac{d\rho_{10}^*}{dt} = -i\{(\mathscr{W}_{00} - \mathscr{W}_{11}) \cos \omega t \, \rho_{01} + \mathscr{W}_{01} e^{-i(\omega_0-\omega)t}(\rho_{11} - \rho_{00})\}.$$

(b) Consider the case when $\mathscr{W}_{11}$ and $\mathscr{W}_{00}$ vanish by symmetry. Prove that the expressions

$$\rho_{11} = \frac{|\mathscr{W}_{01}|^2}{\Gamma^2} \sin^2(\Gamma t/2),$$
$$\rho_{01} = \frac{\mathscr{W}_{01}}{\Gamma^2} e^{-i(\omega_0-\omega)t}$$
$$\times \{(\omega - \omega_0) \sin(\Gamma t/2) + i \cos(\Gamma t/2)\} \sin(\Gamma t/2)$$

where $\Gamma = \{(\omega_0 - \omega)^2 + |\mathscr{W}_{01}|^2\}^{1/2}$ satisfy the resulting differential equations as well as the initial conditions, $t = 0, \rho_{11} = \rho_{01} = 0$.

16. Prove the following two results:

(a)

$$\lim_{x \to 0} \frac{\sin^2(ax)}{x^2} = a^2.$$

(b) The full width at half-height d of the main peak at $x = 0$ of $x^{-2} \sin^2(ax)$ is

$$d = 2\frac{\eta}{a}$$

where η is the limit as $n \to \infty$ of the following sequence of compositions of sines:

$$\eta = \lim_{n \to \infty} \underbrace{\sin(\sqrt{2}\sin(\sqrt{2}\sin(\sqrt{2}\sin(\sqrt{2}\sin\cdots(\sqrt{2}\sin 1)))))}_{n\text{-fold}}$$
$$\approx 1.391557378\ldots.$$

Bibliography

[1] Daan Frenkel and Smit Berend. *Understanding Molecular Simulation: From Algorithms to Applications*. Second. Vol. 1. Academic Press, 2002.

[2] Nicholas Metropolis et al. “Equation of state calculations by fast computing machines”. In: *Journal of Chemical Physics* 21.6 (June 1953), pp. 1087–1092. ISSN: 0021-9606. DOI: https://doi.org/10.1063/1.1699114. eprint: https://pubs.aip.org/aip/jcp/article-pdf/21/6/1087/18802390/1087_1_online.pdf. URL: https://doi.org/10.1063/1.1699114.

[3] M. H. Kalos and P. A. Whitlock. *Monte Carlo Methods*. New York NY: Wiley, 1986.

[4] C. N. Yang and R. L. Mills. “Conservation of isotopic spin and isotopic gauge invariance”. In: *Physical Review* 96 (1 Oct. 1954), pp. 191–195. DOI: https://doi.org/10.1103/PhysRev.96.191. URL: https://link.aps.org/doi/10.1103/PhysRev.96.191.

[5] Sheldon Lee Glashow. “Partial symmetries of weak interactions”. In: *Nuclear Physics* 22 (1961), pp. 579–588. URL: https://api.semanticscholar.org/CorpusID:48093421.

[6] F. Englert and R. Brout. “Broken symmetry and the mass of gauge vector mesons”. In: *Physical Review Letters* 13 (9 Aug. 1964), pp. 321–323. DOI: https://doi.org/10.1103/PhysRevLett.13.321. URL: https://link.aps.org/doi/10.1103/PhysRevLett.13.321.

[7] Peter W. Higgs. “Broken symmetries and the masses of gauge bosons”. In: *Physical Review Letters* 13 (16 Oct. 1964), pp. 508–509. DOI: https://doi.org/10.1103/PhysRevLett.13.508. URL: https://link.aps.org/doi/10.1103/PhysRevLett.13.508.

[8] Steven Weinberg. “A model of leptons”. In: *Physical Review Letters* 19 (21 Nov. 1967), pp. 1264–1266. DOI: https://doi.org/10.1103/PhysRevLett.19.1264. URL: https://link.aps.org/doi/10.1103/PhysRevLett.19.1264.

[9] Abdus Salam. “Weak and electromagnetic interactions”. In: *Conf. Proc. C* 680519 (1968), pp. 367–377. DOI: https://doi.org/10.1142/9789812795915_0034.

[10] K. Grotz and H. V. Klapdor. *The Weak Interaction in Nuclear, Particle and Astrophysics*. Taylor & Francis, 1990. ISBN: 9780852743133. URL: https://books.google.com/books?id=KaBfQgAACAAJ.

[11] Lochlainn O’Raifeartaigh and Norbert Straumann. “Gauge theory: historical origins and some modern developments”. In: *Reviews of Modern Physics* 72 (1 Jan. 2000), pp. 1–23. DOI: https://doi.org/10.1103/RevModPhys.72.1. URL: https://link.aps.org/doi/10.1103/RevModPhys.72.1.

[12] W. Pauli. “Relativistic field theories of elementary particles”. In: *Reviews of Modern Physics* 13 (3 July 1941), pp. 203–232. DOI: https://doi.org/10.1103/RevModPhys.13.203. URL: https://link.aps.org/doi/10.1103/RevModPhys.13.203.

[13] S. Weinberg. *The Quantum Theory of Fields Volume 1*. Cambridge University Press, 1996.

[14] Mark Srednicki. *Quantum Field Theory*. Cambridge University Press, 2007.

[15] M. H. Kalos, D. Levesque, and L. Verlet. “Helium at zero temperature with hard-sphere and other forces”. In: *Physical Review A* 9 (1974), p. 2178. DOI: https://doi.org/10.1103/PhysRevA.9.2178. URL: https://doi.org/10.1103/PhysRevA.9.2178.

[16] Hao Jiang et al. “Ar_nHF van der Waals clusters revisited: II. Energetics and HF vibrational frequency shifts from diffusion Monte Carlo calculations on additive and nonadditive potential-energy surfaces for $n = 1$–12”. In: *Journal of Chemical Physics* 123.5 (2005), p. 054305. DOI: https://doi.org/10.1063/1.1991856. eprint: https://doi.org/10.1063/1.1991856. URL: https://doi.org/10.1063/1.1991856.

[17] Anne B. McCoy. “Vibrational excited states by diffusion Monte Carlo”. In: *Advances in Quantum Monte Carlo*. American Chemical Society, 2006. Chap. 11, pp. 147–164. DOI: https://doi.org/10.1021/bk-2007-0953.ch011. eprint: https://pubs.acs.org/doi/pdf/10.1021/bk-2007-0953.ch011. URL: https://pubs.acs.org/doi/abs/10.1021/bk-2007-0953.ch011.

[18] P. Håkansson et al. “Improved diffusion Monte Carlo propagators for bosonic systems using Itô calculus”. In: *Journal of Chemical Physics* 125.18 (Nov. 2006), p. 184106. ISSN: 0021-9606. DOI: https://doi.org/10.1063/1.2371077. eprint: https://pubs.aip.org/aip/jcp/article-pdf/doi/10.1063/1.2371077/15390514/184106_1_online.pdf. URL: https://doi.org/10.1063/1.2371077.

https://doi.org/10.1515/9783111610207-018

[19] Charlotte E. Hinkle and Anne B. McCoy. "Characterizing excited states of CH5+ with diffusion Monte Carlo". In: *The Journal of Physical Chemistry A, Molecules, Spectroscopy, Kinetics, Environment, & General Theory* 112.10 (2008). PMID: 18251525, pp. 2058–2064. DOI: https://doi.org/10.1021/jp709828v. eprint: https://doi.org/10.1021/jp709828v. URL: https://doi.org/10.1021/jp709828v.

[20] E. Curotto and M. Mella. "Quantum Monte Carlo simulations of selected ammonia clusters (n = 2–5): isotope effects on the ground state of typical hydrogen bonded systems". In: *Journal of Chemical Physics* 133 (2010), p. 214301.

[21] M. Mella. "Higher order diffusion Monte Carlo propagators for linear rotors as diffusion on a sphere: development and application to O2@He(n)". In: *Journal of Chemical Physics* 135 (2011), p. 114504.

[22] K. Roberts, R. Sebsebie, and E. Curotto. "A rare event sampling method for diffusion Monte Carlo using smart darting". In: *Journal of Chemical Physics* 136.7 (2012), p. 074104. DOI: https://doi.org/10.1063/1.3685453. eprint: https://doi.org/10.1063/1.3685453. URL: https://doi.org/10.1063/1.3685453.

[23] Andrew S. Petit and Anne B. McCoy. "Diffusion Monte Carlo in internal coordinates". In: *The Journal of Physical Chemistry A, Molecules, Spectroscopy, Kinetics, Environment, & General Theory* 117.32 (2013). PMID: 23410209, pp. 7009–7018. DOI: https://doi.org/10.1021/jp312710u. eprint: https://doi.org/10.1021/jp312710u. URL: https://doi.org/10.1021/jp312710u.

[24] M. Mella and E. Curotto. "Quantum simulations of the hydrogen molecule on ammonia clusters". In: *Journal of Chemical Physics* 139 (2013), p. 124319. DOI: https://doi.org/10.1063/1.4821648. URL: https://doi.org/10.1063/1.4821648.

[25] S. Wolf, E. Curotto, and M. Mella. "Quantum Monte Carlo methods for constrained systems". In: *International Journal of Quantum Chemistry* 114 (2014), pp. 611–625.

[26] J. B. Anderson. "A random walk simulation of the Schroedinger equation: ${H^+}_3$". In: *Journal of Chemical Physics* 63 (1975), p. 1499. DOI: https://doi.org/10.1063/1.431514. URL: https://doi.org/10.1063/1.431514.

[27] Kenta Hongo et al. "Diffusion Monte Carlo study of para-diiodobenzene polymorphism revisited". In: *Journal of Chemical Theory and Computation* 11.3 (2015). PMID: 26579744, pp. 907–917. DOI: https://doi.org/10.1021/ct500401p. eprint: https://doi.org/10.1021/ct500401p. URL: https://doi.org/10.1021/ct500401p.

[28] E. Curotto. "Ion-Stockmayer clusters: minima, classical thermodynamics, and variational ground state estimates of Li^+ $(CH_3NO_2)_n$ (n = 1–20)". In: *Journal of Chemical Physics* 143 (2015), p. 214301. DOI: https://doi.org/10.1063/1.4936587. URL: https://doi.org/10.1063/1.4936587.

[29] E. Curotto and Massimo Mella. "On the convergence of diffusion Monte Carlo in non-Euclidean spaces. II. Diffusion with sources and sinks". In: *Journal of Chemical Physics* 142.11 (2015), p. 114111. DOI: https://doi.org/10.1063/1.4914516. eprint: https://doi.org/10.1063/1.4914516. URL: https://doi.org/10.1063/1.4914516.

[30] E. Curotto and Massimo Mella. "On the convergence of diffusion Monte Carlo in non-Euclidean spaces. I. Free diffusion". In: *Journal of Chemical Physics* 142.11 (2015), p. 114110. DOI: https://doi.org/10.1063/1.4914515. eprint: https://doi.org/10.1063/1.4914515. URL: https://doi.org/10.1063/1.4914515.

[31] Michel Caffarel et al. "Using CIPSI nodes in diffusion Monte Carlo". In: *Recent Progress in Quantum Monte Carlo*. American Chemical Society, 2016. Chap. 2, pp. 15–46. DOI: https://doi.org/10.1021/bk-2016-1234.ch002. eprint: https://pubs.acs.org/doi/pdf/10.1021/bk-2016-1234.ch002. URL: https://pubs.acs.org/doi/abs/10.1021/bk-2016-1234.ch002.

[32] H. M. Christensen, L. C. Jake, and E. Curotto. "Smart darting diffusion Monte Carlo: applications to lithium ion-Stockmayer clusters". In: *Journal of Chemical Physics* 144 (2016), p. 174115. DOI: https://doi.org/10.1063/1.4948562. URL: https://doi.org/10.1063/1.4948562.

[33] Massimo Mella and E. Curotto. "Assessment of the effects of anisotropic interactions among hydrogen molecules and their isotopologues: a diffusion Monte Carlo investigation of gas phase and adsorbed clusters". In: *The Journal of Physical Chemistry A, Molecules, Spectroscopy, Kinetics, Environment, & General Theory* 121.26 (2017). PMID: 28616991, pp. 5005–5017. DOI: https://doi.org/10.1021/acs.jpca.7b03768. eprint: https://doi.org/10.1021/acs.jpca.7b03768. URL: https://doi.org/10.1021/acs.jpca.7b03768.

[34] E. Curotto and M. Mella. “Diffusion Monte Carlo simulations of gas phase and adsorbed D_2-$(H_2)_n$ clusters”. In: *Journal of Chemical Physics* 148.10 (2018), p. 102315. DOI: https://doi.org/10.1063/1.5000372. eprint: https://doi.org/10.1063/1.5000372. URL: https://doi.org/10.1063/1.5000372.

[35] J. B. Anderson. “Quantum chemistry by random walk: higher accuracy”. In: *Journal of Chemical Physics* 73 (1980), p. 3897. DOI: https://doi.org/10.1063/1.440575. URL: https://doi.org/10.1063/1.440575.

[36] G. E. DiEmma, S. A. Kalette, and E. Curotto. “Classical and quantum simulations of a lithium ion solvated by a mixed Stockmayer cluster”. In: *Chemical Physics Letters* 725 (2019), pp. 80–86. ISSN: 0009-2614. DOI: https://doi.org/10.1016/j.cplett.2019.04.007. URL: http://www.sciencedirect.com/science/article/pii/S0009261419302805.

[37] A. D. Stringer and E. Curotto. “An ergodic measure for diffusion Monte Carlo ground state wavefunctions: application to a hydrogen cluster with an isotopic impurity”. In: *Chemical Physics Letters* 734 (2019), p. 136728. ISSN: 0009-2614. DOI: https://doi.org/10.1016/j.cplett.2019.136728. URL: http://www.sciencedirect.com/science/article/pii/S0009261419307092.

[38] Victor G. M. Lee, Lindsey R. Madison, and Anne B. McCoy. “Evaluation of matrix elements using diffusion Monte Carlo wave functions”. In: *The Journal of Physical Chemistry A, Molecules, Spectroscopy, Kinetics, Environment, & General Theory* 123.20 (2019). PMID: 31021632, pp. 4370–4378. DOI: https://doi.org/10.1021/acs.jpca.8b11213. eprint: https://doi.org/10.1021/acs.jpca.8b11213. URL: https://doi.org/10.1021/acs.jpca.8b11213.

[39] Nick S. Blunt and Eric Neuscamman. “Excited-state diffusion Monte Carlo calculations: a simple and efficient two-determinant ansatz”. In: *Journal of Chemical Theory and Computation* 15.1 (2019), pp. 178–189. DOI: https://doi.org/10.1021/acs.jctc.8b00879. eprint: https://doi.org/10.1021/acs.jctc.8b00879. URL: https://doi.org/10.1021/acs.jctc.8b00879.

[40] Lena Jake and E. Curotto. “On diffusion Monte Carlo in spaces with multi-valued maps, boundaries and gradient torsion”. In: *Chemical Physics Letters* 762 (2021), p. 138167. ISSN: 0009-2614. DOI: https://doi.org/10.1016/j.cplett.2020.138167. URL: https://www.sciencedirect.com/science/article/pii/S0009261420310769.

[41] Ryan J. DiRisio, Fenris Lu, and Anne B. McCoy. “GPU-accelerated neural network potential energy surfaces for diffusion Monte Carlo”. In: *The Journal of Physical Chemistry A, Molecules, Spectroscopy, Kinetics, Environment, & General Theory* 125.26 (2021). PMID: 34165989, pp. 5849–5859. DOI: https://doi.org/10.1021/acs.jpca.1c03709. eprint: https://doi.org/10.1021/acs.jpca.1c03709. URL: https://doi.org/10.1021/acs.jpca.1c03709.

[42] Ryan J. DiRisio et al. “Using diffusion Monte Carlo wave functions to analyze the vibrational spectra of $H_7O^+{}_3$ and $H_9O^+{}_4$”. In: *The Journal of Physical Chemistry A, Molecules, Spectroscopy, Kinetics, Environment, & General Theory* 125.26 (2021), pp. 7185–7197. DOI: https://doi.org/10.1021/acs.jpca.1c05025. URL: https://doi.org/10.1021/acs.jpca.1c05025.

[43] Ryan J. DiRisio, Jacob M. Finney, and Anne B. McCoy. “Diffusion Monte Carlo approaches for studying nuclear quantum effects in fluxional molecules”. In: *WIREs Computational Molecular Science* 12.6 (2022), e1615. DOI: https://doi.org/10.1002/wcms.1615. eprint: https://wires.onlinelibrary.wiley.com/doi/pdf/10.1002/wcms.1615. URL: https://wires.onlinelibrary.wiley.com/doi/abs/10.1002/wcms.1615.

[44] Kevin Ryczko, Jaron T. Krogel, and Isaac Tamblyn. “Machine learning diffusion Monte Carlo energies”. In: *Journal of Chemical Theory and Computation* 18.12 (2022). PMID: 36317712, pp. 7695–7701. DOI: https://doi.org/10.1021/acs.jctc.2c00483. eprint: https://doi.org/10.1021/acs.jctc.2c00483. URL: https://doi.org/10.1021/acs.jctc.2c00483.

[45] A. R. Zane and E. Curotto. “Electrolyte clusters as hydrogen sponges: diffusion Monte Carlo simulations”. In: *Physical Chemistry Chemical Physics* 24.42 (2022), pp. 26094–26101. DOI: https://doi.org/10.1039/D2CP03658D. URL: http://dx.doi.org/10.1039/D2CP03658D.

[46] Holly G. Hixson et al. “Experimental and simulated vibrational spectra of H2 absorbed in amorphous ice: surface structures, energetics, and relaxations”. In: *Journal of Chemical Physics* 97.2 (1992), pp. 753–767. DOI: https://doi.org/10.1063/1.463240. eprint: https://doi.org/10.1063/1.463240. URL: https://doi.org/10.1063/1.463240.

[47] Cancan Huang and Brenda M. Rubenstein. “Machine learning diffusion Monte Carlo forces”. In: *The Journal of Physical Chemistry A, Molecules, Spectroscopy, Kinetics, Environment, & General Theory* 127.1 (2023). PMID: 36576803, pp. 339–355. DOI: https://doi.org/10.1021/acs.jpca.2c05904. eprint: https://doi.org/10.1021/acs.jpca.2c05904. URL: https://doi.org/10.1021/acs.jpca.2c05904.

[48] Gabriella E. Ravin and E. Curotto. “A potential energy surface of spectroscopic accuracy for a lithium ion–hydrogen clusters”. In: *Chemical Physics Letters* 834 (2024), p. 140951. ISSN: 0009-2614. DOI: https://doi.org/10.1016/j.cplett.2023.140951. URL: https://www.sciencedirect.com/science/article/pii/S0009261423006565.

[49] Marie S. Corrie and E. Curotto. “Diffusion Monte Carlo and molecular dynamics on atlases”. In: *Chemical Physics Letters* (2025), p. 142277. ISSN: 0009-2614. DOI: https://doi.org/10.1016/j.cplett.2025.142277. URL: https://www.sciencedirect.com/science/article/pii/S0009261425004178.

[50] V. Buch, S. C. Silva, and J. P. Devlin. “Rotational spectrum of a quantum rotor adsorbed on a rough and disordered surface: para-H2 and ortho-H2 on amorphous ice”. In: *Journal of Chemical Physics* 99.3 (1993), pp. 2265–2268. DOI: https://doi.org/10.1063/1.465237. eprint: https://doi.org/10.1063/1.465237. URL: https://doi.org/10.1063/1.465237.

[51] C. J. Umrigar, M. P. Nightingale, and K. J. Runge. “A diffusion Monte Carlo algorithm with very small time-step errors”. In: *Journal of Chemical Physics* 99.4 (Aug. 1993), pp. 2865–2890. ISSN: 0021-9606. DOI: https://doi.org/10.1063/1.465195. eprint: https://pubs.aip.org/aip/jcp/article-pdf/99/4/2865/13421024/2865_1_online.pdf. URL: https://doi.org/10.1063/1.465195.

[52] V. Buch and J. P. Devlin. “Preferential adsorption of ortho-H2 with respect to para-H2 on the amorphous ice surface”. In: *Journal of Chemical Physics* 98.5 (1993), pp. 4195–4206. DOI: https://doi.org/10.1063/1.465026. eprint: https://doi.org/10.1063/1.465026. URL: https://doi.org/10.1063/1.465026.

[53] Massimo Mella. *Sviluppi ed applicazioni dei metodi stocastici nella determinazione della struttura elettronica atomica e molecolare*. Universita’ degli Studi di Milano, Ph. D. thesis, Oct. 1996.

[54] Dario Bressanini et al. “Analytical wavefunctions from quantum Monte Carlo simulations”. In: *Recent Advances In Quantum Monte Carlo Methods*. World Scientific, 1997, pp. 1–19.

[55] Carl Eckart. “Some studies concerning rotating axes and polyatomic molecules”. In: *Physical Review* 47 (7 Apr. 1935), pp. 552–558. DOI: https://doi.org/10.1103/PhysRev.47.552. URL: https://link.aps.org/doi/10.1103/PhysRev.47.552.

[56] T. F. Miller III and D. C. Clary. “Torsional path integral Monte Carlo method for calculating the absolute quantum free energy of large molecules”. In: *Journal of Chemical Physics* 119 (2003), p. 68.

[57] M. F. Russo and E. Curotto. “Stereographic projections path integral in S^1 and $(S^2)^m$ manifolds”. In: *Journal of Chemical Physics* 118 (2003), p. 6806.

[58] José Zúñiga et al. “On the use of optimal internal vibrational coordinates for symmetrical bent triatomic molecules”. In: *Journal of Chemical Physics* 122.22 (June 2005), p. 224319. ISSN: 0021-9606. DOI: https://doi.org/10.1063/1.1929738. eprint: https://pubs.aip.org/aip/jcp/article-pdf/doi/10.1063/1.1929738/15366918/224319_1_online.pdf. URL: https://doi.org/10.1063/1.1929738.

[59] M. F. Russo and E. Curotto. “Stereographic projections path integral for inertia ellipsoids: applications to Ar_n – HF clusters”. In: *Journal of Chemical Physics* 120 (2005), p. 2110.

[60] M. W. Avilés and E. Curotto. “Partial averaging and the centroid virial estimator for stereographic projection path-integral simulations in curved spaces”. In: *Journal of Chemical Physics* 122 (2005), p. 164109.

[61] E. Curotto. “A reweighted random series method for stereographic projection path integrals”. In: *Journal of Chemical Physics* 123 (2005), p. 134102.

[62] Behrooz Hashemian, Daniel Millán, and Marino Arroyo. “Charting molecular free-energy landscapes with an atlas of collective variables”. In: *Journal of Chemical Physics* 145 (2005), p. 174109. DOI: https://doi.org/10.1063/1.4966262. URL: https://doi.org/10.1063/1.4966262.

[63] M. W. Avilés, P. T. Gray and E. Curotto. “Stereographic projection path-integral simulations of $(HF)_n$ clusters”. In: *Journal of Chemical Physics* 124 (2006), p. 174305.

[64] F. Langley et al. “Rigid quantum Monte Carlo simulations of condensed molecular matter: water clusters in the $n = 2$–8 range”. In: *Journal of Chemical Physics* 126 (2007), p. 084506.

[65] E. Curotto, David L. Freeman, and J. D. Doll. “A stereographic projection path integral study of the coupling between the orientation and the bending degrees of freedom of water”. In: *Journal of Chemical Physics* 128 (2008), p. 204107.

[66] Edgar Bright Wilson, J. C. Decius, and Paul C. Cross. *Molecular Vibrations: The Theory of Infrared and Raman Vibrational Spectra*. New York NY: Dover, 1955.

[67] M. W. Avilés, M. L. McCandless, and E. Curotto. “Stereographic projection path integral simulations of $(HCl)_n$ clusters ($n = 2$–5): evidence of quantum induced melting in small hydrogen bonded networks”. In: *Journal of Chemical Physics* 128 (2008), p. 124517.

[68] P. E. Janeiro-Barral, M. Mella, and E. Curotto. “Structure and energetics of ammonia clusters $(NH_3)_n$ ($n = 3$–20) investigated using a rigid-polarizable model derived from *ab initio* calculations”. In: *The Journal of Physical Chemistry A, Molecules, Spectroscopy, Kinetics, Environment, & General Theory* 112 (2008), p. 2888.

[69] M. Ragni et al. “Orthogonal coordinates and hyperquantization algorithm. The NH_3 and H_3O^+ umbrella inversion levels”. In: *The Journal of Physical Chemistry A, Molecules, Spectroscopy, Kinetics, Environment, & General Theory* 113.52 (2009). PMID: 19757778, pp. 15355–15365. DOI: https://doi.org/10.1021/jp906415m. eprint: https://doi.org/10.1021/jp906415m. URL: https://doi.org/10.1021/jp906415m.

[70] E. Curotto. *Stochastic Simulations of Clusters: Quantum Methods in Flat and Curved Spaces*. Boca Raton, FL: CRC, 2010.

[71] S. Wolf and E. Curotto. “Ring polymer dynamics for rigid tops with an improved integrator”. In: *Journal of Chemical Physics* 141.2 (2014), p. 024116. DOI: https://doi.org/10.1063/1.4887460. eprint: https://doi.org/10.1063/1.4887460. URL: https://doi.org/10.1063/1.4887460.

[72] Zhaojun Zhang, Fabien Gatti, and Dong H. Zhang. “Full dimensional quantum mechanical calculations of the reaction probability of the H + NH3 collision based on a mixed Jacobi and Radau description”. In: *Journal of Chemical Physics* 150.20 (May 2019), p. 204301. ISSN: 0021-9606. DOI: https://doi.org/10.1063/1.5096047. eprint: https://pubs.aip.org/aip/jcp/article-pdf/doi/10.1063/1.5096047/9651933/204301_1_online.pdf. URL: https://doi.org/10.1063/1.5096047.

[73] Kemal Oenen, Dennis F. Dinu, and Klaus R. Liedl. “Determining internal coordinate sets for optimal representation of molecular vibration”. In: *Journal of Chemical Physics* 160.1 (Jan. 2024), p. 014104. ISSN: 0021-9606. DOI: https://doi.org/10.1063/5.0180657. eprint: https://pubs.aip.org/aip/jcp/article-pdf/doi/10.1063/5.0180657/18287665/014104_1_5.0180657.pdf. URL: https://doi.org/10.1063/5.0180657.

[74] B. S. DeWitt. “Dynamical theory in curved spaces. I. A review of the classical and quantum action principles”. In: *Reviews of Modern Physics* 29 (1957), p. 377.

[75] James K. G. Watson. “Simplification of the molecular vibration-rotation Hamiltonian”. In: *Molecular Physics* 15.5 (1968), pp. 479–490. DOI: https://doi.org/10.1080/00268976800101381. eprint: https://doi.org/10.1080/00268976800101381. URL: https://doi.org/10.1080/00268976800101381.

[76] James D. Louck and Harold W. Galbraith. “Eckart vectors, Eckart frames, and polyatomic molecules”. In: *Reviews of Modern Physics* 48 (1 Jan. 1976), pp. 69–106. DOI: https://doi.org/10.1103/RevModPhys.48.69. URL: https://link.aps.org/doi/10.1103/RevModPhys.48.69.

[77] Frederick O. Meyer and Rogers W. Redding. “Molecular symmetry and motions of the Eckart frame”. In: *Journal of Molecular Spectroscopy* 70.3 (1978), pp. 410–419. ISSN: 0022-2852. DOI: https://doi.org/10.1016/0022-2852(78)90179-0. URL: https://www.sciencedirect.com/science/article/pii/0022285278901790.

[78] L. S. Schulman. *Techniques and Applications of Path Integration*. New York: John Wiley & Sons, 1981.

[79] D. Marx and P. Nielaba. “Path-integral Monte Carlo techniques for rotational motion in two dimensions: quenched, annealed, and no-spin quantum-statistical averages”. In: *Physical Review A* 45 (1992), p. 8968.

[80] D. Marx and M. H. Müser. "Path integral simulations of rotors: theory and applications". In: *Journal of Physics Condensed Matter* 11 (1999), R117.

[81] G. N. Mil'shtejn. "Approximate integration of stochastic differential equations". In: *Theory of Probability and Its Applications* 19.3 (1975), pp. 557–562. DOI: https://doi.org/10.1137/1119062. eprint: https://doi.org/10.1137/1119062. URL: https://doi.org/10.1137/1119062.

[82] K. W. Schmid et al. "Beyond symmetry – projected quasi – particle mean fields: a new variational procedure for nuclear structure calculations". In: *Nuclear Physics A* 499 (1988), 63–92.

[83] David O. Harris, Gail G. Engerholm, and William D. Gwinn. "Calculation of matrix elements for one-dimensional quantum-mechanical problems and the application to anharmonic oscillators". In: *Journal of Chemical Physics* 43.5 (Sept. 1965), pp. 1515–1517. ISSN: 0021-9606. DOI: https://doi.org/10.1063/1.1696963. eprint: https://pubs.aip.org/aip/jcp/article-pdf/43/5/1515/18839366/1515_1_online.pdf. URL: https://doi.org/10.1063/1.1696963.

[84] Seung E. Choi and J. C. Light. "Use of the discrete variable representation in the quantum dynamics by a wave packet propagation: predissociation of NaI -> NaI -> Na(2S) + I(2P)". In: *Journal of Chemical Physics* 90.5 (Mar. 1989), pp. 2593–2604. ISSN: 0021-9606. DOI: https://doi.org/10.1063/1.455957. eprint: https://pubs.aip.org/aip/jcp/article-pdf/90/5/2593/18974776/2593_1_online.pdf. URL: https://doi.org/10.1063/1.455957.

[85] James T. Muckerman. "Some useful discrete variable representations for problems in time-dependent and time-independent quantum mechanics". In: *Chemical Physics Letters* 173.2 (1990), pp. 200–205. ISSN: 0009-2614. DOI: https://doi.org/10.1016/0009-2614(90)80078-R. URL: https://www.sciencedirect.com/science/article/pii/000926149080078R.

[86] Daniel T. Colbert and William H. Miller. "A novel discrete variable representation for quantum mechanical reactive scattering via the S-matrix Kohn method". In: *Journal of Chemical Physics* 96.3 (Feb. 1992), pp. 1982–1991. ISSN: 0021-9606. DOI: https://doi.org/10.1063/1.462100. eprint: https://pubs.aip.org/aip/jcp/article-pdf/96/3/1982/11261858/1982_1_online.pdf. URL: https://doi.org/10.1063/1.462100.

[87] Hua Wei and Jr. Carrington Tucker. "Discrete variable representations of complicated kinetic energy operators". In: *Journal of Chemical Physics* 101.2 (July 1994), pp. 1343–1360. ISSN: 0021-9606. DOI: https://doi.org/10.1063/1.467827. eprint: https://pubs.aip.org/aip/jcp/article-pdf/101/2/1343/19070921/1343_1_online.pdf. URL: https://doi.org/10.1063/1.467827.

[88] María Fernanda González et al. "Quantum trajectories from a discrete variable representation method". In: *The Journal of Physical Chemistry A, Molecules, Spectroscopy, Kinetics, Environment, & General Theory* 111.41 (2007). PMID: 17696411, pp. 10226–10233. DOI: https://doi.org/10.1021/jp072237o. eprint: https://doi.org/10.1021/jp072237o. URL: https://doi.org/10.1021/jp072237o.

[89] Aurel Bulgac and Michael McNeil Forbes. "Use of the discrete variable representation basis in nuclear physics". In: *Physical Review C, Nuclear Physics* 87 (5 May 2013), p. 051301. DOI: https://doi.org/10.1103/PhysRevC.87.051301. URL: https://link.aps.org/doi/10.1103/PhysRevC.87.051301.

[90] T. S. Chihara. *An Introduction to Orthogonal Polynomials*. Dover Books on Mathematics. Dover Publications, 2011. ISBN: 9780486479293. URL: https://books.google.com/books?id=IkCJSQAACAAJ.

[91] Charles F. Dunkl and Yuan Xu. *Orthogonal Polynomials of Several Variables*. 2nd ed. Encyclopedia of Mathematics and its Applications. Cambridge University Press, 2014.

[92] Claude Cohen-Tannoudji, Bernard Diu, and Franck Laloë. *Quantum mechanics*; 1st ed. Trans. of: Mécanique quantique. Paris: Hermann, 1973. New York, NY: Wiley, 1977. URL: https://cds.cern.ch/record/101367.

[93] Ramamurti Shankar. *Principles of quantum mechanics*. New York, NY: Plenum, 1980. URL: https://cds.cern.ch/record/102017.

[94] Jun John Sakurai. *Modern Quantum Mechanics*; rev. ed. Reading, MA: Addison-Wesley, 1994. URL: https://cds.cern.ch/record/1167961.

[95] E. P. Wigner. *Group Theory: And Its Application to the Quantum Mechanics of Atomic Spectra*. Pure and Applied Physics: A Series of Monographs and Textbooks. Elsevier Science, 1959. ISBN: 9780127505503. URL: https://books.google.com/books?id=IOJEAAAAIAAJ.

[96] Joseph A. Gallian. *Contemporary abstract algebra. Undefined/Unknown*. Ninth edition. Cengage Learning, 2017. ISBN: 978-1-305-65796-0.

[97] Gilmore Robert. *Lie Groups, Lie Algebras, and Some of Their Applications*. Dover Publications, 2006. ISBN: 0486445291.

[98] Dwight E. Neuenschwander. *Tensor Calculus for Physics*. Johns Hopkins University Press, 2014. ISBN: 9781421415659. URL: https://doi.org/10.56021/9781421415642.

[99] Dwight E. Neuenschwander. *Emmy Noether's Wonderful Theorem*. Johns Hopkins University Press, Baltimore, Md., 2011. ISBN: 978-0-8018-9694-1.

[100] Rodney Loudon. *The Quantum Theory of Light*. Oxford University Press, 2000. ISBN: 9780198501763.

[101] R. P. Feynman. "Space-time approach to non-relativistic quantum mechanics". In: *Reviews of Modern Physics* 20 (1948), p. 367.

[102] H. Kleinert. *Path integrals in Quantum Mechanics, Statistics, Polymer physics, and Financial Markets*. Singapore: World Scientific, 2002.

[103] P. V. O'Neil. *Advanced Engineering Mathematics*. Thomson Brooks/Cole, 2003. ISBN: 9780534401306. URL: https://books.google.com/books?id=oOcuKAAACAAJ.

[104] O. A. Nieves. *Gaussian Integrals and their Applications*. 1st. Chapman and Hall/CRC, 2024. ISBN: 9781003501329. URL: https://doi.org/10.1201/9781003501329.

[105] Jens Peder Dahl and Michael Springborg. "The Morse oscillator in position space, momentum space, and phase space". In: *Journal of Chemical Physics* 88.7 (Apr. 1988), pp. 4535–4547. ISSN: 0021-9606. DOI: https://doi.org/10.1063/1.453761. eprint: https://pubs.aip.org/aip/jcp/article-pdf/88/7/4535/18969239/4535_1_online.pdf. URL: https://doi.org/10.1063/1.453761.

[106] W. D. Curtis and F. R. Miller. *Differential Manifolds and Theoretical Physics*. Pure and Applied Mathematics. Academic Press, 1985. ISBN: 9780080874357. URL: https://books.google.com/books?id=sWXcAyR37acC.

[107] David L. Freeman and Jimmie D. Doll. "A Monte Carlo method for quantum Boltzmann statistical mechanics using Fourier representations of path integrals". In: *Journal of Chemical Physics* 80.11 (June 1984), pp. 5709–5718. ISSN: 0021-9606. DOI: https://doi.org/10.1063/1.446640. eprint: https://pubs.aip.org/aip/jcp/article-pdf/80/11/5709/18948477/5709_1_online.pdf. URL: https://doi.org/10.1063/1.446640.

[108] Charusita Chakravarty. "Path integral simulations of atomic and molecular systems". In: *International Reviews in Physical Chemistry* 16.4 (1997), pp. 421–444. DOI: https://doi.org/10.1080/014423597230190. eprint: https://doi.org/10.1080/014423597230190. URL: https://doi.org/10.1080/014423597230190.

[109] D. M. Ceperley. "Path integrals in the theory of condensed helium". In: *Reviews of Modern Physics* 67 (2 Apr. 1995), pp. 279–355. DOI: https://doi.org/10.1103/RevModPhys.67.279. URL: https://link.aps.org/doi/10.1103/RevModPhys.67.279.

[110] Kurt R. Glaesemann and Laurence E. Fried. "Improved heat capacity estimator for path integral simulations". In: *Journal of Chemical Physics* 117.7 (Aug. 2002), pp. 3020–3026. ISSN: 0021-9606. DOI: https://doi.org/10.1063/1.1493184. eprint: https://pubs.aip.org/aip/jcp/article-pdf/117/7/3020/19009181/3020_1_online.pdf. URL: https://doi.org/10.1063/1.1493184.

[111] A. Korzeniowski et al. "Feynman-Kac path-integral calculation of the ground-state energies of atoms". In: *Physical Review Letters* 69.6 (1992), p. 893.

[112] A. Sarsa, K. E. Schmidt, and W. R. Magro. "A path integral ground state method". In: *Journal of Chemical Physics* 113.4 (2000), pp. 1366–1371.

[113] Javier E. Cuervo, Pierre-Nicholas Roy, and Massimo Boninsegni. "Path integral ground state with a fourth-order propagator: application to condensed helium". In: *Journal of Chemical Physics* 122.11 (2005).

[114] Javier Eduardo Cuervo and Pierre-Nicholas Roy. "Path integral ground state study of finite-size systems: application to small (parahydrogen) N (N = 2–20) clusters". In: *Journal of Chemical Physics* 125.12 (2006).

[115] E. Vitali et al. “Path-integral ground-state Monte Carlo study of two-dimensional solid He 4”. In: *Physical Review B, Condensed Matter and Materials Physics* 77.18 (2008), p. 180505.

[116] Steve Constable et al. “Langevin equation path integral ground state”. In: *The Journal of Physical Chemistry A, Molecules, Spectroscopy, Kinetics, Environment, & General Theory* 117.32 (2013), pp. 7461–7467.

[117] Yangqian Yan and D Blume. “Path integral Monte Carlo ground state approach: formalism, implementation, and applications”. In: *Journal of Physics B, Atomic, Molecular and Optical Physics* 50.22 (2017), p. 223001.

[118] R. P. Feynman. “Space-time approach to non-relativistic quantum mechanics”. In: *Reviews of Modern Physics* 20 (2 Apr. 1948), pp. 367–387. DOI: https://doi.org/10.1103/RevModPhys.20.367. URL: https://link.aps.org/doi/10.1103/RevModPhys.20.367.

[119] I. R. Craig and D. E. Manolopoulos. “Quantum statistics and classical mechanics: real time correlation functions from ring polymer molecular dynamics.” In: *Journal of Chemical Physics* 121 (2004), pp. 3368–3373. DOI: https://doi.org/10.1063/1.1777575. PMID:15303899.

[120] Mariana Rossi, Michele Ceriotti, and David E. Manolopoulos. “How to remove the spurious resonances from ring polymer molecular dynamics”. In: *Journal of Chemical Physics* 140.23 (2014).

[121] Liang Zhang et al. “Ring polymer molecular dynamics approach to quantum dissociative chemisorption rates”. In: *Journal of Physical Chemistry Letters* 14.31 (2023), pp. 7118–7125.

[122] Ian R. Craig and David E. Manolopoulos. “Chemical reaction rates from ring polymer molecular dynamics”. In: *Journal of Chemical Physics* 122.8 (2005).

[123] Ian R. Craig and David E. Manolopoulos. “A refined ring polymer molecular dynamics theory of chemical reaction rates”. In: *Journal of Chemical Physics* 123.3 (2005).

[124] Thomas F. Miller and David E. Manolopoulos. “Quantum diffusion in liquid water from ring polymer molecular dynamics”. In: *Journal of Chemical Physics* 123.15 (2005).

[125] Thomas F. Miller and David E. Manolopoulos. “Quantum diffusion in liquid para-hydrogen from ring-polymer molecular dynamics”. In: *Journal of Chemical Physics* 122.18 (2005).

[126] Bastiaan J. Braams and David E. Manolopoulos. “On the short-time limit of ring polymer molecular dynamics”. In: *Journal of Chemical Physics* 125.12 (2006).

[127] Yu V. Suleimanov, Joshua W. Allen, and W. H. Green. “RPMDrate: bimolecular chemical reaction rates from ring polymer molecular dynamics”. In: *Computer Physics Communications* 184.3 (2013), pp. 833–840.

[128] Scott Habershon et al. “Ring-polymer molecular dynamics: quantum effects in chemical dynamics from classical trajectories in an extended phase space”. In: *Annual Review of Physical Chemistry* 64.1 (2013), pp. 387–413.

[129] Artur R. Menzeleev, Franziska Bell, and Thomas F. Miller. “Kinetically constrained ring-polymer molecular dynamics for non-adiabatic chemical reactions”. In: *Journal of Chemical Physics* 140.6 (2014).

[130] Gregory A. Voth. “Path-integral centroid methods in quantum statistical mechanics and dynamics”. In: *Advances in Computational Fluid Dynamics* (1996), pp. 135–218.

[131] Seogjoo Jang and Gregory A. Voth. “A derivation of centroid molecular dynamics and other approximate time evolution methods for path integral centroid variables”. In: *Journal of Chemical Physics* 111.6 (1999), pp. 2371–2384.

[132] Seogjoo Jang. “Path-integral centroid dynamics for general initial conditions: a nonequilibrium projection operator formulation”. In: *Journal of Chemical Physics* 124.6 (2006).

[133] Timothy D. Loose, Patrick G. Sahrmann, and Gregory A. Voth. “Centroid molecular dynamics can be greatly accelerated through neural network learned centroid forces derived from path integral molecular dynamics”. In: *Journal of Chemical Theory and Computation* 18.10 (2022), pp. 5856–5863.

Index

https://doi.org/10.1515/9783111610207-019

www.ingramcontent.com/pod-product-compliance
Lightning Source LLC
Chambersburg PA
CBHW080703220326
41598CB00033B/5294
* 9 7 8 3 1 1 1 6 1 0 0 8 5 *